D0936446

The Moonlandings

The Soviet–American race to land the first man on the Moon was a technical challenge unlike anything in recent human history. Reginald Turnill, the BBC's Aerospace Correspondent, covered the entire story first-hand, and his reports were heard and seen by millions around the world. With unparalleled access to the politicians, scientists and technicians involved in the race to the Moon, Turnill got to know all the early astronauts – Alan Shepard, John Glenn, Neil Armstrong, Buzz Aldrin – as they pioneered the techniques that made the moonlandings possible. He became a friend of Dr Wernher von Braun, the German rocket pioneer and mastermind behind it all. This unique eyewitness account of one of the most thrilling adventures of the twentieth century is written in a lucid style, packed with action and drama, and is a fascinating read for all those interested in the story of the race to the Moon.

REGINALD TURNILL started work in Fleet Street at the age of 15, and by 19 he was covering the national news as a Press Association staff reporter. After joining the BBC in 1956 he covered the launch of Sputnik 1 and found it so exciting that he made space reporting his speciality. As the BBC Aerospace Correspondent, Turnill spent the rest of his career covering all the manned space missions as well as planetary missions such as Mariner, Pioneer, Viking and Voyager. Since leaving the BBC staff, Turnill has continued to broadcast and write about space, and he created the first spaceflight directory. Turnill is the only non-American to have been presented with NASA's Chroniclers Award for contributions to public understanding of the space programme.

Chimp 'Ham', first primate in space (the Russians had sent dogs) receiving an apple with a big grin after 6 minutes of weightlessness on a suborbital flight to an altitude on 253 km in a Mercury spacecraft on 31 January 1961. Though he looked happy, when shown a spacecraft later, he made it clear he had no wish to go again. His survival in spite of malfunctions gave the seven Mercury astronauts the confidence to start manned flights four months later. (NASA)

The Moonlandings

An Eyewitness Account

REGINALD TURNILL

Foreword by **Dr Buzz Aldrin**

PUBLISHED BY THE PRESS SYNDICATE OF THE UNIVERSITY OF CAMBRIDGE
The Pitt Building, Trumpington Street, Cambridge, United Kingdom

CAMBRIDGE UNIVERSITY PRESS
The Edinburgh Building, Cambridge CB2 2RU, UK
40 West 20th Street, New York, NY 10011–4211, USA
477 Williamstown Road, Port Melbourne, VIC 3207, Australia
Ruiz de Alarcón 13, 28014 Madrid, Spain
Dock House, The Waterfront, Cape Town 8001, South Africa

http://www.cambridge.org

First published 2003

Printed in the United Kingdom at the University Press, Cambridge

Typeface Trump Medieval 9.5/15 pt *System* QuarkXPress™ [SE]

A catalogue record for this book is available from the British Library

Library of Congress Cataloguing in Publication data

Turnill, Reginald.
The moonlandings : an eye witness account / Reginald Turnill.
p. cm.
Includes bibliographical references and index.
ISBN 0 521 81595 9
1. Project Apollo (U.S.) 2. Space flight to the moon–History. I. Title.
TL789.8.U6 A5844 2002
629.45′4′0973–dc21 2002023374

ISBN 0 521 81595 9 hardback

Why 'Apollo'?

The spacecraft was named for one of the busiest and the most versatile of the Greek gods. Apollo was the god of light and the twin brother of Artemis, the goddess of the Moon. He was the god of music and the father of Orpheus. At his temple in Delphi, he was the god of prophecy. Finally, he was also known as the god of poetry, of healing, and of pastoral pursuits. (NASA)

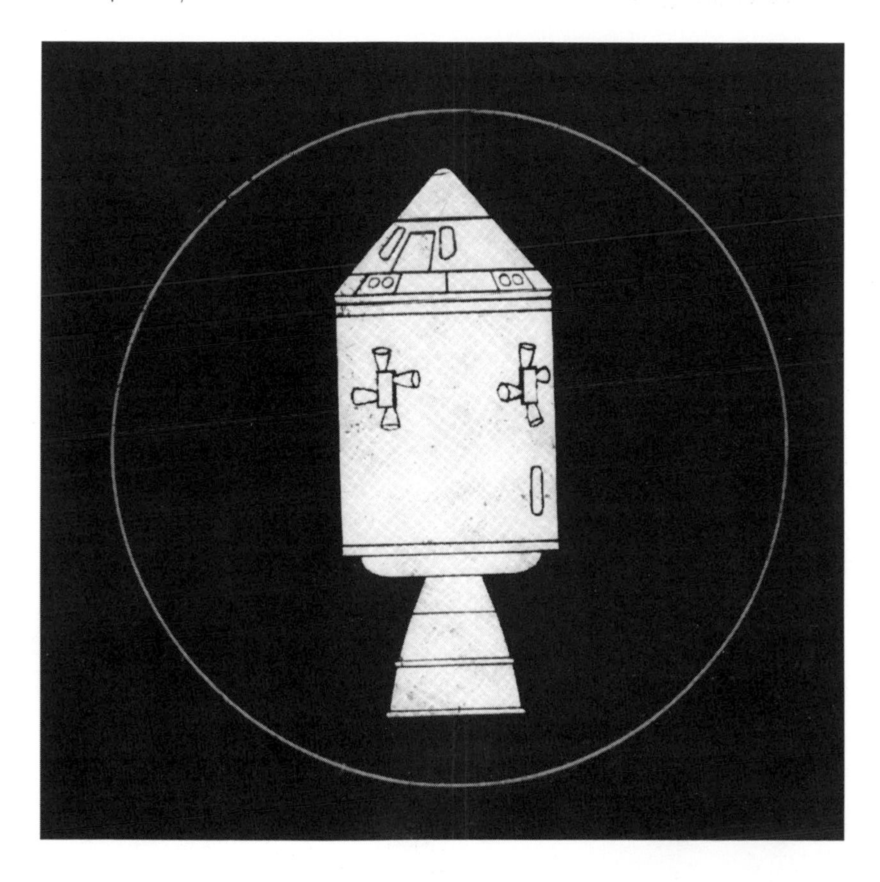

The Chroniclers

Presented by the National Aeronautics and Space Administration
John F. Kennedy Space Center

REGINALD TURNILL

In recognition of your many contributions to the public understanding of the space program through your work at the Kennedy Space Center. Working in concert with communicators in every medium, the excellence of your career has provided knowledge, ideas and inspiration world wide.

Your name has been permanently inscribed on the 'roll of honor' titled 'The Chroniclers' at the KSC Press Site.

Hugh W. Harris
Director Public Affairs Office

Jay F. Honeycutt
Director Kennedy Space Center

[The Author is the only non-American to be awarded this certificate.]

'We indeed have come a long way, and we are fortunate that there have been men like Reg Turnill on hand to document our progress': the late GEORGE LOW, NASA Deputy Administrator, who had had overall responsibility for the Gemini and Apollo programmes, in a 1978 Foreword to the *Observer's Spaceflight Directory*

Contents

Foreword

DR BUZZ ALDRIN

Neil Armstrong and Buzz Aldrin made history on 20 July 1969 when, in Apollo 11, they became the first humans to land on the Moon.

Who wants a journey into space?

I am always being asked whether I am disappointed because, 33 years after I stepped on to the Moon, humans have not gone on to explore Mars, nor even returned to the Moon. That would be like saying I am disappointed with life, and that certainly is not true. At 72, as one of six members of President George W. Bush's Commission on the Future of the Aerospace Industry I am finding plenty of stimulation looking 100 years ahead, as well as making proposals for the short-term exploration of space.

It is true, however, that I would like to see more general understanding and endorsement of our exciting concepts for the future. For their own good reasons, Neil Armstrong and Mike Collins, with whom I went to the Moon in July 1969, are not involved in these concepts. But it would be good to hear more vocal support from some of the other space pioneers who won glory in the early days.

The story of how we got to the Moon has been written many times, but this one is a very different and personal story. As we progressed through the three stages of our journey to the Moon, through Projects Mercury and Gemini to Apollo, all of us in NASA were well aware of the enthusiastic reporting coming out of the BBC; and one thing I can say about Reg Turnill is that he has been a persistent supporter from the very early hours and days of the space program right up to today, and all of us appreciate that effort.

This story starts with a European look at the early days of Dr Wernher von Braun. My early days were spent in a different world. As the son of an aviation pioneer – my father was Edwin Eugene Aldrin, an aviation pioneer in his own right after studying under America's great rocket inventor, Robert Goddard – I was steeped in aeronautics and astronautics from my childhood. I graduated from West Point in 1951, and was then caught up in the Korean War, during which I flew, and thankfully survived, 66 combat missions. When I was selected by NASA as one of the early astronauts in 1963, I knew Dr Wernher von Braun as a brilliant man with great vision, and it is fascinating to see his plan for manned bases on the Moon and Mars, of which he gave a copy to Reg Turnill on the day I stepped on the Moon, properly published at last.

If it had been carried out, von Braun's plan would have got humans to Mars in the 1980s. But the fact is that, stimulated by the politics of the US-Soviet race to be first on the Moon, Neil and I stepped on its surface well before mankind at large was ready for such planetary adventures.

But now we are ready. During those 33 years since Apollo 11, American, Russian and European astronauts have learned to live and survive in Earth orbit – and the performance of the astronauts and cosmonauts now constructing the International Space Station (ISS) has been almost flawless.

The Presidential Commission on Aerospace, which will report about the time this book is published, is looking at two short-term space interests: privatising the Space Shuttle and the ISS, and bringing together NASA and the US Department of Defense on the definition of evolutionary, reusable launch vehicles. NASA would like to see definitions of both reusable boosters and a successor to the present Space Shuttle by about 2009–11.

A feature that I would like to see included is a multi-purpose crew container – an ejectable pod, which could be used for aborts any time during launch into orbit until the vehicle returns to land on the runway. It would serve as a space taxi between low Earth and lunar orbits and even be adapted as a wheeled lunar lander. I see a potential for making use of things we have already developed, and modifying them so that we get reusability and reliability. We need prudent objectives, with NASA working closely with Defense, so that we can economise on needs, and advance from some of the scars of revolutionary approaches like the X33.

The crew pod I propose would initially carry 9 or 10 astronauts – the sort of number that NASA would like to have in the next-generation space Shuttle. But I would like to see versions of it adapted to carry passengers – you might get as many as 30 people in a pod. I would like to move rapidly towards getting private citizens into space, first in the present Space Shuttle and then in its successor. With dual launches of the pods, we might get 60–80 people per launch going into orbit. The Russians started the concept of adventure travel for a high price, to help solve some of the economic problems of their space program. Some of us would like to see progression towards some kind of national or international lottery selection. 'Who Wants to be a Millionaire' might even become 'Who Wants a Journey into Space?'

Those are short-term possibilities to involve and enthuse ordinary people and demonstrate that, before long, space travel need not be confined to highly specialised astronauts and scientists. But at the same time we need to be developing the ability to protect Earth from potential asteroid impacts and work towards what I call Survival of Species Settlements off the Planet Earth.

These would not necessarily be on the Moon or Mars. The Moon is not too attractive for such settlements, and space stations by themselves require too much support. The latest thinking is to build such settlements on Earth-crossing or near-Earth asteroids, which would have the necessary materials. We would have to nudge them nearer Earth so that they could be easily visited and in the process commercially exploited, and at the same time demonstrate the ability to divert asteroids threatening to impact Earth.

To get there we need to select some strategies and integrate them as we move through this century, with targets in 2010, 2020, 2030 and so on, so that we can measure our progress and set objectives. We do of course need to go back to the Moon, establish a base there, and go on to

Mars – landing first on the Martian moons, and going from there to the surface. Some people think we would use the Moon as a jumping-off point to Mars, but I do not mean that at all. Our expeditions would depart from low Earth orbit to high Earth orbit, with some fuel being added, perhaps from the Moon or from LO_1 [the Libration Point 36 000 miles on the Earthside of the Moon]. These taxi spacecraft would then intercept cycling spaceships in perpetual orbits between Earth and Mars.

I have been much encouraged with the progress some US universities are making in improving and shortening 'cycling orbit concepts' for going to Mars, and some encouraging reports are likely to emerge from them. We need to start with a heavy launcher, able to put large wet and dry volumes into 28.5° orbits, rather than the International Space Station's 56° orbit established to meet Russian requirements. The 28° orbit can be reached from Florida and Kourou – and we need to develop a third launch site in Brazil. I have made a personal visit there to get a sense of the possibilities of Alcantra, which is close to São Louis, in the mouth of the Amazon and about 2.5° south of the Equator. In that orbit we can experiment with artificial gravity and large volume support, followed by adventure travel or tourist hotels.

Of course I get asked for costs and time frames. I wish I could do that. But as I see it my task is to try to point the way, and leave others to figure out ways of making it all economically attractive. The day may come when the diversion of just one asteroid on a collision course with Earth will make it all look very economical indeed!

Buzz Aldrin

[Dr Aldrin has told his own story of Project Apollo in two autobiographies, *Return to Earth* (Random House, 1973) and *Men from Earth* (Bantam Press, 1989).]

Growing Up with Space – Sources and Acknowledgments

There was a rare moment of job satisfaction for me in the late 1970s during a visit to a Government space centre in southern England (*not* the one at Cheltenham) involved in military communications, when the Director shook my hand and said: 'I'm most interested to meet the author of the *Observer's Spaceflight Directory*. There are things in there I am not allowed to talk about.' I did not ask, nor did he tell me, what they were.

While NASA used to boast, with some justification, about the openness of both their manned and unmanned space activities, at the Pentagon the Department of Defense was usually spending at least as much, sometimes twice as much, on military space as NASA was on 'civil' activities. The Soviet–US space race, which started in earnest in the 1950s, was a major part of the cold war, and the military on both sides were paranoid about concealing from each other what they were doing, what they were actually achieving, and how much – or, very often, how little – they knew about the opposition.

For a dedicated space correspondent this was a source of amusement rather than frustration. Exploring what is going on in space must be much less fun nowadays when, if you want to find out what the Russians are doing, you just ask them! Baiting the military in those days was all too easy. Their problem was that, if they denied a space correspondent's assessment of what was happening, they might disclose how much they knew. The most they could do was to condemn it as 'idle speculation'!

All the same, it was a matter of professional honour to get things right, and as the Soviets moved through their series of Vostok and Voskhod manned flights, surrounded by dozens of unmanned flights in very varied orbits, the purposes of which were camouflaged under the cover-all label 'Cosmos' and attributed to scientific research, I stumbled upon a way to do it.

During my visits to Washington I came across Dr Charles Sheldon, tall, white-haired and erudite, a man of infinite courtesy. He headed the US Library of Congress, and his principal task soon became the requirement to produce written briefs explaining the meaning of the space race to senators and congressmen. His problem was that he was prohibited from including any 'classified' information in his reports, and he fought a constant war with the Pentagon. When, just as we tried to do, he analysed the likely purpose of Soviet Cosmos launches, they accused him of releasing classified information. He had to prove to them that the information was 'in the public domain'.

In other words, if the BBC or any other knowledgeable space correspondent produced a likely explanation or analysis, that analysis was in the public domain and Dr Sheldon could discuss it in his reports to Congress. And in those days British aerospace correspondents were at least as well informed as their American counterparts, and far less inhibited in what they published.

There was an international requirement even then that a nation launching a spacecraft must announce it, together with basic details of the orbit and inclination, and the Russian section of the BBC's monitoring service at Caversham soon became adept at picking up these announcements. They would either pass them on to the BBC Newsroom, or ring me direct at home. And I soon realised that Soviet scientists had higher standards than their Government, and almost never lied. Thus one could trust the facts they had released, consider the possible uses of that particular orbit, and think about what they had *not* said. Many a time, rung up in the small hours, I got a sub-editor to read to me exactly what the Russians had announced, and then recorded a brief news report interpreting what it meant, without waking my wife by turning on the bedside light.

This became possible because a small group of specialists began to gather and collate these announcements. A key figure in Britain was Dr Desmond King-Hele, a principal scientific officer in the Space

Dr Desmond King-Hele, one of the last Farnborough 'boffins', observing satellites in 1967 in his Hampshire garden. His deckchair observations of the first 200 man-made satellites advanced our knowledge of Earth's shape and atmosphere. (Royal Aircraft Establishment)

Department at what was then the Royal Aircraft Establishment at Farnborough. His 1966 book *Observing Earth Satellites* still makes astonishing reading. By then there were already a thousand artificial satellites in orbit, and he spent his nights in a deck chair observing about 200 of them through a pair of binoculars.

King-Hele was not alone, for satellite tracking attracted enthusiasts in the same way that aircraft spotting had done. King-Hele compared his observations with those of amateur Russian satellite trackers through Dr Alla Massevitch of the Astronomical Council of the USSR Academy of Sciences, and in the US with Dr Fred Whipple, Director of the Smithsonian's Astrophysical Observatory. If Soviet intelligence chiefs were aware of these exchanges they no doubt regarded them as those of harmless amateurs. Their own main interest was in reaching the Moon, which these amateurs regarded, in King-Hele's words, as 'a villain; an unwanted light that can't be turned off'.

In fact the observations by these early satellite trackers of the way in which natural forces changed the orbits of man-made satellites led to revolutionary advances in our knowledge of the Earth's shape and upper atmosphere; and in 1966 King-Hele became the first Fellow of the Royal Society to be elected for contributions to space science.

Long before that, I became aware of King-Hele's activities when I awoke to the value of his Satellite Tables, issued monthly by the RAE. The information in those, like Caversham's monitoring reports, was somewhat esoteric for the layman; but for us it brought valuable facts instantly 'into the public domain'. Whether the news was of US or Soviet launches, I would ring up Charles Sheldon, and tell him what I had learned. Sometimes, especially in the case of the Russians, it was news to him; often it seemed likely that he already knew of the latest US and Soviet launches, but could not say so because the Pentagon had 'classified' them. Once the facts were published elsewhere he could justify including them in his own reports and even discuss them in a limited way.

I would tell him I was proposing to do a radio or TV news report in a few minutes' time and was thinking of saying the latest 'Cosmos' was a rehearsal for a new manned 'spectacular' – or perhaps that a US launch was a particular spy satellite. 'What makes you think that?' he would ask. 'I suppose', I would then speculate, 'it might be something quite different'. 'That seems more likely,' he might reply. I would then do my broadcast with confidence, and Caversham would sometimes tell me there had been sarcastic references to it on Moscow radio. But it usually turned out to be right, and, queuing up one day to pass through 'Checkpoint Charlie' into East Berlin, I discovered with satisfaction that I was listed by the Russians as 'a troublemaker'.

Later there was another valuable source of informed speculation. Geoffrey Perry, the science master at Kettering Grammar School in Northamptonshire, formed some of his sixth formers into a space observing group, and with very basic instruments they soon became adept at monitoring satellite launches and orbits. They even listened in to the exchanges between the cosmonauts and Moscow's mission control – and were encouraged in this way to learn some Russian.

Perry became famous when his group identified Plesetsk as a new launch site 17 years before the Soviets admitted it and 34 years before I got an official admission from the Pentagon that they had worked it out for themselves just as quickly. The American magazine *Aviation Week and Space Technology* – with whom I also exchanged information as a result of a long-standing friendship with the Editor-in-Chief Robert Hotz – gave Perry a contract and a grant for new equipment.

When Charles Sheldon died, all too early, from spinal cancer, his assistant Marcia S. Smith (whose help during many years since I should also acknowledge) took over the work of producing space reports and analyses for the Congressional Research Service, and got Perry to write large sections of her reports because no one could suggest that his knowledge had been obtained from classified sources. Geoff Perry was another great loss when he died suddenly in 1999.

Nicholas L. Johnson became yet another invaluable source. As an advisory scientist working for the US space contractor Teledyne Brown Engineering, he began producing his own annual *Soviet Year in Space*, and in 1987 I was proud of the fact that I persuaded Jane's to publish his first major book *Soviet Military Strategy in Space*. Nowadays Nick is a senior scientist with NASA at Houston, and acknowledged as probably the world's leading authority on the problems of space debris.

One result of all this was that I never quoted any sources of information in my reports to ensure that no one got into trouble; and in my books I made only general acknowledgments. A problem now is that I cannot identify many of the 'quotes' in my text; but they are all to be found in the hundreds of reports and documents that I have acquired during the last 50 years! Many are listed in the bibliography at the back of this book. Typical of the most useful have been Wernher von Braun's own *History of Rocketry and Space Travel* (1966), NASA's

This New Ocean: A History of Project Mercury (1966), and numerous reports led by Marcia Smith and produced by the Science Policy Research Division of the Congressional Research Service such as 'US Civilian Space Programs 1958–1978' and 'Soviet Space Programs 1967–75'. Some of these even have snippets of information originating from me.

The people who have helped and encouraged me during nearly 50 years of space writing and reporting are a vast blur of friendly and knowledgable faces. Many of course have died. One or two stand out, like Bill O'Donnell of NASA and USIS, and Jesco von Puttkamer, also of NASA, who would talk non-stop for four hours about 'putting the dweam back into space'. The successive 'Voices of the Cape' have always been major sources of information, from the original Col. 'Shorty' Powers of Mercury days, through Jack King of Apollo days, down to Hugh Harris of Space Shuttle days, to whom I am especially indebted for help and friendship in recent years.

I ought to begin the formal acknowledgments with the late Tom Maltby, who became Head of the BBC Correspondents in the 1960s, because he was the man who turned an unwilling Industrial Correspondent into the Aerospace Correspondent. For 20 years after that I should thank the BBC for overcoming its financial reluctance when pressurised into allowing me to commute between London, Washington, Cape Canaveral, Houston and many other space centres.

Once there I benefited from the willingness of NASA's launch and flight directors and countless other officials who presented themselves to the space correspondents for questioning after every event and incident, no matter how tired they were.

The NASA News Centers provided the necessary bases for covering the missions. There was and always has been a small core of knowledgable 'public affairs' officers in these places upon whom one could rely for more accurate guidance than was contained in the official handouts! It was the contractors' press officers for companies like Rockwell, Boeing, MacDonnell Douglas, Grumman, Lockheed,

Martin Marietta, Morton Thiokoll and numerous others over the years to whom one turned for technical help when crises occurred.

In a competitive world a wise reporter always works closely with one or two colleagues as an insurance against missing a breaking story. In my case it was Ronnie Bedford, Science Editor of the *Daily Mirror* (in those days highly esteemed for its factual accuracy), and Angus Macpherson, aerospace correspondent first of the *News Chronicle* and then of the *Daily Mail*. During crises like that of Apollo 13, they would generously whisper the latest developments into my ear while I was doing live broadcasts.

There are countless others whose generosity and helpfulness enabled me, to my eternal surprise, to survive as BBC Aerospace Correspondent until compulsory retirement from the staff on my 60th birthday – since when they have continued to help me as a self-employed correspondent. I feel guilty about not naming them all, but the blotchy memory of an octogenarian makes that an impossibility. But there was, and thankfully still is, another colleague who was always more skilled at extracting the bits of information that give one's reports colour and depth, and that is Margaret Turnill. She just happens to be my wife of 60+ years!

Illustration acknowledgments

Most of the illustrations in this book are culled from NASA's voluminous files, which they have always generously made available to all the media. The others are gratefully acknowledged in individual captions where appropriate. The photograph on p. vi is by Margaret Turnill.

1 The Context: A Twentieth [illegible]

1 The Context: A Twentieth-century Faust

Five days after the fall of the Berlin Wall, I was talking to young East German space scientists who had never heard of Peenemunde. There was vague recognition of the name 'Dr Wernher von Braun', but they knew nothing about him.

Forty-five years of silence and censorship east of the Berlin Wall meant that by October 1990 knowledge of one of the twentieth-century's most remarkable men had been almost obliterated. For von Braun not only designed and built the rockets that landed the first men on the Moon; the drive and determination with which he led the missions changed man's perception of himself and added to his language in a way not experienced since Shakespeare wrote his plays and sonnets. [Women had tried but failed to penetrate either the ranks of the astronauts or the top echelons of NASA in the moonlanding years of the 1960s and 1970s.]

My own encounters with von Braun started in the late 1950s. The impact in 1944 in Sydenham, south-east London, of one of his first V2 rockets had hastened the arrival of my younger son, and for two years I could not bring myself to shake his hand. After that I was surprised to find quite a warm professional friendship developing between us.

'He was like Faust!' My own impression, formed years earlier, that this German rocket engineer had mentally sold his soul to the Devil in exchange for the facilities and the money to fulfil his ambition to build a rocket capable of sending men to the Moon, was finally confirmed in a Dresden cafe during the 1990 International Astronautical Congress.

I had tracked down Hans Endert, by then aged 71, who had been one of von Braun's engineers at Peenemunde during the final developments of the V2s in 1943–45, and asked him to describe the man for whom he worked. I had not expected that my own view of von Braun would be confirmed with the use of the same simile – especially as Endert had had no opportunity to observe the man's extraordinary development after arriving in the United States.

At first von Braun's execrable English made it impossible for the BBC to broadcast interviews with him; and his mouth full of metal teeth was quite revolting for the viewer. One suspected the Faustian pact as much as US technology for his transformation, within a few years, into a gleamingly fit and handsome middle-aged man with near-perfect English flowing from between a set of pearl-white teeth.

Endert, both a victim and beneficiary of von Braun's quest for the Moon, had just been retired from his job as Director of East Germany's space agency. West Germany had taken it over. And for the first time in 45 years he was free to sit and talk about his experiences. He filled in the gaps in my earlier knowledge of the development of the V2 rockets.

The 'V' did not stand for 'Victory' as was generally believed by the recipients of the one-ton warheads which the rockets delivered in England. The V stood for 'Vergeltungswaffen' or reprisal weapons. Most major countries, including Britain, France, Germany, Japan and the US, were already studying rockets – amid much secrecy – as potential weapons, before the 1939–46 war.

The story of Germany's rocket development and test range at Peenemunde, an island on the Baltic Coast, and of the way that Western intelligence, helped by Sweden's recovery of a V2 that veered off course and fell in their territory, gradually learned what was happening, is a matter of history.

For Hans Endert the story started with a raid by the RAF on Peenemunde on 17 August 1943. Of 571 Lancaster and Halifax bombers, 40 were shot down as they returned to England in bright moonlight, at a cost of more than 300 RAF lives. Most bombs seem to have missed the main target – the adjacent installations at which both the V1 flying bombs and the V2 rocket bombs were being built and tested. But the bombs did hit the Germans' living quarters, killing 735, including Dr Walter Thiel, responsible for developing the V2 rocket engine. Nearby Zinnowitz however, with a comfortable hotel at which von Braun and his top associates spent much time – and which still survives – escaped serious damage.

Following that raid, the ranks of the German Army were hurriedly combed for engineers capable of replacing the dead men; and Endert, busily employed maintaining radio communications on the Russian front, was pulled back and rushed to Peenemunde. 'A great stroke of luck', he said many times. 'It saved my life.'

Although production of the V-weapons was moved to the Harz Mountains following the raid, development flights continued at Peenemunde until the Soviet and Allied forces converged upon it in 1945. Von Braun, whose loyalty and motives were doubted – perhaps with some reason – by the Gestapo and SS, was at one time arrested and imprisoned, but released within a few days when his friends, who included General Dornberger, protested to Hitler that without von Braun there would be no V-weapons.

Von Braun's daring and determination were equal to the final test when it seemed that Soviet forces would reach Peenemunde first. Taking advantage of conflicting orders, from Berlin to evacuate, and from local commanders to stand firm and fight to the last, he bluffed his way 250 miles south-west to Bleicherode past military roadblocks and Gestapo checkpoints. It was an epic journey with a trainload of documents, drawings and papers to enable him, somewhere at some time, to continue his exploration of rocket technology. With him went about 5000 engineers and their families. His own *History of Rocketry* records how the huge convoy of railroad cars, trucks and other vehicles was emblazoned with red and white signs reading 'Vorhaben zur besonderen Verwendung', or 'Project for Special Disposition' – an entirely mythical project.

While he hid his documents in mountain caves, from which American intelligence later recovered this treasure trove, von Braun sent his brother Magnus, who spoke better English, to make contact with the Americans. Magnus told a somewhat astonished Private (First Class) Fred P. Schneiker of the 44th Infantry Division that 150 top

Dr Wernher von Braun, then aged 33, suffering from an arm broken in a car accident, centre, 'surrenders' with his rocket team to the Americans at Reutte in May 1945. It was more a negotiation than a surrender. The most confident man I ever met, von Braun was well aware of the value of what he had to offer the US. His brother Magnus von Braun, who had made the first contacts, is on the right. From left: Charles L. Stewart, US counter intelligence agent; Lt-Col. Herbert Axter, on Gen. Dornberger's staff; and extreme right: Hans Lindenberg. (NASA)

German rocket personnel wished to join the Americans to continue their rocketry work. It was a tricky situation, for by then they were hiding in an inn behind the German lines from an SS general who wanted to use them as bargaining hostages.

Again, the story of how it all worked out, as von Braun and General Dornberger planned it, is a matter of history. The only casualty of his epic 'long march' seems to have been von Braun himself: a photograph of his group finally surrendering to the Americans at Reutte in May 1945 shows him wearing a rather smug smile and a huge plaster caste on his left arm. For years I thought this was just camouflage and yet another detail of his great escape plan; but later I was told that it was the result of a motor accident, and that the injury was so bad that amputation was discussed. When I knew him, however, he did not appear to be handicapped in any way.

Less important members of his entourage, such as Hans Endert, fell into the hands of the Russians, and it was three months before they learned that von Braun had gone over to the US. 'Then, knowing that the Americans knew everything,' Endert told me, 'I had no scruples about helping the Russians, because they offered me a decent salary and food rations which I could get nowhere else.' He was well treated and sent off to an establishment near Leningrad.

'Did you like that?' I asked – provoking him to throw his hands in the air and reply: 'Only an Englishman could ask such a question!'

But, as I suspected, there were compensations for him. Just before that he had fallen in love with a German girl who was bilingual in Russian, so that she went with him and acted as interpreter between the German and Soviet scientists. Three years later they were given one week's notice that with 175 others they were being sent back to East Berlin. There they were given priority housing, and for the rest of his career Endert was employed by the East German Space Agency on their numerous co-operative Soviet space projects.

When I met him he had just been back to Peenemunde, and was astonished to find little but bog and trees growing where Test Stand No.7 had stood – a large hall 50 × 40 metres, and 40 metres high. Having first removed all the equipment they could, the Soviets had blown it up with great thoroughness. Endert and his colleagues were planning to use a bulldozer to regain access to the site, uncover what remained, and establish a museum telling the story of Peenemunde and its contribution to space history. It did not seem to be the appropriate time to express my hopes that such a museum would include the names of the 300 RAF personnel who had lost their lives in a raid that foiled Germany's last desperate bid to reverse the course of the war in their favour.

Von Braun himself entered my personal life during that unhappy period for America in 1961 when the Soviets had launched Gagarin and

Titov into orbit, and NASA was struggling to match those achievements against national indignation about being second in the space race. The US had in fact been concentrating on production of intermediate and long-range military missiles rather than on bigger and more powerful rockets capable of lifting men into low Earth orbit.

And so far as Earth-orbiting satellites and men-in-space were concerned, the effort had also been handicapped by the bitter rivalry between the three Services to claim these glamorous activities. The US Navy, with the most powerful political lobby, won Government and Congress support for their Vanguard rocket – which, however, kept blowing up during tests. The Air Force, which thought and still thinks that space belongs to them, maintained that their early Atlas rockets would be best; while the US Army, politically the poor relation of the three Services, and despite its corps of German rocket engineers, had its claims repeatedly turned down that it could do the job with the rockets developed from the V2.

Watching from Cape Canaveral the string of failures which accompanied these political struggles, the growing corps of international space correspondents, led by the British media, learned how von Braun's team, still conducting their meetings in German, refused to give up. He finally won the inter-Service battle for the US Army on 31 January 1958 with the successful launch of Explorer 1. At last the US had a satellite in orbit – and only four months after Russia's historic Sputnik 1. It was done with a rocket called Juno 1, a four-stage development of Jupiter C, which itself had been developed from Redstone, which in turn was directly evolved by von Braun from his wartime V2.

From that day the US Army's von Braun team was unassailable. Redstone was used to launch Alan Shepard and Gus Grissom on the first two US 'spaceflights', which were actually suborbital up-and-down space lobs, in the tiny Mercury capsules. The US Air Force's Atlas rockets had to be used – with misgivings in view of their early history of unreliability – for the subsequent four Mercury orbital flights, and the Titan, also based on an Air Force missile, for the 10 two-man Gemini flights. But it was von Braun who was building the Moon rockets.

With unlimited money and manpower – just as in his Peenemunde days – he had the time to go on developing Juno and Jupiter into a rocket big enough to fulfil President Kennedy's 1961 programme aimed at landing men on the Moon and returning them safely to Earth before the end of that decade.

'Saturn' was selected as the name for the family of rockets needed for the moonlandings because that huge planet came next after Jupiter. Space correspondents like myself were inevitably sceptical about the proposals for the 3000 ton Saturn 5 when we were still watching and reporting Atlas and Centaur failures, with robot Ranger spacecraft regularly crashing on the Moon without sending back the close-up pictures that were essential before men could be sent.

But von Braun, having moved rapidly from prisoner-of-war to US citizen and Director of the Marshall Space Flight Center – as the US Army Ballistic Missile Agency at Huntsville, Alabama, was soon more respectably named – quickly proved himself to be a superb politician as well as leader and inspirer of the rocket design team. His private demeanour and enthusiasm convinced both President Kennedy and Vice President Lyndon Johnson, while publicly he was unerringly effective, briefing the media at news conferences and doing radio and TV interviews.

Once I got past the Public Affairs 'minders', von Braun always performed for me, well aware that it gave him access to the BBC's world-wide audience; and when his compatriot, Dr Kurt Debus, became Director of the Kennedy Space Center at Cape Canaveral, NASA's manned spaceflight programme was effectively being run by Germany's rocket refugees.

Two days before the lift-off of Apollo 7, the first manned flight on a Saturn rocket in October 1967, von Braun was shaking his head dubiously, and capturing headlines around the world by pointing to the recent unmanned Soviet flight around the Moon, and telling us that 'at best' the actual moonlanding would be a 'photo finish' between the two Super Powers racing to be first. In an interview with me he stressed the immense prestige value of winning the race. 'After all', he said, 'who remembers the second man to fly the Atlantic Ocean?'

He was wrong there of course, since Colonel Lindbergh's flight came seconds after that of Britain's Alcock and Brown, and it is Lindbergh's solo flight that is remembered. But such esoteric details did not worry the American public; what did worry them was the threat that they might lose the moonrace. And, thus aroused, those worries ensured that the prospect of wounding financial cuts to the Apollo programme died away.

The Soviets on their side provided von Braun with a steady flow of ammunition for his repeated warnings that the slightest US hesitation could cost them the race. Cosmonaut Shatalov's boastful assertion at one critical point that when the first American stepped on the Moon Soviet cosmonauts would be there to greet him added hundreds of millions of dollars to NASA's budget!

Inevitably, as this story tells, once the Apollo 11 crew had achieved their landing in July 1969, the remainder of von Braun's life was all anti-climax. President Nixon started cutting back the programme as soon as he and his Administration had extracted the maximum benefit from it.

'We've built a railway line to the Moon, and we're only going to run one train along it,' mourned von Braun in another interview with me that day. In fact, seven were run, six of them successfully. The one Apollo failure had nothing to do with the Saturn launcher, which ended its career uniquely with a faultless performance. The Skylab space station missions in 1973, and the Apollo–Soyuz manned link-up of 1975, provided some use for left-over Saturn hardware, and then its life was over.

von Braun's master plan for landing men on Mars by 1982, of which he gave me a copy the day Apollo 11 landed, remained unread in Washington. The White House was worried about ending the Vietnam War and the media was obsessed with Senator Edward Kennedy's car crash off Chappaquiddick Bridge in which his woman passenger drowned.

Apart from that, von Braun never knew failure, nor the bitterness of the recriminations evoked by re-awakened consciences during the last two decades of the twentieth century. Fifteen of his Saturns placed

45 Americans in space without the loss of a single life. Three giant Saturn 5s, originally intended to carry Apollos 18, 19 and 20 to more remote parts of the Moon (and which would have solved the problem of whether it contains any frozen water!), remained unused to inspire awe in thousands every day as they are displayed like dinosaurs at the Kennedy, Johnson and Marshall Space Centers.

von Braun's last five years were spent as technical and development vice-president of Fairchild Industries, and he finally met his destiny in that role in 1977, aged only 65. He had achieved what he set out to do, but I was saddened to see him wasting away as cancer consumed him. In what was probably his last interview he told me, shortly before he died: 'I just envy the youngsters who have a chance of going on where we leave off.'

I used that quote to dedicate a book I wrote for young people called *Space Age* to the man whose hand I had for so long refused to shake. Since then I have had many second and third thoughts, and wondered whether, after all, I should have shaken it.

The doubts were re-awakened when in October 1993 my wife and I drove to Nordhausen, just inside what used to be East Germany, and visited Dora-Mittelbau, the mass production factory in tunnels beneath the mountains in which Dr Arthur Rudolph was directing the mass production of V2s when the war ended. Had the war gone on for another six months London would have been rendered uninhabitable.

Out of 60000 European slaveworkers brought there from the Buchenwald concentration camp and elsewhere, 20000 were worked and starved to death in the catacombs. The production and reliability of the V2s was hindered by the determined sabotage of those workers, and there were frequent hangings, sometimes outside Rudolph's office, when the culprits were caught.

We saw where the corpses of hundreds were incinerated each week. We shivered in the rain where the workers had been paraded for hours, often naked, before and after their 12-hour shifts in the tunnels.

The relevance here of all this is that von Braun, while working at Peenemunde, was a frequent visitor to Dora after production was transferred there, and could not have been ignorant of the sufferings

Arthur Rudolph, when manager of the Saturn 5 programme office at Huntsville under Wernher von Braun. Not until after his retirement did questions begin to be asked about his past at Dora. (NASA)

imposed upon those whose work did not in the end win Hitler's war, but which did make space travel possible soon after it. And when Germany fell, Arthur Rudolph was included in the team von Braun took to America.

Having worked with von Braun in Germany and America for 38 years, Rudolph was awarded NASA's Distinguished Service Medal in 1969 – but was forced to abandon his American citizenship and flee back to Germany when at last the media began to investigate what happened at Dora. By then von Braun himself was dead. The twentieth-century Faust himself had escaped such final retribution. Having died before the collapse of the Soviet Union, he was buried with his reputation, and even bigger and better American honours, untarnished.

Was this a monstrous injustice? Or should we console ourselves that, had von Braun not been allowed to continue his work, there might

German engineer Dr Hans Endert, who said Wernher von Braun, under whom he worked at Peenemunde, was 'like Faust'. He is pictured standing on the overgrown site of the test facility, which was blown up by the Russians in 1945. (Hans Endert)

still be no human footprints on the Moon? And how far, if at all, should we blame von Braun for the awful happenings in the Harz Mountains?

Albert Speer, Hitler's Minister of Armaments and War Production, whose idea it had been to transfer V2 production from Peenemunde to the bomb-proof tunnels, explained how it all came about after his arrest as a war criminal. He was one of the few Nazis to admit publicly to a guilty conscience:

> Basically, I exploited the phenomenon of the technician's often blind devotion to his task. Because of what seems to be the moral neutrality of technology, these people were without any scruples about their activities.

28. ...Encore des morts...

Part of the price for landing men on the Moon: Life and death at Dora, the mass production factory for V2 rockets in the Harz Mountains, 1943–45. Of 60 000 slave workers, 20 000 were worked or starved to death. Shortly after his release, an artist survivor, Maurice de la Pintiere, depicted its horror in 35 sketches, ending with the camp's busy crematorium. (Maurice de la Pintiere)

13. ...Où chaque tronc d'arbre leur semblait une trop lourde croix

24. ...Tandis que résonnaient les plaisanteries macabres des gardes-chiourmes

2 Preparing for Manned Spaceflight

It was late September 1960, flying over the Arctic Circle, when I read what I still recall as the most exciting book I ever encountered – even more exciting than Wynwood Reed's *Martyrdom of Man*, which had so much influenced my youth. It was slim and free, entitled *NASA–Industry Program Plans Conference July 28–29 1960*.

It contained the complete texts of papers presented at the two-year-old NASA's first such conference, outlining their 10-year plan for that decade. It described Project Mercury, which aimed to achieve the first orbital flight of an astronaut in 1961. An unmanned flight around the Moon and unmanned reconnaissance flights to Mars were scheduled for 1964, with a manned flight round the Moon for 1965–67, and a manned moonlanding and return to Earth 'beyond 1970'.

It was all set out with such calm and logical confidence that I believed in it instantly. It became clear that everything that was already happening – the launching of scientific satellites by military missiles, the plans for military reconnaissance and meteorological satellites, the talk of probes to send back pictures of the Moon, Mars and Venus, were all leading to one thing – manned spaceflight within my journalistic career.

Although I had been covering more and more space stories, I am afraid I had until then picked up and reflected the slightly amused and patronising attitude of the media towards rocketry. We had after all grown up with inbred mental clichés, associating men on the Moon with the 'mad boffin' stories of Verne and Wells, and rockets with explosions and wild inaccuracies. The successful development of guidance systems for the East–West rocket arsenals had been cloaked with military secrecy; pronouncements by rival service chiefs about the precision with which they could aim at and hit their potential targets thousands of miles away on different continents was suspiciously like rhetoric intended to frighten the enemy. Many of us, myself included, had failed to take very seriously the application of these marvellous new techniques to worthwhile civil space exploration.

Looking down upon the frozen wastes from our cruising height of about 20000 feet between reading chapters of my NASA book, I decided to become part of this story – and space was to be the dominant interest of my journalistic life for the next 40-odd years. NASA – the National Aeronautics and Space Administration – had been created out of the original NACA – National Advisory Committee for Aeronautics – following US alarm at the launching of the first Soviet spacecraft.

Seven days after the creation of NASA, Project Mercury had been born. This was the first phase of America's manned flight programme, aimed at placing men in Earth orbit. Now, after only two more years of research, came this comprehensive plan to develop the technology to send men to the Moon and back.

Behind it all, of course, was Dr Wernher von Braun, the German rocket pioneer, who had made the lives of my wife Margaret and myself miserable in the closing stages of the war with his V2 missiles. But in those days the US was still downplaying the contribution of the wartime German *émigrés* to their space programme, and the NASA booklet included no chapter by von Braun. There was, however, a significant reference to him in the section *Introduction to Launch Vehicle Programs* by Major General Don Ostrander: '. . . the Marshall Space Flight Center under the direction of Dr Wernher von Braun was transferred to NASA on July 1 of this year and is our principal field of activity for the conduct of our launch vehicle programs.' von Braun had at last escaped from the US Army.

The NASA booklet actually detailed five projects in the moon-landing programme; a sixth, Project Gemini, was to be inserted later between Mercury and Apollo. While the ability of astronauts to live and work in space, plus development of the rockets to get them there, was included in Projects Mercury and Apollo, three unmanned projects

(Ranger, Lunar Orbiter and Surveyor) were to be conducted in parallel, to explore and photograph the Moon so thoroughly that astronauts could arrive with maps of the terrain, showing possible landing places with their contours and altitudes in more detail than is available about most Earthly surfaces.

What was a bit baffling as I read about NASA's avowedly civil Project Apollo was that I was travelling in a 40-seater Convair provided by the US Military Air Transport Service for the exclusive use of 17 British air and defence correspondents, all graded as 'honorary major generals' on a three-weeks defence tour of the US. The main message being conveyed to us during this tour turned out to be that US Air Force (USAF) generals took the view that 'the man on the high ground dominates the world', and that the highest ground available was likely to be, quite soon, the Moon.

More specifically the US and British governments had apparently laid on this spectacular occasion to get over to their respective electorates, as well as to the Soviets, their joint determination to produce an airborne ballistic missile called Skybolt. It was a time of unprecedented panic amid Western defence, whose intelligence experts had decided that there was a huge 'missile gap': that the Soviets had hundreds of operational intercontinental ballistic missiles (ICBMs) compared with a miserable 12 in the West. With such a commanding lead it was felt that the Soviets could well be tempted, and maybe were already planning, to overwhelm the US and its allies with a pre-emptive nuclear strike. The US and Britain were still in the final stage of the 'free-falling bomb era' – dependent upon the ability of their long-range B52 and V-bombers to break through the enemy's rocket defences and drop free-falling nuclear bombs.

The US was just introducing 'Hound Dog', a stand-off bomb – really an unmanned aircraft carrying a nuclear bomb – which could be released from a bomber 500 miles from the target to find its own way there by the magic of its own guidance system. Britain was a year away from completing its own stand-off bomb, called Blue Steel, to be launched from the V-bombers. The snag was, according to current intelligence wisdom, that by the time Hound Dog and Blue Steel were operational in sufficient numbers, the Soviets would almost certainly be able to intercept them with homing missiles. But Skybolt, being manufactured by the US at enormous cost, would overcome all these problems, because not only would it have a range of over 1000 miles, but it was a ballistic missile, attacking its target by travelling in a great arc before descending from the edge of outer space, thus making it invulnerable for many years until the invention of an anti-missile missile. If Russia succeeded in inventing that first, we were assured at a particularly grim briefing, she would have decisively broken the nuclear stalemate in her own favour.

The US, with no 'national' Press like that of Britain, had no group of journalists comparable with the British air and defence correspondents. Our tour was to persuade us to write stories and articles sounding the alarm about the missile gap and the need for Skybolt, so that those stories would be picked up and repeated around the world – including of course the American Press.

In the case of the BBC, an international audience was assured, because my broadcasts were routinely picked up and repeated by External Services in my own voice in their English programmes, and in translation in many other languages. Thus I was 'No.1' among a very motley collection of honorary major generals, and never had to fight, like many of the others, for a single bedroom during our night stops. Arriving desperately tired in new time zones – 'jetlag' was yet to be invented – there would be bitter altercations at reception desks at hotels and military bases. One member of the team used to take the receptionist aside and say he must have his own room because he was homosexual, and usually got away with it, since the receptionist could not know that he was perhaps the randiest womaniser amongst us.

I fear this extraordinary collection of journalists caused our hosts problems that they could never have anticipated. Two, including the erudite *Daily Telegraph* man grieving over the recent defection of his wife with their young son, were advanced alcoholics. The real doyen of the group, and the man most feared by his colleagues, was Harry

Chapman Pincher, of the *Daily Express*. We all had our own seats and desks in the VIP Convair, and Pincher sat there lost in thought, seldom talking to his colleagues either on the ground or during the flights – and never asking questions at the briefings. He waited until they were over and sought private interviews, so that no one ever knew what sort of story he was proposing to send. The others talked endlessly among themselves trying to identify what the story was, and then hammered out the details by comparing their interpretations of the briefings.

I was usually an anxious participant in these discussions. Transmitting similar stories had many advantages. If everyone sent different stories, Editors always complained that they would have preferred the story used by an opposition paper; they also had an uneasy feeling that their man's story was not really worthwhile unless other newspapers gave prominence to it as well. Pincher's ability to remain aloof from all this was partly because he was a director of the *Daily Express*, and partly because he had an unerring sense of timing which I could never match. I was frequently much too soon with a story the importance of which was not yet fully apparent to the newsdesks. Pincher had the gift of dropping it into his newspaper, perhaps weeks later, just as the subject was about to break the political surface, so that everyone hailed it as a momentous scoop!

For 17 journalists there were no fewer than six 'escorts and observers', whose interests and attitudes conflicted a good deal when it came to their vain efforts to curb our worst journalistic excesses. Major Bob Spence, a USAF public relations man, thought it was important that the tour should provide an opportunity for all to have a good time and explained that he had laid on plans to that effect. He anticipated that there would be only limited and irregular opportunities for rest periods, but displayed his preparations for overcoming any such problems in the form of a mobile medicine chest. It contained a wide range of pills to deal with all possible stomach disorders, plus an impressive array of wake-up, keep-going, and go-to-sleep pills. There was to be a heavy demand for them.

These preparations were heartily endorsed by Lt-Colonel

Sammy Lohan, deputy head of public relations at Britain's Ministry of Defence, who was several years later to hit the headlines after causing Prime Minister Wilson much irritation about alleged 'leaks' during expense-account lunches with Chapman Pincher. Sammy was a loud and colourful figure in his fifties, with luxuriant flowing white moustache, and his ready flow of 'background information' about Britain's defence policies resulted in his company being eagerly sought by those defence correspondents whom he favoured. They rarely included me, because Sammy knew that I could get the story on the BBC so quickly that the newspapermen would be exposed to abuse from their Editors which would rebound upon him, and he in turn cherished their favours. (After a few demonstrations of my ability to overtake their stories, my fellow defence correspondents learned to co-operate with me, and I would then hold my story back for perhaps 12 hours, so that it went on the air in the BBC's early morning news bulletins at the same time as the newspapers hit the breakfast table. My closest friends and colleagues, Ronald Bedford of the *Daily Mirror* and Angus Macpherson of the *News Chronicle* – later the *Mail* – both took the view that my brief BBC reports were no more than sound bites, whetting the interest of their more leisurely readers.)

Sammy was seeking to avoid such routine and sordid rows on this occasion; and his expectation that the tour would provide opportunities for fleshly pleasures – he confided that his ambition was to add a beautiful black girl to his tally of bedmates – was either shared or treated with amused tolerance by the rest of the party. Larry Moe, from US Information Services, was another warm, larger-than-life character from Texas, with much more rigid standards of probity and morality. He moved around the world's Embassies, and was to cause a sensation over dinner in a crowded restaurant at Cape Canaveral when he ended an argument with a national newspaperman among us by roaring: 'Why, you're nothin' but a Fleet Street whore!'

Plans to allow us to witness the test launch of a Titan, the newest ICBM, were frustrated when the attempt had to be 'aborted' at minus 2 minutes 40 seconds in the countdown – the first of scores of aborted countdowns that were to dog my future years. But this was brilliantly compensated for as our Convair was flying us from the Cape to Washington two days later. Mysterious delays in our take-off were explained when our climb was held at 1500 ft and we were told to look below at the launchpads about 10 miles north of us. Aircraft are rarely allowed so near at launchtime, so we had a unique grandstand view of mushrooming golden flame as the 2 300 000 lb-thrust first-stage ignited, and then watched the vehicle soar up towards and past us on a 6000-miles trajectory to its target area in the Pacific Ocean.

Generals and nuclear scientists had briefed us at the annual three-day Air Force Association Convention in San Francisco, and at that moment we were on our way to more briefings at the Pentagon and State Department. Dr Edward Teller, 'Father of the H-bomb', denounced those who maintained that Russia would learn nothing of military value by her imminent space probes towards Mars and Venus. 'That's all very well', he declared with passion, 'but what I want to know is: what KIND of nothing!'

Having instructed us very firmly on the superior wisdom and practicality of maintaining the nuclear deterrent with the airborne Skybolt rather than the seaborne Polaris missile, the Air Force reluctantly handed us over at Santa Monica to the US Navy for 36 hours aboard the USS *Kearsarge* (35 000 tons) one of 15 USN aircraft carriers then operating in the Pacific. The Air Force, however, effectively irritated a clutch of top admirals by landing us at the wrong place, so that the admirals had to cool their heels while we were bussed to the right place for their cocktail party. Admiral Raborn, Director of Special Naval Projects, also known as 'Mr Polaris', who was bulldozing the 16-missile Polaris submarines into service to ensure that the Navy recovered its nuclear deterrent supremacy from the Air Force's B52 bombers, opened his briefing in the Pentagon with a dramatic photograph of mountainous seas in the North Atlantic, and nothing else visible. It was captioned: 'A Polaris submarine on patrol'.

For Britain at that time, it was very much a case of Skybolt *or* Polaris, and when we asked the Admiral for his view of Polaris's suitability for Britain he replied: 'It's sort of like Marilyn Monroe: it can do

The Russian dog Laika was the first living creature to be launched into space on 3 November 1957, preceding America's chimpanzee Ham by just over three years. Unlike Ham, Laika was not recovered. She died after seven days in orbit when her oxygen ran out. (Novosti)

something for everybody.' When the US Air Force recovered possession of us they riposted: 'Polaris is a good system. What a pity the Navy has oversold it!'

Between the briefings I was able to send back a flow of broadcasts about the intensifying Soviet–American space race. General Donald Flickinger, who was making an international name for himself on space stragegy ('Have you read Flickinger?' the stuffier defence correspondents used to inquire portentously of those of us they suspected had not), misled us grievously though possibly unintentionally during the Air Force Convention. He assured us that there was strong evidence that one or two Soviet cosmonauts had been killed when recovery procedures failed in a first attempt to launch men into space, and that subsequently two more had been successfully launched and recovered in the sort of space capsule which had earlier been used to carry dogs. It was argued that these missions were intended to provide Khrushchev, about to visit New York, with a 'space spectacular' to announce during his visit. (Subsequently his famous loss of temper at the United Nations, accompanied by banging his shoe on the table, was attributed to irritation at his inability to make any such announcement.)

The Mercury Seven, the first astronaut team, dressed for space and looking incredibly old-fashioned. Their selection was announced on 9 April 1959, six months after NASA was formally established. Left to right, front: Walter Schirra, Donald Slayton, John Glenn and Scott Carpenter. Back row: Alan Shepard, Virgil 'Gus' Grissom, and Gordon Cooper. (NASA)

A less spectacular but more accurate story was provided within a few days when what should have been NASA's Pioneer Six, intended to orbit the Moon, blew up just after launch.

While we were being so comprehensively educated on the military importance of space, and the right of the Services to share in these glamorous activities, we recalled that NASA had been set up on the basis – as the US had proposed to the United Nations in January 1957 – that it would pursue the use of outer space for peaceful purposes only. NASA's leaders believed from the organisation's inception that they must assert and maintain their independence from the military – culminating in the Apollo moonlanding plaque that 'we came in peace for all mankind'. But from those early Mercury days it has always been a position that was extremely difficult to sustain, both in theory and practice.

There were two glaringly obvious reasons. First, all the seven Mercury astronauts announced in April 1959, and most of those selected later for Gemini and Apollo, were of necessity military test pilots; since they were only seconded from their various Services, that was where their first allegiance lay.

Secondly and even worse, the rockets and launchpads used for the Mercury and Gemini missions were supplied and operated by the US Air Force. Even when NASA had built its own manned spaceflight centre (later to become the Kennedy Space Center) to launch their own Saturn rockets from launchpads on Merritt Island just behind the Cape's military pads, the Administration remained, and still remains, dependent upon the Air Force for the use and operation of the Eastern Test Range facilities. Every launch involves thousands of military personnel plus patrolling recovery ships and aircraft. And finally the Destruct Officer was and is a military man, sitting at a console with a finger hovering over the red destruct button, with the unchallengable right to blow up and destroy any launch vehicle, manned or unmanned, if it veers off course and in his judgment poses a threat to any populated area – not only Florida but touchy neighbours like Cuba.

3 Gagarin Puts Russia Ahead

With my newly acquired background knowledge I was able during 1960 and early 1961 to wedge what I hoped were well informed pieces on the development of the Soviet–American space race between the more pressing day-to-day coverage of British progress with vertical take-off aircraft, the development of radar, and the 'scandal' of the Queen's aircraft being involved in a near-miss with a military plane over Europe.

US manned spaceflight moved nearer with the launch and recovery in January 1961 of Ham, a lively 37 lb male chimpanzee, on a 'space lob'. The fact that Ham survived many malfunctions was reported to have given the seven Mercury astronauts confidence that by using their sharpened test-pilot wits on such flights their chances of survival would be excellent. However, Wernher von Braun and Kurt Debus, the two Germans mainly responsible for the performance of 'Old Reliable', as their Army-built Redstone rocket (a descendant of the V2 through Jupiter) was known, were far from satisfied that what they called their 'Wee-hickle' could yet be regarded as 'man-rated'.

Ham had to sit strapped in his Mercury capsule for more than four hours while solutions were found to problems as diverse as the gantry elevator getting stuck and internal spacecraft temperatures rising to three times the proper level. Ham's suit, however, remained in the comfortable mid-60s. When he was at last launched, the rocket's thrust was higher than planned, and he had to endure 17*g* – being seventeen times his normal weight – during lift-off. The capsule reached a speed of 5800 mph instead of 4400 mph, and a height of 157 instead of 114 miles. After six minutes of weightlessness Ham splashed down in the Atlantic 16 minutes later, 60 miles from the nearest recovery ship. The peak re-entry *g* was 14.7, 3*g* greater than planned. Launch malfunctions had left an inlet valve open so that, as the capsule wallowed in the sea awaiting a recovery helicopter, it was slowly filling with water. When Ham and his spacecraft were winched up in the nick of time, there was 800 lb of sea water on board.

During all that, however, Ham had performed almost faultlessly. He – and five other potential chimp astronauts – had been taught on the same principles that public schoolboys used to be taught in British schools; errors brought physical punishment. Ham was seated in front of a dashboard with two lights and two levers that required 2 lb of effort to depress, and failure to do so when required resulted in a series of mild electrical shocks to the soles of the feet. The right-hand lever, cued by a white warning light, had to be depressed within 15 seconds, and the left-hand lever, cued by a blue light about every two minutes, had to be depressed within five seconds to avoid punishment. Ham pushed the right lever about 50 times, and received only two shocks for bad timing; with the left lever his performance was perfect. His recorded reaction time was 0.82 of a second, compared with his pre-flight performance of 0.80 of a second.

Back aboard the recovery ship Ham appeared reasonably happy, and readily accepted an apple and half an orange. But an internal NASA memo recorded that some time later, when Ham was shown the spacecraft, 'it was visually apparent that he had no further interest in co-operating with the spaceflight program'. Most of these details I learned much later. At the time I was merely able to report that Ham's mission meant that NASA had accomplished the last stage but one in their efforts to put a man in space.

The Redstone rocket used for Ham and due to launch the first US astronauts on 15-minute up-and-down space lobs was not powerful enough to place a man in orbit. For that job NASA had selected a modified version of the Atlas intercontinental missile. Preparations to place a Mercury capsule carrying a robot astronaut on an Atlas as a rehearsal for the first manned orbital flight were well advanced when the Soviets got there first.

One reason was that Wernher von Braun had insisted on inserting an extra Mercury/Redstone test to ensure that when a man was sent

Major Yuri Gagarin, 27, ready to become the first man in orbit on 12 April 1961. (Novosti)

up the booster did not over-accelerate him as had happened to Ham. This extra test was successfully performed on 25 March 1961 – but the next day the Soviets announced the successful launch and recovery of their fifth 'Korabl Sputnik', containing a dog. Soviet scientists preferred dogs rather than chimpanzees as stand-ins for human beings; and now they were able to announce that on three out of five missions the dogs had been successfully recovered. While some still like to argue that there was no 'space race', NASA's official history of Project Mercury records that following this news, 'at the beginning of April 1961 . . . feverish activity pervaded Hangar S and the service structure' at Cape Canaveral.

Nevertheless this first stage of the space race was lost. On 12 April 1961 Moscow announced that Major Yuri Gagarin had successfully orbited the Earth in a 108-minute flight in a 5-ton Vostok spacecraft. While US intelligence services were disappointed but not surprised, it was a shattering blow to the prestige-conscious American public. Colonel John ('Shorty') Powers, the Cape's public information officer, whom later I came to know well, became instantly famous when he played into the hands of the media by replying, when woken by a newsman at 4am for a comment on the Soviet flight: 'We're all asleep down here.' He became much more inventive later, and that too was to be his undoing.

Gagarin at the height of his fame in 1966, with daughters Lena and Galia. (Novosti)

I abandoned some follow-up stories I was doing about reports that the US Air Force was to close down one of its 15 British bases by sending home 10000 servicemen and their families from an area of Norfolk whose continued prosperity depended upon their presence. That day the BBC's news programmes had no room for anything but the Soviet achievement and the threat it posed to the West. By the end of the day I was pointing out that amid all the rejoicing there was a lot we did not know about Gagarin's flight. We had heard his voice from space – but it was a recording, fed to us from Moscow after his brief single orbit. The BBC's listening station at Tatsfield in Surrey had not succeeded in discovering the frequencies on which Gagarin had been reporting until Vostok was re-entering the Earth's atmosphere. We did not know where he had been launched nor whether he had come down on land or splashed down in the sea, as US astronauts would be doing. In a two-way transatlantic discussion with Robert Hotz, Editor of *Aviation Week* on the Ten O'Clock programme, we speculated that his flight might have been made a few days earlier, with communications being recorded, and released only when Gagarin had been recovered uninjured.

It was six years after Gagarin's flight before the Russians put the Vostok rocket which launched him on public display. This picture was taken when the rocket appeared at the 1967 Paris Air Show. The Soyuz rocket, with its system of clustered engines, used ever since for manned launches, still looks remarkably similar. (Margaret Turnill)

Next day I was interviewing 77-year-old Lord Brabazon, who had qualified for Britain's first pilot's licence 51 years earlier, on what he made of it all, when there were urgent messages to get back to Broadcasting House. Gagarin was to hold a news conference in Moscow two days later, and I must somehow be there. It was a great chance to find out what had really happened. There was no regular BBC Moscow correspondent at that time, because the Khrushchev regime refused to issue a residential visa – though Reuters had been allowed to base a correspondent there, and Paul Fox heading a BBC TV crew had been allowed in for a few days to cover the imminent May Day procession intended to intimidate the West with spectacular displays of long-range nuclear rockets.

For once there were no holdups in obtaining a visa. By the time I got to Moscow there were about 600 of the world's reporters and photographers scrambling to get seats for the Gagarin press conference in the Academy of Sciences. Russian behaviour was at its worst; the officials were determined to humiliate us as part of their demonstration of international superiority. I spent all the hours I had between my arrival and the conference at the Foreign Office pleading for a ticket of admission. No doubt they intended to give the BBC one in the end, but this was not apparent as they insisted that all tickets had been issued, I was not entitled to one, and so on and so on.

Having finally got a ticket, and spent a few exhausted hours in bed, I arrived early at the Academy to ensure that I got in. My ticket was waved aside as if it were of no importance and, as the conference was due to begin, I was still outside. Plenty of other people were being admitted, however. Crocodiles of nurses from local hospitals, Ministry officials, and 'representatives of the workers' of course, elbowed their way past. The hall appeared to be full when somebody somewhere must have given a signal. Just when we had lost hope, and were imagining what it would be like telling our Editors that we had reached Moscow but failed to gain admission to the conference hall, a clamouring group of distraught journalists, all dignity lost, were admitted amid nudges, winks and grins from the officials and troops who had been holding us back. I managed to locate an odd empty seat amid the nurses, and forced my way into it past feet and thighs determined not to give way. There was not the slightest chance of making contact with a Visnews cameraman assigned to provide something for television news; but radio news was still much more important than TV.

My fears that it would all be in Russian and I would not have the slightest notion of what was going on were quickly allayed. Boris Belitsky, whom I was to encounter many times over the years, was on the platform, translating in good English. If newsmen wanted to ask questions, he said, the questions must be written and sent up to the platform. Of course every reporter wanted to be able to write about his personal question. On a sheet torn from my notebook I wrote in capitals: WHEN SHALL WE BE ABLE TO SEE THE ROCKET AND SPACESHIP, and pushed out of the seat again, down to the front, and dropped it in a crude cardboard letter box.

An hour and a half of preliminary speeches followed. Academician Keldysh, President of the Academy, and others talked endlessly, meaninglessly and humourlessly in the Soviet way, using precious minutes on their formal openings alone. At least there was time to study Yuri Gagarin, a 27-year-old Major in the Soviet Air Force. I described him, flanked as he was by his nation's most distinguished academicians and scientists, as slight, fair-haired and incredibly composed and confident. His ready smile, displaying a perfect set of gleaming teeth, was shortly to become familiar around the world when he was sent off to make public appearances to persuade the 'non-aligned nations' that they would be wise to link their futures to the all-powerful Soviet Union rather than to the space-backward West. The long speeches had the advantage for him that he could thumb through the questions – presumably Boris Belitsky had translated them – and take advice on his answers from Keldysh beside him.

'Major Gagarin's press conference was a masterpiece of good-humoured evasion,' I reported. 'When it was over, many of the questions being asked by Western scientists about the details of his achievement remained unanswered.'

The English translation of his answers was frequently drowned in applause and laughter at our expense. For instance, when asked whether he landed in the capsule or had ejected and finished his descent on the end of a parachute, he replied: 'The landing proceeded successfully and my presence here demonstrates the success of the systems.' Then came my question.

The Russian reply to this brought cheers and applause so violent that the substantial nurses on each side of me were bouncing in their seats, frustrating my notetaking attempts, and turning to thrust their jeering faces into mine. I heard little of the translation, but its effect was that the last thing the Soviets intended to do was to reveal the secrets of their super rockets to the wicked Western warmongers. Weightlessness, Gagarin assured us, would not affect man's ability to work in space, adding: 'I want to do a lot more. I want to go to Mars and Venus, and do some real flying.' And he appeared to hint that spacecraft were already being designed for flights to the Moon.

It was a real loss that this attractive personality was not in fact to make any further spaceflights. By the time his masters brought him back from wining and dining around the world, promoted him to Colonel and appointed him Commander of the Soviet Cosmonauts' Detachment, he was reported to be much overweight. He did become backup to Cosmonaut Komarov as Commander of Soyuz 1, and had he replaced Komarov he would also have been the first man to die in space in April 1967. But Gagarin died with another pilot in a jet trainer crash in the following March, which remained suspiciously unexplained for years. The Soviet failure to provide an explanation led to speculation, almost certainly unfair, that Gagarin's taste for the good life had made him less sharp as a pilot.

David J. Shayler finally solved the 30-year mystery in his book *Disasters and Accidents in Manned Spaceflight* (Praxis Publishing Ltd), in 2000. Gagarin was on a final training flight in a two-seater MiG-15 fighter in preparation for his return to spaceflight, and had completed a spin manoeuvre when another MiG-15 fighter shot past him far too close, as a result of flight control failures. Gagarin's aircraft crashed, it is thought, as a result of being caught in the ensuing turbulence.

The irony of Gagarin's death is that if he had lived he might have become a foolish old man; instead he remained for ever a 34-year-old hero of mythological proportions.

4 The Moon and How to Get There

It was only 18 days after Gagarin became the first human in orbit that President Kennedy announced, in May 1961, that the United States proposed to land a man on the Moon and bring him safely home before the end of that decade. He said that they would do it, not because it was easy, but because it was 'hard'!

Too right, thought NASA's top managers! At that time the youthful National Aeronautics and Space Administration had only vague theories as to how such a landing could be accomplished. Despite the confident 10-year programme which had so impressed me and many others, their scientists and technicians had actually achieved only one 15-minute manned space log; and while Project Apollo had been announced 10 months earlier, its stated aim was merely to fly men around the Moon – 'a circumlunar mission' – without landing.

The President's 'deadline' led to some rather desperate planning. Sending men to the Moon was relatively easy; the difficult part was bringing them back again. Two Lockheed engineers proposed that an astronaut should be sent on a one-way trip and left there, with food, oxygen and other supplies being rocketed to him for several years while methods and equipment were devised for bringing him back. This solution was still being advocated in June 1962 by Bell Aerospace engineers, who pointed out that while he was waiting the astronaut could perform valuable scientific work. It would be a hazardous mission, they conceded, but 'it would be cheaper, faster, and perhaps the only way to beat Russia.' NASA's historians say there is no evidence that their administrators ever took such a plan seriously; but they did listen to it, and it is recorded.

NASA had inherited from the US Air Force a general assumption that 'direct ascent' was the way to get to the Moon. As explained

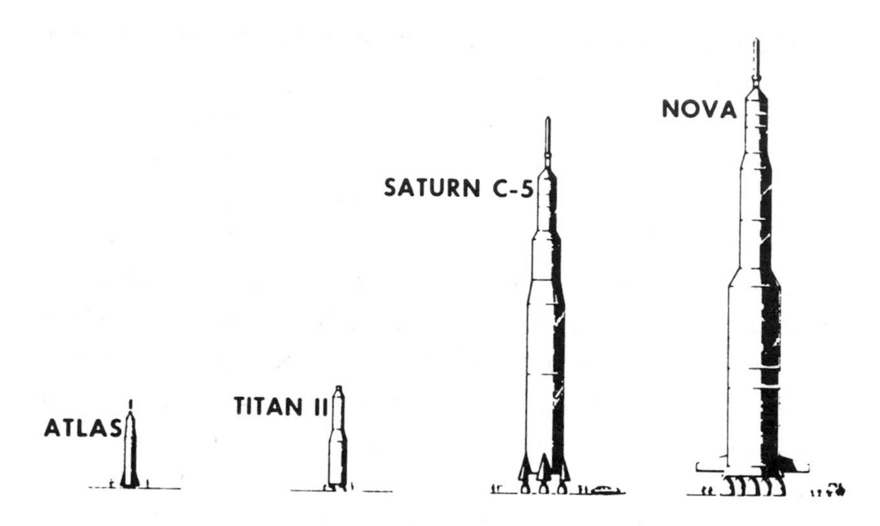

Launchers: Comparative sizes of NASA launchers for manned spaceflight. Nova, right, would have been needed for a direct-flight landing on the Moon. It was decided that using Saturn 5 and the lunar-orbit rendezvous technique, although slightly more risky, would cost $1.5 billion less and enable landings to be made 6–8 months sooner. (NASA)

earlier, the USAF had decided some years before that a manned base on the Moon was desirable for defence reasons, and had been working on a plan for a lunar expedition called Lunex since 1958. They thought they could send three men there and back in a huge three-stage rocket called Nova, providing an initial thrust of 12 million lb – almost twice as big as the projected Saturn 5.

Nova was the largest of a series of rocket designs proposed by Dr Wernher von Braun and his team of German rocket engineers. Von Braun had always supported direct approach as the best way to get men to the Moon. Although rendezvous and docking techniques in Earth or lunar orbit were much discussed, practical tests were a long way off, so no one was sure that they would work. The weakness of direct approach, on the other hand, was that a huge weight – the whole third stage of the rocket – had to be slowed down for the lunar landing, still carrying enough propellant to re-launch part of itself and its crew on the return journey to the Earth.

To lessen the landing weight, lunar surface rendezvous had been proposed. For that an unmanned tanker vehicle would be sent first – but then the problem was that the manned lander must touch down near enough for the astronauts to transfer the tanker's fuel. That in turn required that the final landing would have to be controlled by the onboard astronauts. But how would they be able to see the surface from the pointed top of the rocket, with their sloping windows looking skywards? Mirrors, periscopes, TV and even hanging porches were proposed, and a lot of time was wasted on this concept until it was finally agreed that it would not work. It was also felt that developing a rocket as large as Nova would take far too much time.

Assembling rockets and spacecraft in either Earth orbit or lunar orbit were repeatedly proposed as solutions, because this could be done by multiple launches of one or more of seven alternative variations of von Braun's proposed Saturns. But von Braun himself was still describing any rendezvous proposals as 'premature' at meetings in February 1962.

John Houbolt, assistant chief of the Dynamics Load Division at the Langley Research Center, had been arguing the case for LOR, as lunar orbit rendezvous soon became known, quite passioniately since 1960. He maintained that if a simple spacecraft could be dropped off to land two astronauts on the Moon and then bring them back to the parent craft waiting in lunar orbit, enormous weight savings would be achieved. It would no longer be necessary to take the heavy Apollo craft, with its heatshield and fuel for the return flight to Earth, down to the lunar surface. Lowering all that weight and lifting it off again consumed many tons of propellant which could all be saved.

But the disadvantages – that such a lightweight ferry could place only a small payload on the Moon, and, worst of all, if its lift-off were less than perfect it would miss its rendezvous with the parent craft and doom the astronauts to a slow death – meant that this option was not seriously considered.

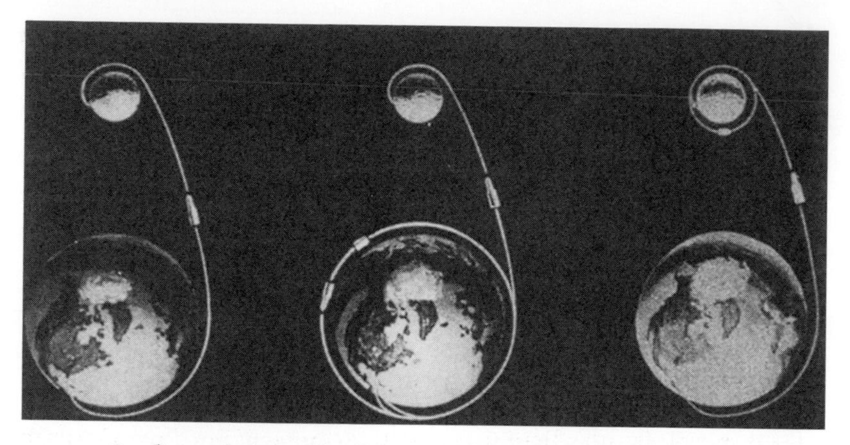

Landing techniques: the three contending methods for the manned lunar landing. From left, direct ascent, Earth-orbit rendezvous, and lunar-orbit rendezvous. (NASA)

By contrast, a missed rendezvous in Earth orbit would merely mean a failed mission, with the astronauts being brought safely home. So Houbolt's arguments that LOR was much simpler than EOR, and that his plan meant taking 7000 lb (3200 kg) instead of 150 000 lb (68 000 kg) down to the lunar surface, were at first discounted.

Slowly, however, the Manned Spacecraft Center at Houston, led by Brainerd Holmes, who was brought in to head the programme after successfully completing the then RCA's Ballistic Missile Early Warning System (during which air and defence correspondents like myself had been immensely impressed by his abilities), were won over to LOR. Its over-riding advantage was that only one Saturn 5 rocket would be needed for a complete moonlanding mission instead of two for EOR, and the savings in time and cost were enormous. It soon became clear that it was the only way in which a moonlanding could be accomplished within the decade.

But the Marshall Space Flight Center at Huntsville stubbornly adhered to its view that EOR was the way to go. Brainerd Holmes decided that von Braun must be won over. A shrewd negotiator, he realised that LOR would mean a substantial loss of work for the rocket centre, so he arranged for his deputy, Joseph Shea, to invite von Braun to

Washington to point out to him that, if EOR were chosen, Houston would be overloaded with work. 'It just seems natural to Brainerd and me that you guys ought to start getting involved in the lunar base and the roving vehicle, and some of the other spacecraft stuff.'

NASA's historians say that Wernher, who was known to have wanted for a long time to get into spacecraft design and not be confined to launch rockets, 'kind of tucked that in the back of his mind and went to Huntsville'.

Two months later came the conversion. At an all-day conference in June, when a final decision was desperately overdue, all the presentations by von Braun's lieutenants still favoured EOR. Their German leader sat listening and making notes for six hours. Then he got up and made a 15-minute speech which shocked his staff but finally settled the issue. 'Our general conclusion', he said, 'is that all four modes [under discussion for reaching the Moon] are technically feasible and could be implemented with enough time and money.' He then listed what he called 'Marshall's preferences': 1) lunar orbit rendezvous; 2) Earth orbit rendezvous, using the refuelling technique; 3) direct flight with a Saturn 5, using a lightweight spacecraft and high energy propellants; and 4) direct flight with a Nova or Saturn C8 rocket.

His staff listened open-mouthed while von Braun said he readily admitted that when first exposed to the LOR proposal they were 'a bit sceptical', but so was the Manned Spacecraft Center at Houston. It had taken quite a while to substantiate the feasibility of the method and finally endorse it. So it could be concluded that the issue of 'invented here' or 'not invented here' did not apply to either of the centers; both had actually embraced a scheme suggested by a third source!

Shea's headquarters staff then costed the four contending modes of approach to the Moon, and reached the satisfying conclusion that LOR would cost almost $1.5 billion less than either EOR or direct flight – $9.5 billion versus $10.6 billion. On 11 July 1962 the media was told at a news conference that the NASA centers were unanimously of the opinion that a moonlanding was to be accomplished by means of a lunar orbit rendezvous. Not for the first time, nor the last, the abrupt change of policy came as a shock to space correspondents like myself. In this case we had been subjected to innumerable briefings stressing the hazards of such an approach. But Brainerd Holmes told the American Rocket Society a few days later: 'Essentially we have now "lifted off" and are on our way.' Events proved that he was right.

5 The Seven Story Begins

Mercury 3

Astronaut Alan Shepard, then 37, and a US Navy Commander, was selected to pilot the first space lob on May 1961, and thus become America's first man in space. He named his spacecraft Freedom 7 because it so happened that his Mercury capsule was the seventh to be built, would be launched by the seventh Redstone rocket, and was intended to be the first of a series of at least seven flights for seven Mercury astronauts. Suggestions that anyone was superstitious were vehemently denied, but with everything meticulously numbered the figures attracted their own significance – just as they did later on Apollo 13.

The right of the astronauts to name their own spacecraft had long roots. Test pilots at Edwards Air Force Base who either did not apply or were not selected for Project Mercury spoke scathingly of the seven Mercury astronauts as 'spam in a can'. Despite the hair-raising risks taken by test pilots like Chuck Yeager, they did to a large extent have their lives in their own hands. When things went wrong in vehicles like Dynasoar travelling at several thousands miles an hour the pilots could and often did survive by their own skill and improvisation. In the Mercury capsule, they sneered, the astronauts would be no better than the chimpanzees, orbiting the Earth in a 'free fall' mode, contributing little or nothing to the mission and unable to control their own destinies.

Thus challenged, the Mercury men insisted from the start that they must always be able to 'fly' their spacecraft – thus adding to the problems of the scientists and engineers, whose instinct was to retain control of their experiments from the ground. When Alan Shepard named Freedom 7, it was confirmation of their insistence on full participation in the design and development phase. They were, after all, highly-qualified scientists in their own right, and their contributions, as the rocketry men later acknowledged, were to play a major part in final success. Spacecraft naming was not possible when the Shuttle missions started, but the emphasis on individual achievement survived in the crews designing their own mission patches.

'Freedom' of course was meant to emphasise the difference between the rival Soviet and American programmes. Much play was given to the secrecy surrounding the Soviet missions: they gave no advance information of what was intended, nor when it would happen. Thus announcements could be confined to successes – initially when the mission was completed; later after the spacecraft was safely in orbit.

By contrast, boasted America, their space programme was open to public scrutiny, its objectives made public at the time that funding for the mission was sought. Most important of all, the media could report the missions 'live', as they happened. Looking back with fuller knowledge of what went on, I have modified my view considerably. At the time, however, I fully accepted this picture, and gave it much play in my broadcasts.

It was always a relief to leave Moscow, and after covering the Gagarin mission I devoted the return flight to drafting a piece for the *Today* programme. I called it 'War and Peace in Moscow' and described how the Soviet capital was both more friendly and more frightening that on my previous visit in 1959. There had been smiles and salutes from Red Army men instead of glares, and suddenly everyone accepted tips. The Intourist 'guides' were happy, or at least resigned, to let you wander off on your own.

> Then suddenly, at 11 pm, the picture changed. We were having dinner at a hotel near Red Square, and as we talked the whole world started rumbling and roaring. Pulling back the curtains unveiled an astonishing picture: Huge tanks, armoured cars, sixty-foot missiles

shrouded in canvas covers – hundreds of them. The scene was hazy behind a thick pall of dust and blue smoke. Because it was late, many of the street lights were out – but not the huge red stars glowing above the Kremlin. It was a nightmare – was it WAR? No, it wasn't war, it was a rehearsal for May Day. Whichever road our taxi took on the drive back to our own hotel, Red Army men with whistles flagged us down as yet another convoy thundered past, slotting into its allotted position for the May Day Parade.

So this is the Russian enigma in 1961: in London the smiling Soviet Consul gave me a visa in ninety minutes – a lesson in speed and efficiency for any Consulate. In Moscow I saw a public parade of military might such as you'll see nowhere else in the world.

Having recorded that piece as soon as I got back to London, I met my wife – having no doubt kept her waiting as usual – for a pre-arranged lunch at Shirreff's. This was a pleasant, medium-priced restaurant near Broadcasting House, and much favoured by BBC journalists. It was a mistake to take one's wife there. We had barely ordered when Tony Wigan, the Foreign Editor, wandered up to the table. 'Glad I found you, Reg. You did well in Moscow, and we want you to go straight on to Washington and the Cape to cover the American spaceflight.' Margaret's face fell and her lunch was spoiled. She and our sons had seen all too little of me for months and years past. But while I was both guilty and unhappy about that, I was also eager to cover man's next step in the space race.

There were still no direct flights from Britain to Florida, and I did not welcome them when they did become available. One could do incredible things in those days getting airline tickets rerouted on the basis that you were flying no more miles, and I always travelled via New York and Washington. I would contact friends and make personal arrangements with the resident BBC correspondents and engineers; call at NASA headquarters, where once you were known astonishingly frank briefings could be obtained, and visit the offices of the aerospace contractors competing to get into the lucrative space business.

On this occasion I arrived in Washington just in time to pick up the story of the failure of the first attempt to place an unmanned Mercury spacecraft in orbit on top of an Atlas rocket. The range safety officer hit the 'destruct button' 40 seconds into the launch as the vehicle refused to roll over into its correct trajectory – a failure of the inertial guidance system.

'NASA Fails Again' was the inevitable media reaction. It was twelve days after Gagarin. I am glad to say that my reports did emphasise the significance of the fact that the escape tower worked perfectly, lifting the Mercury capsule clear of the rocket before it was blown up, and enabling it to coast up to 24 000 ft with its robot astronaut, then descend safely into the Atlantic for recovery undamaged. One step forward instead of two could not fairly be written off as failure. I have an uneasy feeling, however, that my broadcast report, transmitted and recorded from the Washington studio, was cut before that part was reached.

Alan Wheatley, whom I had been instrumental in bringing across from the Press Association to succeed me as industrial correspondent, was by then claiming the lead in *Radio Newsreel* and the bulletins with his coverage of a massive and bitter dock strike. No one was to know, least of all NASA, that that was to be the last major failure in the Mercury programme – so long as one did not count delays and postponements. On television I pointed out that, after starting late, America had by then orbited 39 unmanned satellites against the Soviet's 14, and added: 'Now she's rushing on with the enormous Saturn launcher, and within ten years it could easily be that the first man on the Moon will be NOT Russian, but American.'

In Florida, along with scores of other newsmen, I date-lined all my pieces 'Cape Canaveral', although we actually saw little of the Cape and its launchpads, and spent our time 20 miles south at Cocoa Beach. That was then more like a tropical seaside village hacked out of jungle clearings, with about six newly established motels.

Cocoa Beach grew up, its foundations funded by visiting newsmen, because of military restrictions surrounding the Cape. It was half-way between the Cape and Patrick Air Force base to the south, where sand dunes and mudbanks between the Banana River and the Atlantic widened out enough for a runway and living accommodation. The Air Force had naturally placed these well away from their highly volatile missile activities, but had built a good road – though still narrow – along the sandbanks linking Patrick to the Cape nearly 40 miles north – where once again swamps widened out and had been built up to provide sites for missile pads.

Nature provided excellent security. The wide Indian and Banana Rivers cut the whole area off from the mainland; and the South Atlantic provided 5000 miles of uninterrupted ocean for missile tests – assuming their direction was accurate. The Air Force had been building up this missile range since 1947, and the Redstone launchpads – Nos 5 and 6 – for the Mercury 'space lobs', and Pad No. 14 for the subsequent Atlas/Mercury orbital flights, were very much in Air Force territory. USAF military men inevitably took much pleasure in demonstrating to NASA who was the boss by enforcing security restrictions in the area to a pathological degree. Newsmen – and especially foreigners like myself – were normally only allowed near the pads in the final stages of a countdown, and were bussed in and out, heavily badged and escorted, under constant threat of having one's badge 'pulled' for the smallest deviation from the rules.

Since the newsmen could not go to NASA's temporary buildings inside the Cape, NASA had to come to us. For me the system had many advantages – the greatest being that it saved me at least two hours' driving every day. A week before the countdown started for major missions, NASA established a News Centre in one of the half-dozen motels which had already sprung up on Cocoa Beach – using a different one each time so that this lucrative business was fairly shared among the growing local community. Regular news conferences were held, and NASA administrators, scientists and flight directors, interrupting long days sorting out their technical and political problems at the Cape, had to drive to Cocoa to talk to us – something they never seemed to resent. Like most Americans they loved the limelight – and knew that their futures depended upon public support.

Between news conferences, 'Public Affairs Officers', as the

public relations people were called, were required to find out the answers to our questions and come back to us with the information when we invoked NASA's constitution laying down that everything being done was unclassified and we were entitled to know about it. They also arranged escorted trips for photographers and cameramen, and later when the BBC began to send me a staff camera crew instead of relying on NASA and news agency film I could get round some of the security by insisting that I must accompany them to direct the filming.

Staying in the same motel as the News Centre was luxury indeed. One could stroll from a swim in the warm ocean, and just slip on a shirt to attend the news conferences. At the Holiday Inn there was an extra bonus. In those early days the astronauts also stayed there. There was a strict protocol which we welcomed as much as they did, since it protected the regular space correspondents against the activities of marauding tabloid reporters: No one approached the astronauts for interviews and autographs. Maybe you chatted to them in passing as fellow hotel guests – as I did on my first visit, when I awoke to the fact that I was talking to Alan Shepard himself. At that stage we had just been told that, of the seven astronauts, Shepard, Grissom and Glenn were being considered for the first flight, although they had actually been told the previous January that they would fly in that order. The fact that that was kept secret for five months until the day of the first launch attempt of MR3 (Mercury/Redstone 3) was not only an indication of the limitations placed upon the 'openness' of the programme, but may constitute an unbroken record in the history of US public relations! We had not yet found contacts inside the organisation willing to reveal such harmless facts.

The motels quickly got to know who were the top newsmen in the business, and although there were at least ten demands for every room, they were very selective. Most of all they enjoyed having a flow of incoming and outgoing international phone calls. As always, I benefited from being 'the BBC' – and English. 'Ah LUV to heer yew talk' from every receptionist, barperson and waitress became very tiresome in those early days in Florida. Most of the sparse population had only encountered an Englishman on the cinema screen, and 'Reggie' was soon watched for in all the motels. But it was a price worth paying when most newsmen could only get a room 40–50 miles away in places like Titusville and Melbourne. At the Crossway Inn, which I most often used for many years, a room was kept for me until the eve of the launch without my having to make an advance booking; if I did ring up from England they were quite hurt at the implication that it might be necessary!

When launch day came on Tuesday 2 May, weeks of hot dry weather were broken by thunderstorms and spells of torrential rain. The last stage of the countdown, starting at midnight under the glare of arclights, went ahead, but we were not allowed on to the Press site until 5am, by which time the count was held at minus 2 hours 40 minutes, waiting vainly for a break in the weather. All sorts of restrictive rules had been imposed: radio and television were not allowed to break into their programmes for live coverage until ten minutes before lift-off. As far as I was concerned, although I had written a lengthy report eight months earlier urging advance planning of our coverage of such occasions, the BBC's coverage degenerated (as news organisations usually do!) into hysteria. Most news and current affairs programmes had believed it would never happen, and had taken little advance interest. Now they all wanted to be seen and heard to be major participants in the space game.

To my disgust, BBC Foreign News had decided that it would be cheaper and technically safer to take a 'feed' of NBC's live coverage, with me interweaving our own reports over their lines to New York (actually from the NBC caravan on the Press site) before and after their live commentaries. So when Alan Shepard actually lifted off the following Friday after seven tantalising 'holds' it was Merrill ('Red') Mueller's voice that Britain heard – and still hears when yet another 'Look Back' programme is done. Red Mueller had been a famous BBC voice during the war, when many of his war correspondent despatches had been used. I could not of course have rattled off the commentary in the way he did it – but would have welcomed the chance to develop my own way.

However, there was plenty of scope for me, filling in the

background of a 15-minute flight that attained a speed of 5180 mph, and reached a height of 116.5 statute miles, after which the astronaut had 5 minutes of weightlessness followed by a re-entry which resulted in him becoming 12 times his own weight and emitting 'the sort of drunken groan normally associated with such a workload'.

The man who endured it, US Navy Commander Alan Shepard, appeared to be temperamentally the most carefree of the three finalists for the first flight, and was married with daughters of 9 and 13. Because the 1-ton Mercury capsule had no toilet facilities and he had to wear the equivalent of a baby's nappies inside his spacesuit, he had had to live on a 'low-residue diet' for three days beforehand (thus exposing all those lies that no final choice had been made); his 2.30 am breakfast consisted of four ounces of orange sherbert (or water ice), four ounces of frozen strawberries in syrup, two sugar cookies (small cakes sprinkled with sugar) and eight ounces of skim milk. Not even a cup of coffee!

Dr Wernher von Braun observed carefully that 'it was a good day and a perfect flight'. But James Webb, then NASA's head – or Administrator to give him his correct title – was moaning in a Press statement that having the media present when a launch was planned and then postponed 'placed a serious psychological burden upon the United States'.

But the man really suffering from lack of privacy had been the astronaut. Anyone, I broadcast, who thought he might feel lonely in space could not be more wrong. With sensors attached to pre-selected, tattooed marks on his chest and back, and a thermometer in his anus, his slightest nervous tremor was signalled instantly through the medical sensors and noted with pursed lips by the medical people sitting comfortably at mission control and watching their visual display monitors.

Between transmitting this sort of information down the NBC lines to the BBC New York, where sometimes they patched me straight through to London and into a programme, and sometimes recorded me either in New York or London for use a few minutes later, I had time to stroll around and look at what my fellow newsmen were doing. Years of covering big stories on the Press Association had long since taught me

NASA's decision makers. Left to right: Administrator James Webb, his No. 2, Robert Seamans, and Brainerd Holmes, briefly Manned Spaceflight Director, making the surprise announcement on 11 July 1962 that the lunar-orbit-rendezvous technique would be used for the moonlanding. It was successful exactly seven years later. But all three had been replaced before that. (NASA)

the necessity to keep a tight rein on one's emotions on these occasions. It was reassuring to find, however, that the general hysteria prevailing among BBC editors and engineers was equally prevalent among my competitors. Japanese reporters were especially noticeable amid the babel. However, one good-looking young man was keeping calm. He was walking up and down holding a hand-microphone practising his TV dateline: 'Robert Abernathy, NBC, Cape Canaveral', over and over again. Thirty years later he was grey-haired and famous, still good-looking, and a fellow survivor in a pitiless medium.

President Kennedy, reported by our Washington correspondent as appearing to be 'tired and worried', welcomed Shepard's flight as 'an historic milestone in our exploration into space', and added: 'We have a long way to go in the field of space. We are behind but we are working hard and we are going to increase our efforts.' It was a strong hint of the famous speech which was to follow three weeks later, committing the nation 'to the goal, before this decade is out, of landing a man on the Moon and returning him safely to the Earth'.

At last the BBC news programmes began to heed my year-old

warnings that space was going to provide the story of the decade. Now they wanted me to hurry from the Cape to Washington to cover the celebrations and the 'ticker tape' parade for the Mercury astronauts. But for the only time in my career I refused to go. Margaret's mother, who had provided her – and me – with valiant support right through the war and after during my absences, and lavished more grandmotherly love upon our sons than anyone had the right to expect, had been desperately ill in King's College Hospital for weeks – and I had been most unhappy about the standard of medical attention she was getting there.

She died the day before Shepard was launched, and I insisted that I must return home for the funeral. My BBC bosses expressed a good deal of resentment that I should contemplate returning to base before they had had the full return on their financial investment in my activities. It was then that I first suspected that, unlike the Foreign Correspondents Christopher Serpell, Gerald Priestland and other Oxbridge staffmen like Foreign Duty Editor Keith Bell (a Lord's Taverner usually suffering from post-lunch irritation when I talked to him at my breakfast time) I was regarded then and would be always as among the 'other ranks' rather than the 'officers'. I had observed that when domestic tragedy – and, worse, bouts of alcoholism – struck the officer members of the BBC, there was instant warm sympathy with injunctions to 'drop everything'. In my case I discovered they had expected me to seek Army-style 'compassionate leave' so that they could turn it down with a kindly explanation that too much BBC money had been invested in me to throw away on mere sentiment!

But return home I did – and three weeks later the Mercury spacecraft caught up with me at the Paris Air Show. For more than 30 years, the Soviets and the Americans competed to steal the limelight at this show, usually with their latest space exhibits. In 1961 the Mercury spacecraft, with its scorched heatshield, was successful in drawing away much world attention from Soviet successes. NASA claimed that 650 000 people went to Le Bourget to see it.

The Soviets came back strongly in mid-July, however, when Yuri Gagarin reached Britain on his world tour. Huge crowds turned out to greet him at airports, and during a four-day tour that took him to London, Manchester and elsewhere, Government ministers, senior civil servants and Service chiefs turned out for him as they have never done since for a space occasion. In London he was given an escort of 22 Metropolitan Police motor cyclists, and visited the Queen and Prince Philip, and Prime Minister Macmillan. As one of the few people who had encountered both the Soviet and American spacemen, I was called upon by *Radio Newsreel* to compare their qualities:

> I watched them both as they were transformed overnight from ordinary airmen into immortals of legendary fame. They both remained as calm and as relaxed as they were before. For years past as the separate space programmes have been planned and developed, Russian and American scientists have worried about the sort of qualities a spaceman needs. Both have rejected the devil-may-care bachelor type, ready to risk his life for a whim. Both have picked a family man with family cares and a determination never to take a chance.
>
> But most of all it seems Russia and America have both decided that spacemen must have an instinct for public relations. Shepard's calm broadcasts from his space capsule – and the use of 'A-OK', meaning All-OK, at once earned the nickname A-OK Alan; but it was his ability to face the glare of a television world with a combination of modesty and assurance that really won the hearts of the Americans.
>
> We've just seen Yuri Gagarin do exactly the same thing. Like Shepard, he made us feel there was no reason why we should not be spacemen too. All that's needed is a little knowledge and careful training. Supermen were superfluous.

Mercury 4

After that, a three-week tour of America's Ballistic Missile Early Warning System (BMEWS), which I myself had proposed during the 1960 defence correspondents' tour, and which is described elsewhere, diverted my attention from the space race. But on 21 July 1961, in a repeat of the Shepard mission, Astronaut Virgil (Gus) Grissom, 35 – a US Air Force Lt-colonel – was successfully launched but almost drowned before being rescued. His Mercury craft, which he named Liberty Bell 7, retaining the figure 7, as did all subsequent Mercury astronauts, had been fitted with the more adequate observation window called for by Shepard, as well as an improved manual control system.

The observation window brought Grissom Earth views of such astonishing clarity that they distracted him from using the attitude manoeuvring controls installed upon the astronauts' insistence – though they did work well when he found time to test them. His excitement about how much detail he could see of the Cape and West Palm Beach from a slant range of 150 miles (his maximum altitude was 118 miles) was the first indication that the human eyeball could perform far better in space than was ever expected. But his reports, and those of subsequent astronauts, were for a long period treated with much scepticism, and attributed by earthbound scientists to imagination and other factors. They maintained that they knew better, having long ago defined the limitations of human sight. It was several years before they could accept that astronauts could in fact see details of ships and airports from distances of 300 miles, and in some conditions the contours of ocean floors.

Grissom was also to face scepticism as to whether or not he himself had punched the detonator which blew off his spacecraft's hatch, resulting in near-drowning for himself and the loss of Liberty Bell – the only spacecraft in the Mercury and Gemini programmes not to be recovered. After splashing down only 3 miles from the aiming point, Grissom asked the prime helicopter recovery pilot to delay the pickup for five minutes while he recorded the cockpit panel data. Then he told the helicopter pilots he was ready, removed the pin from the hatch cover detonator and lay back in his dry couch. 'I was lying there, minding my own business, when I heard a dull thud.' The hatch cover blew away, and sea water began swishing into the spacecraft as it bobbed in the ocean. Shipping water and sinking fast, he pushed himself headfirst out of the narrow hatch. The primary helicopter pilot, who from past experience assumed the astronaut in his airfilled suit would float, concentrated on recovering Liberty Bell. But though he succeeded in latching on to it, the spacecraft had filled with water and weighed over 5000 lb – 1000 lb greater than the helicopter's lifting capacity. In danger of his aircraft being dragged into the sea, the pilot was forced to release Mercury, which sank in 2800 fathoms.

The pilot had already called in the backup helicopter to pick up Grissom, who was by then in serious trouble. While making his notes inside the spacecraft he had released a lot of air from his suit because it kept ballooning around his neck and making it difficult to write; now he was being pushed under by the downdraught of two helicopters. Grissom's head was barely above water when the co-pilot, who fortunately had had experience in rescuing both chimpanzee Ham and Alan Shepard, got the lifeline to him. Grissom grabbed the sling, and though he was dunked twice, was finally hoisted, gasping, to safety.

Nothing was ever found that could have caused the hatch to blow without action by Grissom, but he continued to maintain that he did not do it until he lost his life in the Apollo 1 fire nearly six years later. He was able to point out that Glenn, Schirra and Cooper all suffered a slight hand injury when they hit the hatch plunger at the end of their flights; he, Grissom, was the only one who ended without a hand injury.

Despite the loss of Liberty Bell, NASA decided that Grissom's flight was so successful that no more Mercury/Redstone 'space lobs' were necessary; plans for a third were cancelled, so that they could speed up the first Mercury/Atlas orbital flight. But public impatience with NASA's slow progress was exacerbated a month later when Vostok 2, carrying Cosmonaut Gherman Titov, completed a 17-orbit, 25-hour flight – matching what was regarded as an ambitious 18-orbit target for the last of the Mercury missions 18 months later! (Later the last Mercury flight was extended to 22 orbits.) Titov was the first spaceman to experience what is still a regular and familiar problem – space

Monkeys in space: in the early Mercury and Gemini days the cartoonists found a ready source of satire in NASA's use of monkeys to pave the way for men in space. This contemporary cartoon shows Ham briefing the first astronauts to succeed him in space, and saying: 'Then, at 900,000 feet, you'll get the feeling that you MUST have a banana!'

sickness. He complained of feelings 'akin to seasickness', and said he had to be careful not to move his head too swiftly in any direction. NASA's medical advisers noted this with concern, and recommended that the first American in orbit must guard against and watch for 'this peculiar physiological reaction'.

NASA was forced to admit that it would be January the following year before they could hope to place their first astronaut in orbit – a blow to many prestige-conscious Americans who were anxious that, even if America were second to Russia in orbiting the first spaceman, at least the record books would show that both countries achieved it in 1961. In September a Mercury spacecraft was placed in orbit; but it carried only a black box, or 'crewman simulator', on a mission designated Mercury/Atlas 4, increasing confidence in the hitherto unreliable Atlas. That confidence was reinforced at the end of November when MA5 was sent into orbit with Enos, a 37 lb chimpanzee, despite the fact that he had to be brought back to Earth after two orbits instead of three. A faulty environmental system sent Enos' body temperature up to 100.5°, and the attitude control system caused the spacecraft to drift and swing. These faults could have been corrected or compensated for by an onboard astronaut.

Curing the attitude problem was to add to later postponements and was an interesting example of the sort of technical snags that have always dogged spaceflight. It was found that a metal chip in a fuel supply line had cut off the propellant flow to one of the clockwise roll thrusters. This inactive thruster – there were 18 in all – resulted in the spacecraft drifting 30° out of its normal attitude. When it reached that point the automatic stabilisation and control system rolled it back into the correct position, and the sequence started all over again. Each sequence – and it happened nine times – cost an extra one pound of control fuel; hence the need to shorten the flight.

The early re-entry and recovery went so well that NASA considered the whole operation a useful demonstration of the system's versatility – even though on this occasion operating the explosive hatch from outside cracked the observation window and pulled out several bolts.

Criticism, in which I joined, of the use of Enos for this mission, when the astronauts were eager and willing to go and would demonstrably have been more effective, mounted when it was learned that Enos had suffered 79 undeserved punitive electric shocks. He had three levers to operate, and at one time the centre lever malfunctioned, giving him a mild shock when he pulled it, instead of the reverse. Although described all too accurately as 'shocked and frustrated', Enos whose name meant 'man' in Hebrew, continued to operate his levers like a well-trained soldier. During the mission he earned himself 47 measures of water (totalling about one pint) by pulling a lever 20 seconds after a green light appeared; and 13 banana pellets, each of which required him to pull another lever exactly 50 times.

These activities were described as 'voluntary and without penalty', and many of us felt that his performance was better than we could have achieved. Enos died just a year later, aged six, after suffering

Convincing the Presidents: President Kennedy, 2nd left, Major-General MacMorrow, and the then Vice President Johnson, being briefed by von Braun, left, about his Saturn rockets at the Marshall Space Flight Center, Huntsville, Alabama, in 1962. (NASA)

for two months from dysentery, which the Medical Laboratory at Holloman Air Force Base, New Mexico, insisted was 'in no way related to his orbital flight the year before'. Whether that was so or not, Enos had made it possible for John Glenn to become America's first astronaut to make an orbital flight.

There was a sequel to Grissom's flight 38 years later. The third salvage team to search for it found and photographed the missing Mercury capsule 'half a mile deeper than the Titanic' in May 1999 on the floor of the Atlantic 300 miles south-east of Cape Canaveral. A remote-control submarine photographed it, sitting upright, clearly identifiable, with 'United States' still readable on its pyramid-shaped side. Finally, in an operation funded by the cable TV channel Discovery, it was brought to the surface on 20 July 1999 – 38 years less one day after it sank. The hatch which Grissom insisted he did not manually jettison was still missing; and with that mystery unsolved, Liberty Bell was cleaned up and placed on permanent display in a Kansas museum.

6 Glenn Gets There First

Mercury 6

Gus Grissom's flirtation with death at the end of the Liberty Bell mission did much to stimulate media interest in the forthcoming attempt to place John Glenn in orbit. I had little trouble getting myself back to the Cape – or rather Cocoa Beach – about ten days before his launch was expected in the last week of January 1962. I promised my listeners and viewers, as well as my newsdesks, that I would be covering America's biggest-ever week in space. NASA, and the US Air Force, Navy and Army had promised us no fewer than ten missile and space launchings. Not only would John Glenn be placed in orbit, but there was confidence that, after two earlier failures, the unmanned Ranger 3 would be sent off to the Moon to pioneer the way for men.

NASA had also promised accreditation for 400 news people, but Cocoa Beach and the surrounding towns were soon crammed with 800. At my motel they were bedding down in the beach huts.

A major problem for us all, radio and TV as well as newspaper people, was getting our stories over – and for all their short-comings, NASA and the Air Force, with much energetic support from Bell Telephone engineers, had done their best to supplement the saturated telephone and cable facilities across the swamps and rivers. Telstar, the first satellite capable of bouncing some limited television pictures across the Atlantic, was still eight months from launch; and Early Bird, the first of the Intelsat series which made intercontinental TV practical, was still three years away. BBC Television News was growing fast, but still could not match radio.

For Glenn's final countdown and mission I was once again reliant on cadging a few minutes on NBC lines to hook in to the BBC at New York – and as both my requirements and theirs increased, their initial pleasure at providing facilities for the BBC rapidly wore off. Honoured guest during Shepard's launch, I was now a tiresome encumbrance. Most newspaper reporters still cabled their stories, only 'topping them up' by more expensive phone calls as deadlines approached. When I could cadge no circuits I struggled with phone calls, involving delays and variable quality.

American radio men were adept at unscrewing the mouthpiece, removing the microphone and clipping the leads of their microphones to the terminals; their radio stations found the quality acceptable to put on the air. But this practice was illegal in Britain – a breach of Post Office regulations, who jealously pointed out that the BBC was using THEIR lines. BBC engineers and traffic managers started from the premise that telephone quality was not good enough, and instead of working to make this cheap and easily obtained material usable, spent their time moaning unceasingly and depressingly about my own efforts. They thought I should go to a studio in Orlando and swept aside my protests that the round trip of 120 miles on narrow roads would take me four hours, and halve my working time.

When I insisted on my phone recordings being played back over the same phone line they were reassuringly loud and clear – but still brought complaints implying that the poor quality was somehow my fault. I compromised by recording my contributions for *The Eye Witness* and *From Our Own Correspondent* on my tape recorder – not easy, because if I tried it in the open someone or something always made a noise, and in my motel room either my phone or the one next door rang instantly, or a room maid or colleague hammered insistently on my door. Many a time I retreated in desperation to the toilet. News conferences and meals were missed when I made evening and early morning drives to McCoy Air Force Base (not yet called Orlando Airport) or to Melbourne Airport to airfreight tapes to London. Additional time was lost because of the need to cable or telephone the waybill numbers to ensure that TV or radio knew when to collect the material.

Near studio-quality could be obtained by leasing two telephone lines – one for transmission and one for reception – but this was expensive, especially as it needed a BBC engineer to fix it up. For the major Apollo missions many ponderous meetings won approval for this, and I was provided with a crude lashup to which my Uher tape recorder could also be linked, enabling me to send recordings of interviews and Mission Control down the line. My colleagues from all over the world, including downtown US radio stations, all seemed to have the best and latest equipment, and were amazed to see the BBC, with its pretensions to be better than anyone else, muddling along with a tangle of wires spreading like chickweed around and through my typewriter, tape recorder and Press Kits!

Despite the respect accorded to my BBC status, and later to my personal seniority, throughout the Mercury, Gemini and Apollo missions I never ceased to feel like a poor relation among those confident US broadcasters. Cable breakdowns sometimes led to delays of several hours on telephone calls, but if one knew the tricks one could get a generous degree of priority. You asked the local operator for 'an urgent Press call to London, England', doing your best to extend your 'Received Pronunciation' English into the public school version to which you were not entitled; a slight pause to allow the operator to savour the accent, and you followed up with the magic password: 'Please give me Operator 90, White Plains, New York.'

Few Florida operators knew that this was the quick way to get London, and were thankful to be told what to do. 'Operator 90' was a much more knowledgable and sophisticated woman; she usually enjoyed being chatted up, and told how last time she had enabled you to make that marvellous broadcast to millions, because she had given you one of the new transatlantic cable lines, rather than a crackly radio circuit. I tried to keep my voice low during this exercise for fear of alerting pushy competitors on neighbouring phones, fresh from London, and as yet ignorant of the existence of the mysterious 'White Plains'. Their frustrated shouts accompanying their own unavailing attempts to reach their offices provided a covering background to my own calls.

When 'voice pieces' were not insisted upon, I joined my newspaper colleagues in using the excellent cabling facilities, which were then, but not for much longer, cheaper and quicker. There was fierce competition between the two main cable companies, AT&T and Western Union; but Western Union consistently won those battles of the early space years solely on the personality of a little man named Joe Caley. He sought out every arriving newsman likely to give him business, noting his motel room and cable card details so that he or she need not waste time adding them on every cable. If I happened to be in my motel pool when an incoming cable arrived – occasionally a 'herogram' expressing thanks for some timely contribution; more likely a complaint over some alleged failure, or demands for more pieces – Joe would appear at the poolside, and take down my dictated reply while I stayed in the pool.

He also brought in a very pretty and expert teleprinter operator who could beat the phones every time, whom I dubbed 'Miss Western Union' – a title to which she answered proudly for twenty years after. When a story broke during the night Joe would come hammering on my motel door. 'Do you want to send a cable?' he would demand. Ethics forbade him to reveal the subject of other newsmen's cables, but he would hint: 'Everybody else is sending, so I thought you'd want to!' In pyjamas and a silk dressing gown (still in occasional use 40 years later) I would descend to the News Centre to find out what new disaster had struck NASA. Although a skimpy T-shirt showing hairy midriff, plus grubby, very brief shorts went unnoticed, my coverall pyjamas and dressing gown were regarded as either daringly sexy or rather indecent. Thus attired on one occasion I was accosted by a visiting Congressman who shook my hand warmly and said: 'I always wanted to meet a BBC correspondent!'

Joe Caley was the all-American go-getter, and his main rival, Cecil Free, had little chance against him. Cecil, who worked for AT&T, was so British that even at 3 am he always wore immaculately pressed cricket flannels, topped by a blazer and an offended expression because the British newsmen did not automatically give him preference on nationalistic grounds. It was sad, as the years rolled on, and everyone had his own telephone, to see both Joe Caley and Cecil Free struggling for survival as the cable business diminished.

All that, however, was far in the future as I settled down in January 1962 to the longest spell I was ever to have at Cocoa Beach. For a time the postponements of Glenn's launch did not matter too much. There were lots of failures to report: a Polaris submarine missile, intended by the US Navy to demonstrate the doubling of its range to 2500 miles, splashed mightily into the Atlantic when its second stage refused to ignite. Another Navy Project, Composite One – which I called America's first 'Christmas Tree' rocket – joined the Polaris missile at the bottom of the Atlantic when a Thor rocket loaded with five small research satellites failed to develop sufficient thrust to spray them into orbit.

The Ranger 3 moonshot, significant because, like Glenn's flight, it depended upon an Atlas launcher, was almost but not quite a success. It was supposed to crash on the Moon, transmitting close-up pictures in its final moments, but actually missed it by 22 800 miles and ended in solar orbit. Would Glenn suffer a similar fate, we speculated? Ranger 3 transmitted some excellent TV pictures, but unfortunately was pointing the wrong way, and the pictures should reach Sirius in a few more light years.

Between these minor dramas, and postponements of MA6 for reasons like a faulty valve in Mercury's oxygen system, I filled the gaps with plenty of stories. NASA was rushing ahead with Project Gemini (meaning 'Twins'), a two-man spacecraft to keep the astronauts busy in what was expected to be a two-year gap between the end of Project Mercury and the first Project Apollo flight. Gemini was intended to make the first Apollo moonlanding possible three years earlier by developing a capability for spacecraft to rendezvous and dock in space, while orbiting either the Earth or Moon, thus avoiding the necessity for massive vehicles to make direct flights between the two bodies. The rendezvous system offered the exciting prospect that fuel supplies could be parked in orbit, so that the astronauts would not be dependent upon one launch.

Test-firing the escape tower for the Mercury spacecraft – the only time it was ever fired. Escape towers designed to lift Mercury and Apollo spacecraft clear of the launchpad if the rocket below caught fire never had to be used. Gemini spacecraft had ejection seats instead of the escape tower. (R. Turnill)

Rather tiresomely my listeners proved to be more interested in personalised pieces. So I pointed out, after yet another postponement, that I would not be lying in the sunshine upon Cocoa's dazzlingly white beach created by billions of crushed oyster and clamshells. The local residents, mostly golden-skinned youths and gorgeous goodtime girls, drove their jalopies at high speeds along the sands, and it was all too easy to get run over. Nor, I found, could you enjoy a quiet moonlight walk along the sands. Blazing headlights periodically blotted everything out. One night, when all seemed relatively peaceful, a cruising police car coming from the opposite direction passed me, did a U-turn and disconcertingly crept along behind me on dipped lights. After a while it hauled alongside, pacing me, and I could see that it contained four Cocoa Beach police, large, well-armed, and notoriously trigger-happy. One leaned out:

'Say, what's the trouble bud?'

'No trouble.'

'We sure thought you must have trouble.'

'Look, I've had a hard day and I'm trying to have a quiet relaxing walk in the moonlight from the NASA News Centre to my motel.'

The car fell back again while its occupants consulted. Was it possible that this limey really was walking for pleasure? Suddenly the headlights blazed, the car pulled forward, and did an accelerating U-turn, throwing sand all over me. One of the cops leaned out again as the car sped off.

'OK bud, have a NICE walk.'

While 26 ships, 60 aircraft and around 15 000 personnel were moving into position for the first launch attempt, I hired a local camera crew and sent back a filmed 'curtain-raiser' warning that chances of success in the space business seemed at that time about as remote as winning top prize in the football pools. We were getting acquainted – from a respectful distance – with the central figure, sandy-haired John Glenn, who occasionally appeared among us with a ready smile but accompanied it with unwelcome advice. We assembled outside Cocoa Beach's new and expensive-looking Baptist Church one Sunday to watch him enter for morning service and he told us severely: 'You'd do better to come inside and attend the service!'

Glenn was already known as the odd man out among the seven astronauts. They had established the tradition that Cocoa Beach was where they stayed for training, so wives did not go there. This enabled some of the gorgeous girls to bestow their favours upon them; just being an astronaut apparently had an enormous aphrodisiac effect upon these young women, who boasted openly about their tally of astronaut scalps. It was when such activities reached Glenn's ears that he uttered his famous injunction to his colleagues to 'keep your wick in your pants'.

The girls were certainly not after the astronauts' money. Glenn, a colonel in the US Marines, was already 40, with a wife and two teenage children; he had become a hero as a wartime pilot, and after that was the first man to fly supersonically across the US long before becoming an astronaut. All these achievements were earning him the equivalent of about £75 a week – and the astronaut job did not qualify for additional danger money.

At last, at 4 am on Saturday 27 January, a fleet of US Air Force buses began shuttling hundreds of us to the Press site at Pad 14, where the countdown was underway for 'Friendship 7' as Glenn had named his Mercury spacecraft. Bulldozers had pushed up a special mound to give us a view across the scrub to the launchpad about two miles away; in the dark we stumbled among a fantastic tangle of cables and caravans, television and radio equipment. Ninety-nine telephones had been installed in a half circle, and a large brown marquee, normally used for circuses and religious revival meetings, had been hired as a writing room for us. It was not a good idea. The mosquitoes got trapped inside it in large numbers, were very cross about it, and as they tried to find a way out, bit most viciously anyone they encountered. I was thankful to go outside and use the reflected light on the perimeter of the NBC caravan, which was in frantic and continuous use. There was tension in launch control, but it was orderly tension – quite unlike the near-hysteria reigning on the Press mound, where few knew what was happening nor what they were supposed – or able – to do.

Clear skies, and later welcome warming sunshine – Florida nights can be bitterly cold in winter – were soon replaced by a 7000 ft cloud layer. The countdown was held first at T minus 45 minutes, and then, interrupted by a mysterious power failure that stopped the

John Glenn inserts himself into his tiny Mercury spacecraft for the last time before launch. (NASA)

Mercury control room clocks, taken down to T−20 minutes. There it was held; the weather got worse, and I was thankful that I had been able to record a number of a 'filler' broadcasts which the BBC could draw upon during the delays. At last the launch was postponed for at least two days. Glenn, after being strapped in his tiny couch for 5 hours 10 minutes – almost the same time as his planned three-orbit mission would have taken – was pulled out head first, smiling and happy. So were launch control. It had been a splendid dress rehearsal for them – even though the thousands employed in the ships and aircraft, having lost their weekend for nothing, were less than happy.

More delays – 'technical difficulties' with the Atlas which engineers were forbidden to discuss with us – and 'very disturbed, 10-ft high seas' in the Bermuda recovery area were among the series of post-

Launch of Friendship 7 from Pad 14 on 20 February 1962. It was America's first manned orbital flight. John Glenn is aboard the Mercury/Atlas rocket. (NASA)

ponements lasting more than two weeks which made my BBC masters less than happy. Keith Bell became increasingly irritable with me over the necessity to hurry back from his 'lunch' to meet my 10 am calls to discuss whether I should return home. The tabloid newspaper Editors were slightly mollified when their correspondents filled in with a murder story. Early one morning I glanced into the bar of the Starlite Motel, the place where the media nightly exchanged mutual consolation about the unreasonableness of their London offices, to find the big mirror behind the bar shattered and blood-spattered. After we had all gone to bed, our favourite barmaid, a cheerful woman of 31, had been shot four times by a disappointed lover who burst into the bar. Bloodhounds and aircraft were engaged in a hue and cry, concentrating on the dense scrub and swamp of a 10-acre island in the Banana River, which the murderer had reached after a 2-mile swim, reportedly with one shot left in his revolver.

Ronald Bedford, *Daily Mirror* Science Editor, with whom I had developed an enduring friendship since we had shared those defence tours around the US, entertained himself and terrified me by insisting on driving my hired car. Born heavily handicapped with very limited sight – he had a 4-inch focal length for reading with one eye – as well as limited hearing, he could never qualify for a driving licence, but insisted on doing everything that the rest of us did, but much better. In those days the sand dunes lining the Forida coastline had not been linked by bridges and causeways, and there were many quiet, little-used roads. Along these Ronnie drove the car using my eyes and commentary to guide him. When another car approached, and I had to talk him past, like an air traffic controller 'talking down' a disabled aircraft in earlier times, I had to make an effort not to close my own eyes and resort to prayer. When Ronnie, an accomplished pianist, went off with some others visiting night clubs at which he invariably took over the piano from the regular player, I drove myself along lonely tracks into the swamps to enjoy the incredible birdlife – undulating lines of up to 30 slow-cruising pelicans, blue and white herons, and thousands of egrets. For me it magically brought to life the pictures in H.G. Wells' *History of Mankind*.

By 16 February I was cabling London: 'Glenn flight postponed tenth time until earliest Tuesday due clouds and rough seas stop . . . Making provisional arrangements return pending your confirmation.' It was a disastrous mistake, but I had become so worn down with moaning Foreign Duty Editors who seemed to take the view that I was personally arranging the delays, I thought it better to take the initiative rather than hang on until ordered by London to return.

I got back on the 18th, and of course on the 20th Glenn soared into orbit – having done better for breakfast than his predecessors: steak, scrambled eggs, orange juice and coffee. I found myself 'anchoring' the BBC's coverage in the windowless *Radio Newsreel* studio in Broadcasting House, helping to build up the tension during the countdown with my first-hand knowledge. I filled in the gaps when the countdown was held – first the launcher's guidance system proved faulty and had to be changed, then at T−22 minutes a fuel pump valve stuck – with horror stories about the Atlas rocket. Its steel skin was 'no thicker than a sixpence', I explained, and it had been standing on the launchpad exposed to the Atlantic's corroding salt winds ever since the previous December.

After 2 hours 17 minutes of 'holds' and 3 hours 44 minutes inside what Glenn called his 'office', Shorty Powers, the voice of Mercury Control, made the announcement for which America had waited ever since Gagarin's flight: 'Glenn reports all spacecraft systems Go! Mercury Control is Go!'

In defiance of my horror stories and its own dubious past, the Atlas lifted off perfectly as it was to do on three more Mercury flights. Glenn's only criticism, 100 seconds later at the point of maximum acceleration, was: 'It's a little bumpy about here.'

Eleven months after Gagarin, Glenn had reached Earth orbit – with a perigee, (lowest point) of 100 miles, and an apogee (highest point) of 162.2 miles, only 0.05° lower than planned. Mission Control told him that those conditions meant that he could safely stay there for at least seven orbits, while Goddard Space Center, with the most advanced computers, estimated that if necessary he could do almost 100 orbits. Glenn reported that being in a state of zero *g*, or weightlessness, was 'wholly pleasant', and on his much shorter flight he encountered none of the nausea suffered by Titov. Glenn and the world below were intrigued with his reports that at sunrise every 45 minutes his

The pioneers: an early picture of the seven Mercury astronauts. Standing, left to right: Virgil Grissom, Alan Shepard, Scott Carpenter and John Glenn. Seated, left to right: Gordon Cooper, Donald Slayton, and Walter Schirra. Only Shepard got to the Moon. (NASA)

spacecraft was surrounded by what he called 'fireflies', which disappeared in bright sunlight. (Not until the next mission was the mystery solved. Then Scott Carpenter accidentally bumped his hand against the hatch and saw a cloud of the bright particles fly past his window. More taps produced more particles. They came from frost on the outside of the spacecraft, and Glenn's 'fireflies' became Scott Carpenter's 'frostflies'.)

A sticking valve in a jet thruster, similar to the trouble that led to the previous mission by Chimp Enos being reduced to two orbits, enabled Glenn to demonstrate that man could overcome such problems by taking over manual control. Then a telemetry reading started a major alarm by showing that the heatshield upon which Glenn's life would depend during re-entry was no longer locked in position. There was disagreement on the ground as to whether the heatshield really was loose, or whether it was a faulty telemetry reading. Glenn was not told, but guessed there was a problem when site after site, as he passed over them, asked him to check that the retropackage deployment switch was at 'off'. If the heatshield was loose, only the straps of the retropack, which normally would be jettisoned before re-entry, were holding it in place.

On his third orbit, Glenn formally but lightheartedly sent a message to his Marine Commander pointing out that he had completed the required minimum of four hours' flying time per month and had thus qualified for his regular flight pay increment. Then, 4 hours 33 minutes after launch, Glenn initiated the re-entry procedure, and Friendship 7 swung round, so that the heatshield was facing forward, and Glenn facing backward. As the spacecraft plunged back into the atmosphere with the retropackage still in position, and as flaming chunks of it flew past his window, he feared that the heatshield was in fact disintegrating. But all was well. Just as he reached out anxiously to deploy the drogue parachute manually, it shot out automatically at 28 000 ft, and he watched the main 'chute stream out and blossom at 17 000 ft.

Because the re-entry calculations had not taken into account the spacecraft's weight loss in consumables, the spacecraft splashed down in the Atlantic 40 miles short of the predicted area. Even so, the destroyer *Noah* was alongside within 17 minutes. Still inside Friendship 7, Glenn was hoisted aboard. The primary objective of Project Mercury had been achieved.

7 Sequels to the Seven Story

Mercury 7

Glenn's brilliant and colourful description of his mission at the news conference a few days later has always stayed in my mind. I listened with frustration, as it was fed 'live' into the news studio at Broadcasting House. Selecting the best extracts and adding my own interpretation and comment was no consolation for not being present.

That day, 1 March, proclaimed 'John Glenn Day' by New York City, I was certain that Glenn would in time become President of the United States – but not before he became the first man on the Moon. In fact, it was to be 36 years before he flew again – and then only for geriatric research as a payload specialist on the Space Shuttle. He took no further part in the Apollo moonlanding programme. When he was not assigned to a Gemini flight, I was told that he had had a fall in the bathroom, and damaged an ear.

Whatever effect that had on his career, his postflight meetings with the President and politicians in Washington undoubtedly gave him a taste for power and office. Just two years after his flight he resigned from the Marine Corps and NASA to take an executive post in the Royal Crown Cola Company. Selling cola struck most people as an unlikely stepping stone to Washington, but Glenn knew better. Ten years later he became Democratic Senator for Ohio. But his 1983 bid for the presidential nomination was singularly lack-lustre, and what appeared to be his half-hearted support for future space programmes lost him many friends in that area. Gazing down at him from the Senate public gallery in 1991 I thought he appeared to be a sad, grey and burnt-out figure. He came in and out and fiddled at his desk unnoticed while Senator Edward Kennedy, despite being in the news again with more family trouble, was always the centre of a handshaking sycophantic group. I would never have dreamed then that Glenn's career would re-blossom.

After Glenn's flight, and between covering more immediate domestic news, I waited eagerly for the next big spaceshot. Eight months after the Titov flight, the Soviets broke a long gap in their space activities by launching an unmanned research satellite, and I was among those speculating that they might be working on a spacecraft capable of taking two cosmonauts round the Moon and back. Khrushchev, with his instinctive sense of timing, used the occasion to do some 'rocket rattling'. He announced that the Soviets had a 'global rocket' capable of going three-quarters of the way round the world, which was thus able to attack the West through its Antarctic 'back door' – that is, without being detected by the Ballistic Missile Early Warning System watching the Arctic.

About the same time NASA announced that Donald Slayton's nomination as the fourth Mercury astronaut to fly had been cancelled. He suffered from an occasional heart irregularity, medically termed an atrial fibrillation. The astronauts' own physician, then William Douglas, insisted that there was no reason why it should preclude him from spaceflight; but the unfortunate 'Deke' was subjected to exhaustive examinations by the country's top heart specialists in Philadelphia, San Antonio and New York. Inevitably it was finally decided, in Deke's words, 'that the unknown factor in my heart murmer not be added to all the other unknowns for manned space flight'.

The story was of less interest to the public than to space correspondents like myself, who had already developed a warm liking for the laconic but friendly Slayton. He was equally popular with the other six Mercury astronauts, who soon arranged a consolation prize for him: they made him the Chief Astronaut, looking after their interests, and soon the flight selections, as well. Deke remained on the 'active' list of astronauts – no one had the heart to take him off – and his determination to get into space in spite of everything was finally rewarded 13 years later when he was pilot on the 1975 Apollo–Soyuz mission. NASA's long-term memory was better than that of most employers!

Astronaut Scott Carpenter replaced Slayton for what was designated Mercury/Atlas 7 (MA7), and his launch was scheduled for early May. 'Aurora 7', for the 'dawn of manned spaceflight', was Carpenter's choice of name for his Mercury spacecraft. After my MA6 debacle I looked around for some way to get back to the Cape that would not arouse cries of horror about the expense. I was fortunate – I thought. Cunard Eagle Airways had invested £6 million in the purchase of two Boeing 707s, with the intention of competing on the London–New York route, for which they had American approval, but were as usual blocked by the powerful national airlines – BOAC, in this case, who successfully invoked British Government intervention. Right on cue, Cunard Eagle decided to try a Bermuda–Nassau–Miami route instead, and were glad to have the BBC Air Correspondent aboard for the inaugural flight. It was a good story in its own right, and the start of the concept of the luxury holiday – flying first-class one way, and travelling in one of Cunard's Queen liners the other.

Three hours out from London, we overtook the original *Queen Elizabeth*, which was 2.5 *days* out of Southampton for New York. From the 707 flight deck I interviewed Captain Watts, 7 miles below. In his 43 years at sea and 400 Atlantic crossings, he had never flown it. If he had his time over again, would he prefer to command a jetliner? 'I'm quite satisfied with the *Queen Elizabeth*, thank you,' he replied rather primly. I pointed out in a broadcast from the flight deck that just one Boeing 707, with about 150 seats at that time, could carry as many passengers in a year as the *Queen Elizabeth*. The great oceangoing liners taking four days instead of seven hours, and costing much more, were already doomed – and the Jumbo Jets, with over 400 seats, were still to come.

Two days later, I was back at the Cape, filling in time before Scott Carpenter's launch by recording the countdown for a Centaur rocket, intended as a future 'space work-horse' for interplanetary probes and soft landings on the Moon. In the half light of dawn I was peering at my carefully prepared notes, voicing a description of the lift-off against the

1962 and a cold morning at Cape Canaveral. The author (seated) and Ronnie Bedford, *Daily Mirror*, reporting that they have just watched another rocket explode. (Unknown photographer)

background roar of it all, when happily my neighbour shouted: 'It's blown up!' Otherwise, like the unfortunate NASA commentator when Space Shuttle Challenger blew up many years later, I would have gone blithely on. My description of 145 tons of launch rocket, fuelled for the first time with liquid hydrogen, supposed to give 40% more thrust than kerosene, exploding into a fireball above our heads, and leaving behind a black mushroom of smoke as debris rained down into the Atlantic, gave my masters at the BBC little confidence that I would not once again be running up motel bills for some weeks before Mr Carpenter finally took off.

Worse was to come. Two days later my Radio Newsreel piece began:

> It's been a black week here, with four firings and four failures . . . Yesterday a military Pershing missile went off course and was blown up by the range safety officer. The same fate befell a Polaris missile, pride of the US Navy, when launched from a submarine 90 ft below the Atlantic. Now comes the failure of ANNA, one of America's most colourful and controversial projects. She was intended to be a sort of space lightship, sending out high intensity flashes, five at a time, at five second intervals. She was also to be a satellite clock, sending out time signals every 90 seconds.

NASA's proud claims that ANNA would enable scientists to make exact measurements from one point to another on the Earth's surface brought an angry intervention from the Department of Defense (DoD) that this would help the Soviets to pinpoint American military targets. NASA scientists retorted that the Russians already knew the distance to all major US targets to an accuracy of 55 yards. DoD nevertheless declared the project secret, but was persuaded to allow us to watch the launch on the understanding that we would be given only a brief handout, which could not be discussed, and that there must be no Press briefing after the launch. No doubt there was glee at nearby Patrick Air Force base when, after all that, ANNA never got into orbit. Another piece of debris for the South Atlantic.

Meanwhile Scott Carpenter's launch was being postponed for modifications to the parachute-deployment system and then to a temperature control device in the Atlas launcher. With the help of Ronnie Bedford I continued to supply a steady flow of usable pieces – in retrospect, I fear, with increasing desperation. We found that the Winder Aircraft Corporation, based in the centre of Florida, was negotiating to buy the three Saunders-Roe Princess flying boats, mothballed in the Isle of Wight for the past ten years. They were ideal for use as flying test-beds for nuclear-powered aircraft, it was argued. It never happened, of course, but it was a good story at the time!

An interview with Mrs Scott Carpenter, who had brought their four children to the Cape, was followed by an interview with the astronauts' pretty nurse 'Dee' O'Hara, with the unfortunate comment that she saw a lot more of them than their wives did. And a favourite interview, which we have often replayed for our own entertainment over the years, was with the night porter at Cocoa Beach's Starlite Inn. Virgil Blake would now be disparagingly dismissed as 'an Uncle Tom nigger', but was one of the calmest, most well-balanced human beings I have ever met. He had lived on Cocoa Beach for 40 years, and raised 10 children there, making a living by orange picking and road mending until the missile range had brought relative prosperity to the area. For him it meant the ability to earn a dollar and three quarters an hour instead of a dollar and a quarter. He instructed us in the technique of picking up a rattlesnake without getting bitten: 'It's sure death if he hits you.'

His favourite story was of a man, crippled with rheumatism, who had been confined to a wheelchair for 18 years. His family would wheel him to the edge of the sea and leave him there to paddle his feet in the warm salt water while they hunted and fished.

> And come along a bear, a-hunt'n turtle eggs. The bear looked in the chair and saw the old man, and the old man saw the bear. The bear stood up on his two feet and come walkin' towards him. The old man jumped out the chair and ran from here to Williams' Point – must be 22, 23 miles – but he runned it. His family came back, looked and looked, saw a bear track, went home and there he was. He walked till the day he died. He never was down with rheumatism. The bear scared the rheumatism off him.

This story was airfreighted with a 'Note to Mr Bell. This might provide some light relief when things are dull.' But Keith Bell never did appreciate light relief, and I suspected did not listen much to our most serious programme, *From Our Own Correspondent*. To that, one of my favourite outlets, I contributed five minutes on why the US Air Force wanted to continue with nuclear explosion experiments in outer space. NASA was tying up thousands of servicemen recovering their astronauts, complained the USAF, when they ought to be busy on military space projects, such as how to intercept and inspect hostile and possibly nuclear-armed Soviet satellites. President Kennedy was persuaded to underline this issue four months later in a speech on space science at Rice University:

> Whether it will become a force for good or ill depends on man, and only if the United States occupies a position of pre-eminence can we help decide whether this new ocean will be a sea of peace or a new, terrifying theatre of war.

But as many a foreign correspondent has found before and since, my contributions mixing specialist erudition and light relief failed to convince the foreign news meetings back in London that my continued presence at the Cape was really necessary. The programmes used nearly all my pieces, but they had little say in such decisions. So once again I was back in Broadcasting House just two days before the mission was actually launched; once again I listened in the studio to Merrill Mueller, of NBC, doing very graphically (and almost certainly much better) what I had planned to do. But we now had a continuous 'feed' of Mission Control, including their exchanges with the astronaut; so while I filled in the background to the news that others were covering – they included the BBC New York correspondent as well as NBC – I could at least closely follow what was happening.

Carpenter enjoyed his flight too much. Although he was subsequently awarded all the customary honours from Presidential decorations to a hero's welcome at his home town of Boulder, Colorado, NASA executives were less than pleased with his performance. He was the first, but not the last astronaut, not to be offered another flight for that reason. NASA's official history of Project Mercury, which took its title *This New Ocean* from the Kennedy speech quoted above, records what happened:

> During the second orbit, as he had on the first, Carpenter made frequent capsule maneuvers with the fly-by-wire and manual-proportional modes of attitude control. He slewed his ship around to make photographs; he pitched the capsule down 80° in case the ground flares were fired over Woomera; he yawed around to observe

> and photograph the airglow phenomenon; and he rolled the capsule until Earth was 'up' for the inverted flight experiment. Carpenter even stood the capsule on its antenna canister and found that the view was exhilarating . . . Working under his crowded experiment schedule and the heavy manual maneuver program, on six occasions Carpenter accidentally actuated the sensitive-to-the-touch, high-thrust attitude control jets, which brought about 'double authority control', or the redundant operation of both the automatic and the manual systems . . . Ground capsule communicators at various tracking sites repeatedly reminded him to conserve his fuel.

Carpenter made more errors during re-entry preparations – the most important being that he forgot to switch off the manual control system when switching to the fly-by-wire system, so that for 10 minutes fuel was being used from both. He fell behind with his check-list, and Aurora 7 was canted 25° to the right at retrofire time, causing the spacecraft to overshoot the planned impact point by about 175 miles. There was an additional 15 miles of error because he was three seconds late in pushing the button to ignite the first of the three solid fuel retrorockets strapped to the heatshield, which then fired in sequence. Carpenter splashed down safely, but could get no response from the pararescue crew on his radio.

Sweating profusely in the 101 °F temperature inside an alarmingly listing spacecraft, he thought it would be too risky to blow off the main hatch, so instead wriggled his way through the throat of the craft with great difficulty, taking his camera, liferaft and and survival kit. He had to lower himself into the water before he could inflate the liferaft – and then it was upside down. But he managed to flip it upright, crawled on to it, retrieved his camera with its important pictures – and lay there reasonably content for three hours until he was rescued. For the first hour, with communications broken since Aurora 7's re-entry began, the outside world and broadcasters like myself speculated as to whether the US had lost its first astronaut; but I was well briefed on the escape and liferaft procedures, and emphasised the likelihood of survival.

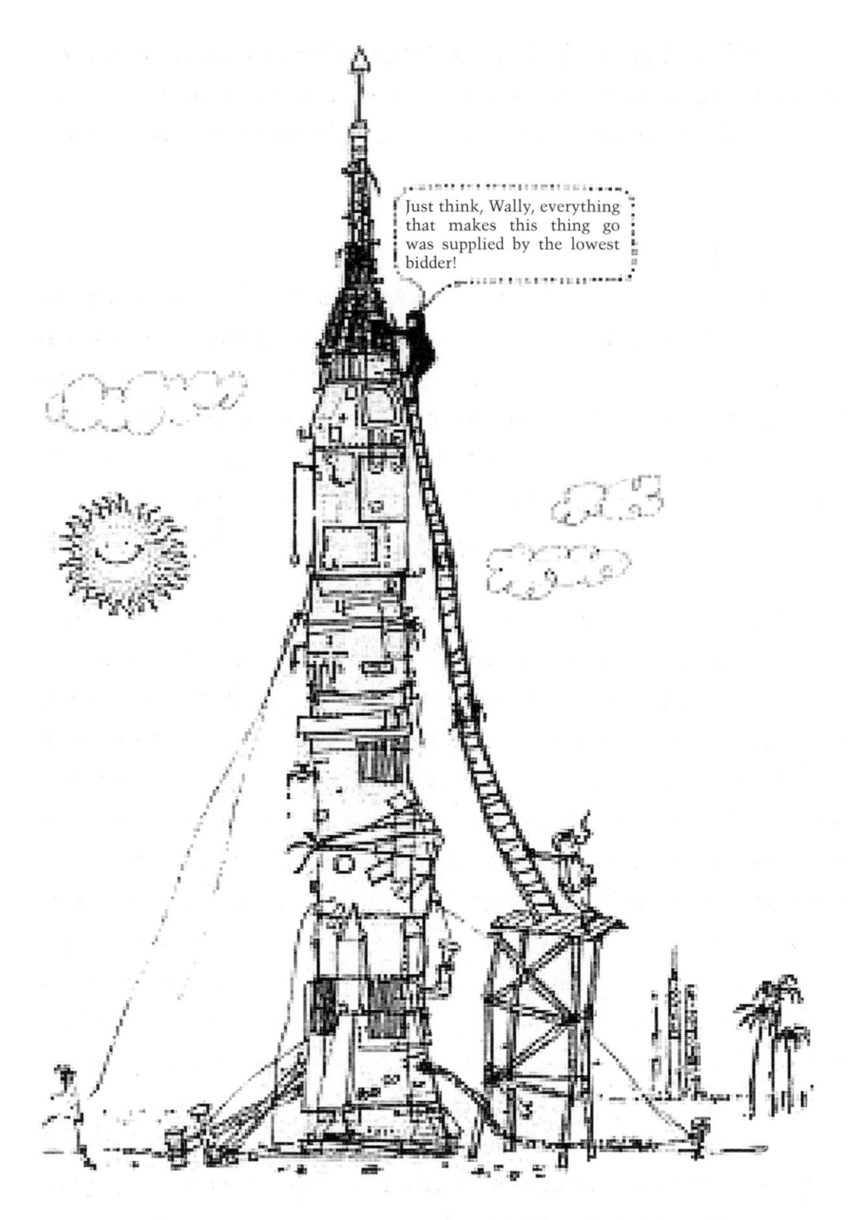

A 1962 cartoon commemorating Astronaut Wally Schirra's observation before his Mercury 8 launch that every part of the rocket and spacecraft was supplied by the lowest bidder. (Raytheon Corporation)

Despite their displeasure with Scott Carpenter's performance, NASA learnt more about Mercury's strengths and weaknesses than they could possibly have done had the mission been carried out to perfection!

Mercury 8

Near perfection was in fact provided in October 1962 by Walter Schirra, a naval pilot who had won the Distinguished Flying Cross for his combat missions over Korea. For his spaceflight, MA8, he named his Mercury craft Sigma 7. Sigma, the engineering symbol for summation, marked the fact that it was to be a six-orbit engineering flight in preparation for a final one-day flight, compared with the exploratory nature of Glenn's flight, and the developmental and scientific nature of Grissom's.

Schirra's most important task was to demonstrate that, although Carpenter had exhausted all his manoeuvring fuel in three orbits, by conserving the fuel it was possible for a Mercury craft to stay in space for a whole day. Schirra achieved this by spending much of his time in the 'free-flight', powered-down mode. He made his re-entry preparations, and fired his retrorocket, with such precision that 90 hours 13 minutes after launch he splashed down only 4.5 miles from the recovery carrier, USS *Kearsarge*. Even after retrofire, his fuel gauges showed more than 50% remaining. A subsequent announcement by the US Air Force that if Schirra's orbit had taken him above 400 miles he could well have been killed by a lingering artificial radiation belt resulting from a US high-altitude nuclear test carried out the previous July attracted little media attention.

The most significant feature of the mission, from the public's point of view, was that communication satellites had begun experimenting with intercontinental television relays, and TV pictures of Schirra's launch were shown in Europe only an hour after it had happened. There was also a small monochrome TV camera inside Sigma, and for two minutes Schirra was interviewed live for US viewers by John Glenn, at Mission Control. 'Great sport', Schirra called it, which in retrospect was surprising, for during the first Apollo test flight he destroyed his excellent chances of getting to the Moon by his refusals to co-operate with demands for TV transmissions. NASA's past public insistence that its space activities were conducted openly rebounded upon them from the moment that live TV coverage became possible. Demands from the expanding and aggressive TV networks for almost continuous live coverage of all manned spaceflight operations grew irresistibly.

Mercury 9

The growing media appetite for TV coverage had an unfortunate effect upon me personally when NASA was at last ready to launch Gordon Cooper. At 36 he was the youngest of the Mercury seven, and his 22-orbit mission offered some unique TV opportunities. Cooper had designated what was to be the last Mercury spacecraft Faith 7, to demonstrate his faith in the mission and in his seven comrades. (A touch of hypocrisy perhaps, for the Seven were fiercely competitive. And the *Washington Post* pointed out that if the spacecraft were lost the world would be reading that America had 'lost Faith in space'!) My problem was that with TV transmissions from the spacecraft planned in advance for a total of 1 hour 58 minutes, BBC TV News wanted considerable coverage direct from me, and was no longer ready to accept versions of my radio pieces.

The geosynchronous TV satellites, which would soon give continuous transatlantic coverage, were being planned, but were still not operational. What was available were two elliptically-orbiting 47° and 42° satellites called Relay 1 and Telstar 2. Their orbits took 186 and 225 minutes respectively, and as they followed one another around, they were 'above the horizon' in relation to both Britain (and much of Europe, of course) and the United States, for periods of 20–30 minutes each. During that time TV signals could be sent up from the US and bounced down to Britain until the satellite dipped below the horizon like a setting sun; as it did so, the viewer's black and white picture began to break up and finally disappear. But use of this facility cost several thousand pounds for only a few minutes of transmission. Television companies soon got together to share the pictures and the cost in what became 'EBU' – the European Broadcasting Union.

Within the BBC there was much excitement about this new facility, and everyone was determined to use it, regardless of cost. The News Division, of course, regarded itself as having priority. But the 'Outside Broadcast' empire, who took over for Royal weddings and funerals, Test matches and other great occasions, felt they were the only people qualified to embark upon such an adventure. They were outraged when, just as they thought they were about to see off News Division, another internal competitor appeared: Science and Features Department. Only they, declared their leaders, Richard Francis and Aubrey Singer, could supply the depth of knowledge required to handle live coverage of such an abstruse subject as manned spaceflight. The possibility of expertise being available in News Division invoked the immediate application of the Nelsonic blind eye.

As this three-cornered contest built up, my masters in News decreed that I must cover Cooper's flight from the BBC's New York office. There I would have access to high-quality radio circuits for insertion into TV and radio programmes, and could keep abreast of what was happening with a continuous feed of Mission Commentary, plus NBC and CBS TV monitors and the ability to dial NASA direct when I felt the need to talk to some of my personal contacts there. It sounded convincing but broke all my personal rules about first-hand reporting.

However, my NASA friends airfreighted copies of the Press Kit and other material to me in London, and I spent every minute of my London–New York flight absorbing the launch, abort and emergency rescue procedures until I felt almost as well informed as Major Cooper. In relative peace, I was fully occupied on 14 May 1963 covering the Cooper countdown, and recording how he crawled into Faith 7 with a 50–50 chance of suitable weather – and how, in accordance with NASA's corny standards of humour, he was handed a 'plumber's friend' to take with him. Decorated with a ribbon, this was a combination of pliers and wrench. We never discovered whether this was a genuine space tool that he might really need, or just a private joke. The weather cleared, and it seemed the flight was on. But then the 11-storey gantry providing access to the Atlas/Mercury stack could not be rolled back because a fuel pump in a diesel engine failed. And when that was changed, Bermuda reported that their radar tracking facilities were 'on the blink'. Together with 22 ships, 100 aircraft and 25 000 personnel, the BBC and I stood down. That was when my personal troubles really began.

Back at the BBC the conflict between Current Affairs and Outside Broadcasts on the one side at Lime Grove, and Science and Features on the other at Kensington House, had at last been resolved. Apparently the Director General had told them they had got to work together – which was really a victory for Science and Features, who could not now be excluded. They were all set to move in with a massive 'TV Special' on the next attempt to launch Carpenter – which came the following day.

Robin Day, newly recruited from Independent Television News to Paul Fox's Current Affairs team, who were the principal customers for Outside Broadcasts facilities, happened to be in New York working on a political contribution to Panorama. He was told to drop everything and stand by to 'anchor' live TV coverage via satellite of the Cooper flight. But Robin knew nothing about space, was the first to admit it, and declined to do it. Nothing would budge him, despite, it was rumoured, direct orders from the Director General. Unmoved by threats and lamentations from London he departed on a flight for his next planned assignment in one of the 'new' African countries with an unfamiliar name.

With no time to fly out another of their big 'names', the new unit was forced to open both eyes and ask News for help. I was ordered to hand over my news coverage from the BBC New York office to a somewhat apprehensive Anthony Wigan when the time came for Cooper's re-entry. My former ally Tony Wigan, a victim of internal strife, had been forced out of his Foreign Editor's post and made to swap positions with John Crawley, now on the fast track to the top of the BBC. I was to do the live commentary on this first major space TV spectacular – and if ever there was a poisoned chalice, this was it. Everyone associated with Current Affairs had a vested interest in ensuring that any contribution from a News correspondent should end in disaster.

In due course I was escorted by a Current Affairs producer based in New York – a good looking young woman – to a seedy area of that city and shut up with her in what would have made a particularly dreary broom cupboard. There were a number of other equally apprehensive correspondents of various nationalities jammed in adjoining cupboards. Two pairs of headphones were placed on my head, and my first reaction of dismay that I must listen to two sources at once changed to horror when I discovered that all four earpieces carried a different source of information and instructions. My protests that I had only two earholes were treated with contempt. There was the familiar NASA mission commentary in Earphone No. 1. Earphone No. 2 carried what later became equally familiar and much better organised: either a NASA or EBU producer announcing periodically that in, say, 30 seconds he would be inserting into this general TV 'feed' some live pictures of space activities, or alternatively, when nothing live was possible, an artist's concept or previously recorded film illustrating the activity that was going on in space, such as re-entry and recovery procedures. This at least one could see on a small TV monitor in a corner of the cupboard.

Earphone No. 3 occasionally – but never when you needed it – carried sound of what the BBC was actually transmitting to the public – a 'programme feed'. Very often this consisted of one's own voice, much louder than anything in the other earpieces, being fed back with up to a second's delay because it had travelled nearly 100000 miles up and down, there and back. Talking against one's own voice is a particularly trying experience. When not required to do that, one was free to keep quiet and decide which of Earphones 1–4 to concentrate upon.

It was essential to make a great mental effort to exclude the programme feed, because when it was not returning one's own voice, what the BBC was transmitting had little relevance to what was being fed over the satellite from NASA via New York. Producers and directors in London never learned that viewers and listeners were more than ready to accept a few seconds' silence in Mission Commentary exchanges, or a temporary lack of intelligible pictures, and would enjoy puzzling it out with the experts. Any such gaps evoked instant producers' panic, and transmission was switched at once to a large assembly of eager 'experts' in the studio, whose gossip bored the viewers to death, but provided a picture which was reassuringly easy for the the producers to handle while they recovered their nerve and waited for NASA to send something that even they understood.

This was where Earphone No. 4 came in. It carried instructions from the said London producer or director, and occasionally from Paul Fox himself. A week or two of rehearsals, plus a little rapport with those originating this medley of information, and the whole thing was just possible if one knew the subject well. Richard Dimbleby was to grow up with the system and master it with an assurance that no one else has achieved. Discovering the medley's various components while attempting to explain what was happening in space to several million people during a live transmission was not a good idea. Even a half-hour briefing would have helped – but that might have rendered the contents of the poisoned chalice ineffective!

Despite all this, for half an hour I enjoyed myself. I was absorbed in the subject and was practised enough to know that the trick when commenting on live pictures is to speak in very short sentences, so that when the pictures change one is not trapped in the middle of a complicated sentence and unfinished idea. 'Paul Fox is very pleased,' said Earpiece No. 4 – and did I detect a note of asperity? But general panic ensued when Cooper's instruments suggested his spacecraft was beginning to fall back to Earth before it was supposed to do so. 'Build up the excitement, Reg!' ordered Paul Fox in Earphone No. 4. Responding, I missed information pouring over from Earpieces 1 and 2, and began commenting on rehearsal film as if it were pictures of what was actually happening. Earphone No. 4 pointed this out.

During the final re-entry I gave up trying to disentangle the simultaneous bombardment of excited and near-hysterical information and instructions from Earphones 1–4. Drawing upon my detailed knowledge of what *ought* to be going on during Cooper's descent, I shut my ears, crossed my fingers, prayed that all was going according to plan, and described that. Robert Robinson, then the Sunday Telegraph TV critic, who was later to have more direct experience of life in a TV studio, panned me heavily for talking over Cooper's own comments during the re-entry. They of course might or might not have been competing for my attention in Earphone No. 1.

'Wind it up, Reg', ordered Earphone No. 4, 'and hand back to Richard Dimbleby in the studio.' It was the first indication I had had that Dimbleby or anyone else was in the studio – another failure on my part to disentangle completely the information medley. Nevertheless, with relief I did so, with what I thought was quite a good peroration, explaining that Cooper's safe splashdown 7000 yards from the recovery ship USS *Kearsarge*, after manually controlling his whole re-entry procedure, was a triumphant end to the five-year Mercury programme intended to pioneer the ability to place men in orbit.

But just as I was about to tear the headphones from sore ears, it all began again. 'Standby Reg!' ordered Earphone No. 4. 'Telstar's gone down, but Relay's come up and they're giving us that for another half-hour.'

When I got back to London my in-tray contained a copy of a memo from Paul Fox to the Foreign Editor. 'We thought Turnill did quite well in the circumstances,' it said, expressing damagingly faint praise for the brief loan of the Aerospace Correspondent. No one mentioned that the following night, so late that few saw it in Britain, I did a live commentary to 17 European countries via Telstar which went without a hitch. But then I had only one pair of headphones, and in front of me in the broom cupboard was a helpful 'approximate timing shot list':

1–30 secs:	Silent shots of carrier.
30–40 secs:	Shots of helicopters in flight.
40–45 secs:	Chute down in water.
40–1:25:	Paramedics on capsule and cutaways of sailors and others on *Kearsarge*.
2:30:	into tape SOUND (band and crowds) comes up briefly . . .

. . . and so on through lifting the capsule aboard the carrier, the hatch being blown open, the astronaut talking with President Kennedy ('SOUND OF CONVERSATION'), to 'astronaut drinks glass of water', and President Kennedy statement from the White House.

I took myself off from New York to Cape Canaveral for Cooper's news conference, careless of the cost, and back to Washington to cover the hero's reception he received at the White House, where the rose garden was packed with 130 000 children given a day off from school. Jackie Kennedy, expecting a baby three months later, appeared very briefly. It seemed churlish to mention amid all the rejoicing that Cosmonaut Titov had completed just such a flight two years earlier; but I did report it, adding with less confidence than was to prove justified, that with the successful completion of the $400 million Project Mercury, American scientists were confident that they would win the race to the Moon before 1970.

I would not have predicted with equal confidence that I myself would continue to survive the BBC's civil wars and be covering that event. Nor would I have predicted that Gordon Cooper, who had done so well despite being rather laconic, would fall out of favour with NASA, and be relegated to the past likc Scott Carpenter.

On 16 November 1963 President Kennedy visited Cape Canaveral to be briefed on the follow-on programmes to Mercury: Projects Gemini and Apollo. Sitting next-but-one to him were Kurt Debus and Wernher von Braun, whose Saturn 5 rockets would fulfil Kennedy's moonrace timetable. A few days after that, Kennedy was assassinated at Dallas. Our pioneering use during MA-9 of the Relay and Telstar satellites proved to have been an invaluable rehearsal for coverage of that shattering event and the drama of the funeral cortege that followed.

1963 was the year when intercontinental TV arrived. By 1964 it was taken for granted.

8 Space Travel: Learning the Rules

Gemini 3

The two-year gap between the last Project Mercury flight in May 1963 and the first two-man Project Gemini mission in May 1965 was a difficult time for NASA. During it the Soviets sent up their first woman (Valentina Tereshkova), then three cosmonauts in one spacecraft, and most sensationally of all achieved the world's first spacewalk. In the US, the steady build-up towards the employment of around 400000 people on the space programme was making good progress – but there was little opportunity to convince the public (and the voters) that this was the case.

I was personally grateful to the Russians for keeping the space stories ticking over during the build-up to the moonrace, for without them interest in spaceflight on BBC and newspaper newsdesks would have been difficult to resurrect. Not that there was any lack of other aerospace stories for me to cover; 1963–65 were traumatic years for British aerospace as a whole – and for me personally as well. I have described that in my autobiography.

But once Project Gemini ('Twins'), with its two-man crew and aim of developing rendezvous and docking techniques, started in March 1965, NASA pushed it along at a breathless pace. They achieved 10 flights in two years – and during that time it was the Soviets who were held up. There was, however, still little confidence among the American public that NASA could beat the Soviets to the Moon. That was clear when two US space engineers, one of whom enjoyed the BBC-like title 'Head of Human Factors' in his company, suggested – as mentioned earlier – that the only way it could be achieved would be to send a volunteer to the lunar surface on a one-way ticket.

He pointed out that it was already technically possible to send

someone to the Moon. He (it was assumed it would be
then be sent supplies to keep him alive for perhaps three y
technologies had been developed that made it possible
back again. It would have been horrendously expensive –
the one period in space history when money did not matter. About 22 cargo rockets of the type then available (early Atlases or Titans) would be needed in the first year alone to establish the one-man moonbase, I reported:

> Robots, including television cameras, would start the operation and the man would follow when pictures came back showing that all was ready. He would have to be protected from radiation emitted by solar eruptions, and that would be done either by lead-shielding delivered by the cargo rockets, or by covering his living quarters 34 ft deep in lunar rubble . . . At present America is hoping to reach the Moon by about 1968. The one-way ticket technique might knock two years off that. But Mr Brainerd Holmes, America's head of manned space policy, tells me that every new step must wait until there's a reasonable chance that the man taking it will return. So unless competition between Russia and America gets a lot fiercer, we're not likely to see the first Robinson Crusoe in space.

Looking forward on New Year's Day, 1965, to the Gemini missions, I disclosed that America's first men on the Moon would have to wear beards. Plans to provide them with battery-driven electric shavers were abandoned because it was thought that weightless whiskers drifting about might clog delicate instruments. For long flights, it was announced, astronauts would be told to start growing a beard a week beforehand, so that they would not feel itchy inside their space helmets. [When the time came, of course, the astronauts improvised their own solutions, and only a few returned unshaven!] Russia, I forecast, would be first to demonstrate a rendezvous capability, and it was known that she intended to be first 'with the next obvious spectacular, sending men on a non-stop flight right round the Moon and back'.

Although the Gemini spacecraft looked very similar and drew

heavily on proven Mercury technology, it was twice the weight and far more complex and versatile than its predecessor. Unlike Mercury before it and Apollo after, there was no escape tower to lift it clear in the event of a launchpad fire; instead the two crewmen were provided with ejection seats, which accounted for much of Gemini's extra weight. It was also the first spacecraft to replace batteries with fuel cells (based on a British development) which provided electrial power through the chemical reaction of oxygen and hydrogen. An enormous advantage was that each cell produced 0.57 litre of drinking water per hour as a by-product. (On an early mission too much water was being produced, and the astronauts found that the only way to dispose of it was to drink it!) It had been hoped that by the start of the Gemini missions techniques would have been developed to enable them to make a land touchdown, thus eliminating the costly and clumsy splashdowns and recovery at sea. But after three years of trials at NASA's Dryden Flight Research Centre on paragliders and Rogallo wings it was ruled out as too risky.

After two unmanned tests, NASA had announced that Gemini 3, the first manned flight, would be 'no sooner than' 23 March. Thus forewarned, the Soviets duly launched a mission one week ahead of it, which made the planned three-orbit Gemini mission look dull indeed. Voskhod 2, with two cosmonauts aboard, stayed up for 17 orbits and 26 hours; and during it chubby Aleksey Leonov, who was to become a General and the most enduring and popular of all spacemen, made the world's first spacewalk. It lasted only 10 minutes, during which he started rotating on his tether at the rate of 10 times per second. But he stayed calm, and the world was astonished by TV pictures carrying his own comments. We called it a 'space swim', which was probably more accurate than either 'space walking' as it later became known, or the ugly NASA 'EVA', the acronym of 'Extravehicular Activity'. Leonov had quite a struggle getting back inside Voskhod 2, because his early spacewalking suit had 'ballooned', and he had to re-enter via an inflatable tube which had been ejected from the spacecraft to provide an airlock. Vague pictures showing this misled us into believing that Voskhod was nose-shaped with a flight deck, instead of being heavy, ball-shaped and crude, and really no more than a slightly modified Vostok.

Leonov and his companion, Pavel Belyayev, also had to overcome a major crisis when their automatic re-entry system failed. I speculated that the spacecraft had overshot the landing area 'by scores, perhaps even hundreds of miles' when there was a four and a half hour silence between re-entry and the announcement that it had landed safely. In fact Belyayev's manual firing of the retrorockets led to a landing in deep snow in a forest near Perm, 1200 miles north of the target. They huddled inside awaiting rescue, and the helicopters did well to reach them only 2.5 hours later.

Pressed by the BBC Newsrooms to match some of the horror stories in the newspapers about the dangers of spacewalking, my NASA friends helped me to come up with this: 'The main danger facing Leonov was the possibility of tearing or damaging his pressurised suit as he left and re-entered the spacecraft. Without its protection he would die instantly; his blood would boil and his body literally burst in the airless emptiness of outer space.'

Rather more prophetic was the comment of Mr Kuzin, a Soviet biologist, who said that Leonov's experiment was extremely important for the building of interplanetary space stations, which would have to be assembled by spacemen manoeuvring outside their spacecraft. Russia, I added, might well consider the building of space stations to be more important than an immediate landing on the Moon. Misled by the inflatable docking tunnel, I then went wrong by adding that Voskhod spacecraft were big enough to take five men, and it was possible that the Soviets would soon launch two five-crew spaceships and try joining them up in orbit.

I had been in Washington only a week earlier and in addition to visiting NASA headquarters had had 'off the record' briefings at the Pentagon, where I was accredited as a 'respectable' defence correspondent, someone having embarrassingly asserted that I was 'a good friend of the United States'. Military reconnaissance and surveillance from space had become so important, said the US Air Force, that they were selecting their own cadre of 16 military astronauts, and planning a series of Manned Orbiting Laboraties (MOLs).

I discovered the details on a subsequent visit to Houston, when, to the horror of my NASA escort, I blundered through a forbidden door. There I saw a mockup of a Gemini spacecraft to be placed in orbit with an adapted upper stage rocket which would be launched separately to provide living and working quarters for up to four weeks at a time. 'PLEASE come out!' my escort pleaded, and the agony in his voice made it clear that it meant the end of his career if my incursion came to light. I backed quietly out before the security men spotted me, and assured him that I had not seen anything at all of interest. MOL was cancelled in August 1969 before being flown; by then it was obvious that unmanned surveillance spacecraft, with infrared and other cameras, telescopes and instruments far superior to the human eye, would be much more versatile and less expensive. Seven of the 16 disappointed MOL astronauts were transferred to NASA, and slowly took over that organisation in the Shuttle era – Richard Truly becoming Administrator, and Robert Crippen becoming Director of the Kennedy Space Center.

Fortified with my background knowledge and an advance copy of the NASA Press Kit, I covered the 5-hour Gemini 3 mission from the studios in London. It was commanded by Virgil Grissom, whom we called a 'veteran' because he had made one previous 16-minute flight, accompanied by an unknown astronaut called John Young, who ultimately became the only person to have made six spaceflights until 61-year-old Story Musgrave matched the achievement in 1996. [Young was still on the 'active' list of astronauts at the turn of the century, which entitled him to fly the T38s. Michael Foale told me, during a visit to Canterbury, how much he and other astronauts had learned from Young, by then a real veteran of 72, when flying with him.]

Gemini 3 went so smoothly that I reported: 'Probably their biggest problem was to choose between Meal A and Meal B – beef pot roast, or instant apple sauce and brownie cubes. They chose the latter.' Grissom and Young, we were told, made their spacecraft 'perform like a ballerina', using their tiny RCS (reaction control system) thrusters to manoeuvre and change their orbit from 100 × 140 miles to 97 × 105 miles – the all-important first step towards the ability to rendezvous and dock with other vehicles in space. Comforted by this, NASA chiefs forecast wrongly that their efforts were likely to be overshadowed by more Soviet 'spectaculars' during Project Gemini; but this mission, together with the fact that the Soviets had nothing to match the 12 000 close-up pictures of the lunar surface recently sent back by Rangers 7 and 8, gave them confidence that before long they would overtake the Russian efforts.

Gemini 4

A fortnight after Gemini 3 I was back at Cape Canaveral, which by then we were having to call Cape Kennedy, renamed – to the indignation of a large portion of the local population – in honour of the assassinated President Kennedy, to see the fulfilment of Arthur C. Clarke's vision of geostationary communications satellites. The US Department of Defense, still courting the goodwill of Britain's defence correspondents, deposited us there for a few hours during a five-day tour. Within a few minutes of its scheduled time, a Thor-Delta rocket launched Early Bird, destined to become the first satellite for what was by 1997 the 138-nation International Telecommunications Satellite Organisation (Intelsat), towards a 21 748 × 22 733 miles orbit.

'Its ability to transmit telephone calls and television programmes around the world will bring nothing but added pleasure to mankind and profit to commerce,' I reported – and then went on to discuss its military significance in a piece for *From Our Own Correspondent*:

> The real thrust behind the Moon race is still the fear by each side that the other might achieve military domination by space techniques. That's why, on my visit to the Cape, among the most impressive sights were the vast new launching facilities now being built for the exclusive use of the US Air Force. With these facilities America will be able, in the next few years, to establish manned defensive bases, orbiting in outer space, in the same way that such bases already exist on land.

Early Bird transformed worldwide interest in the Gemini 4 mission in June 1965. At last millions of people around the world could watch it actually happening. Twelve European nations paid the high price demanded – $22 000 per hour, although you could buy 10-minute segments – to obtain live television coverage via the new geostationary satellite. For the BBC and myself it meant that pre-launch coverage with a film crew sending back rolls of film, followed by live radio coverage, was no longer sufficient. It also meant that my largely unchallenged BBC news coverage was increasingly invaded by cohorts of current affairs commentators, technicians and producers.

On top of that, the new Mission Control Center, built on what had been cattle-grazing land at Nassau Bay, 30 minutes' drive south of Houston, Texas, took over firm control of the flight from the Cape's Launch Control Center as the spacecraft cleared the tower 12 seconds after lift-off. It marked the start of bitter rivalry which has never ceased between NASA's Kennedy and Johnson Space Centers, and was a severe blow to the well-organised comfort of the space correspondents. No longer could one cover the whole mission from the relaxed sunshine of the Cape and the seaside motels of Cocoa Beach. On short missions one had to choose whether to cover from Cocoa Beach, where one got the benefit of pre-launch briefings and a view of the lift-off, but then could only follow what was happening at Mission Control second-hand; or one had to start at Houston, covering the launch second-hand in return for first-hand coverage of the actual flight and splashdown. As the flights got longer, it was necessary to be at the Cape for launch, then make an unseemly dash in a Press-chartered aircraft to cover the rest of it from Houston – usually accompanied by altercations at overbooked hotels when we arrived *en masse* at Nassau Bay.

Gemini 4 I also covered from London, making my bosses happy as I dashed from one studio to another inserting live pieces not only into 'straight' news, but into programmes like *Home This Afternoon* and Jimmy Saville's Radio 2 shows. I kept as close as I could to the mission by wearing headphones which I could plug in, wherever there was a suitable opportunity, to the BBC's 'ring main', on one line of which a continuous feed of mission commentary was being carried. This extraordinary system allowed anyone who understood it to plug in unannounced to listen to incoming pieces from correspondents all over the world as well as to their conversations with newsdesks.

James McDivitt, 35, and Edward White, 34, who had been together at college and at the US Air Force test pilot school, had been selected as the second two-man crew, and were the only pair not to wear a crew patch nor to name their spacecraft. This appealed to me, for in my broadcasts I had consistently avoided using what I thought were the rather childish names selected for their craft by earlier astronauts. McDivitt and White had no doubt been put off by the official disapproval surrounding Grissom's choice of 'Molly Brown' for Gemini 3. Having lost his Mercury 4 spacecraft, Grissom thought it would be a good joke to use the title of the Broadway stage play 'The Unsinkable Molly Brown'. NASA executives thought the name undignified, but hastily withdrew their objections when Grissom proposed to call it 'Titanic' instead. However, all the succeeding Gemini flights were officially referred to only by a roman numeral.

McDivitt and White won a long battle to be allowed to do a spacewalk on their mission. Although Leonov's spacewalk had come as a surprise to the outside world – including the space correspondents – US astronauts had been rehearsing them for months before that, because it was expected that much work would have to done outside spacecraft. Leonov having come to no harm, the Gemini 4 crew pressed to be allowed to catch up with the Russians by tackling NASA's first spacewalk up to a year earlier than planned. Charles 'Chuck' Berry, Gemini's medical director, and a firm favorite with the newsmen because he was always ready to talk to us (among other things, to speculate about the problems of having sexual relations in zero gravity), was worried. He had already demanded that a proposed seven-day mission should be halved because many of his physiologist colleagues thought it might prove fatal. Soviet doctors – partly perhaps to discourage their American rivals – had been uttering many warnings at medical conferences about the physiological dangers of long flights and disorientation during spacewalks. But Chuck withdrew opposition to the Gemini 4 spacewalk after watching, with McDivitt and White, a Soviet filmed interview with Leonov. It included scenes from the spacewalk in which Leonov explained that he had been able to maintain orientation by using the Sun, the spacecraft and the Earth as reference points.

The Gemini 4 mission, launched at 10.16 EDT on 3 June 1965, began with a big disappointment, due to what nowadays seems an astonishing ignorance of orbital mechanics. The first and major task of the mission was to demonstrate the ability to 'rendezvous' with another vehicle in space, as the first step to docking. The idea was to use the second stage of the Titan launcher, which went into orbit with Gemini as its fuel burned out, as a target vehicle. This was described as 'almost as big as a London tube tunnel, and nearly 30 ft long', and I added that the 'military men' hoped to link them up in orbit and use them to assemble big orbiting space stations manned by 10–12 astronauts who would be sent up a few at a time.

As Titan gave Gemini its final push into orbit and then separated from it, McDivitt turned the spacecraft round to look for the trailing vehicle before they drifted too far apart. There it was, venting propellant like a dying monster, and McDivitt estimated its distance at 120 metres, while White thought it was only 75. McDivitt 'braked' Gemini, aimed it, and twice fired his thrusters to close the gap; but the rocket seemed to move away and downwards. He was still trying when the rocket was up to 5 km away, and had used so much fuel in his futile attempts that manoeuvres had to be reduced for the rest of the four-day mission. Reporting NASA's disappointment at failing to beat the Soviets with the first successful rendezvous, I speculated that Gemini would need much more powerful thrusters before it could be achieved.

It took an engineer in the Gemini project office to point out that the crew and everyone else had failed to understand or 'to reason out' the orbital mechanics.

> To catch something on the ground, one simply moves as quickly as possible in a straight line to the place where the object will be at the

right time. As Gemini 4 showed, that will not work in orbit. Adding speed also raises altitude, moving the spacecraft into a higher orbit than its target. The paradoxical result is that the faster-moving spacecraft has actually slowed relative to the target, since its orbital period, which is a direct function of its distance from the center of gravity, has also increased. As the Gemini 4 crew observed, the target seemed gradually to pull in front of and away from the spacecraft. The proper technique is for the spacecraft to reduce its speed, dropping to a lower and thus shorter orbit, which will allow it to gain on the target. At the correct moment, a burst of speed lifts the spacecraft to the target's orbit close enough to eliminate virtually all relative motion between them. Now on station, the paradoxical effects vanish, and the spacecraft can approach the target directly.

The lessons learned were thus at least as useful as if the rendezvous had been a success. And Ed White's spacewalk was so spectacular, and produced such incredible pictures – just as breathtaking to look at nearly 30 years later – that the rendezvous issue was quickly forgotten by the public. It took White much longer to get ready than expected; and when at last he had hoses and umbilical package ready, hand-held manoeuvring unit (which the astronauts shortened to 'zip gun') tethered to his right arm, and chestpack in place, the two of them had great difficulty in getting the hatch unlatched. As he rose at last through the hatch White installed a camera to take the film which we still enjoy, and found that the double-barrelled zip gun worked well until the compressed-oxygen fuel bottle which powered it was quickly emptied.

Trying to work around to look through McDivitt's window from the outside he found himself dangerously near the flaming plumes of gas from the thrusters as McDivitt fired them to steady Gemini. Moving away from them, White enjoyed his walk so much that frantic calls from Mission Control ordering him back inside before the spacecraft passed into darkness went unheeded. With his comment: 'It's the saddest moment of my life' heard by millions via Early Bird, he squeezed back after 21 historic minutes – encountering much the same problems as Leonov had done. The hatch proved even more difficult to

Astronaut Edward White, later to die tragically in Apollo 1, making the first US spacewalk from Gemini 4 in June 1965. It lasted 21 minutes, and he is seen gingerly manoeuvring with an oxygen-powered spacegun. Pictures taken by his Commander, James McDivitt, are still used to symbolise man in space. (NASA)

close, and it was only achieved when White yelled at McDivitt to yank at his legs to give him enough leverage to close the latch.

The new computer failed them when the time came for re-entry, and McDivitt's manual re-entry, one second late, resulted in a heavy splashdown 50 miles from the target. But the mission created far more interest than any succeeding Gemini mission, and its triumphant ending quite muted the news that US troops were entering the fighting in South Vietnam. McDivitt and White proved that the physiologists' fears were misplaced. Chuck Berry could no longer object to flights of 8 and 14 days to prove that men could survive spaceflight long enough for the return trip to the Moon; and I began to talk with confidence of the first lunar mission starting in 1969.

Ed White, a lieutenant colonel in the US Air Force, was the only early astronaut that I never met – even at a formal news conference. The reason of course was that he was to die with Gus Grissom and Roger Chaffee in the Apollo 1 launchpad fire 18 months later, losing not only his life but a strong chance of being one of the first two on the Moon. However, White and I jointly 'starred' in a BBC schools programme transmitted in September 1965 about the difference in measuring weight and mass.

It started with an interview with White, in which he said that everybody's first reaction to weightlessness was to smile, because it was such a pleasant sensation. As for problems, it was more 'a getting-used-to-it situation'. But you did have to be a good housekeeper, making sure that everything was attached or tethered, and remember that 'cookies would crumble' and water would separate into droplets and float around the spacecraft unless care was taken. White was a little patronising: 'It's a pleasure to talk to a group of young people like yourselves . . . you're going to find as you progress along, all the knowledge that you can pick up now when you are young is going to pay very great dividends in future.'

There was no question of my being patronising, for my contribution was one of life's most painful and alarming experiences. The programme's producer asked me to make a recording of what the opposite of weightlessness was like – becoming several times one's own weight. The RAF readily agreed to take me up in a Jet Provost with a enthusiastic young pilot. With my tape recorder in my lap and holding a microphone, this was the result:

> . . . We're really the water in a bucket as we make this very tight left-hand turn. I'm being dragged down with terribly heavy eyes. My body's turned to lead. I'm blacking out, my lips will hardly work. I'm not being dragged down so much now, but my lips feel as if the dentist has frozen some of my teeth. My head is so heavy that it's

> being forced right down into my neck, it must weigh 30 lb. Why I don't go through the bottom of the plane I don't know. I can feel every link in my spine, I'm sweating very heavily. My lips are working a little bit now, but I can't move my arms and legs. I'm feeling a bit sick, partly because I've had to take off my oxygen mask to record this. Being four times one's own weight feels so awful . . . Thank goodness we've levelled off. I just hung on grimly until it's all over, and that's all I could ever do. How pilots manage to keep control is always a mystery to me.

'Sounds uncomfortable, doesn't it?' was the understated comment of the schools presenter on my tortured speech. In fact I was actually suffering at least 5*g* – and there was a moment when I was absolutely certain that my neck was about to snap. My half scream of 'That's enough!' was on the original tape but to my relief was edited out on transmission. The pilot presumably never heard my plea; and while I was surprised and relieved that my neck did not snap, I was left with with some permanent weakness – the result of having to bend my head to talk into the microphone, although the pilot had warned me that it was essential to keep the spinal column straight. Neither Ed White's weight nor mine had actually changed during our opposite experiences, explained the programme: 'Weight depends on *where* it is measured . . . the true measure of the quantity of material in an object is its MASS.'

Twenty years later, when my friend Geoffrey Perry, of Kettering Space Group fame, retired from schoolmastering, he sent me a tape-recording of the programme, which he had been using ever since to demonstrate the point!

Gemini 5

Shortly after that broadcast, Gordon Cooper and Charles (Pete) Conrad made an eight-day flight in Gemini 5 – nearly three days more than the longest Soviet flight at that time, and equal to a return trip to the Moon. They lost track of a 50 lb 'rendezvous evaluation pod' which they ejected soon after orbit because of fuel cell problems, but succeeded later in completing a rendezvous with a 'phantom' target four times. They splashed down 80 miles short of the target point because in programming the computer someone had fed in the Earth's rotation rate as 360° per day, omitting to add the two decimal place numbers .98. It was another lesson in the need for precision in space travel learned by NASA the hard way; and there was also an effort made after this mission to check the hero-worship building up around the astronauts. After landing they disappeared from public view for an 11-day 'debriefing' – analysing the results and applying the lessons for future missions. By the time they held their post-mission news conference, interest had diminished.

The Soviets, displeased at being overtaken, criticised Gemini 5 as a military mission, because one task given to the astronauts was to try to identity the launch of a Minuteman missile; but Cooper and Conrad got no more than a glimpse of it.

Reporting this I also quoted Dr Wernher von Braun, journalistic necessity at last overcoming my reluctance to forgive him for what his V2 rockets had done to Britain and the Turnill family during the war. While the astronauts were demonstrating to sceptical scientists and doctors that they could survive in space, von Braun had just completed ten successful test firings of the engines for the Saturn 5 rockets which he was building for the moonlandings. 'He's so confident in the future of his rockets that he's now looking far beyond the Moon,' I reported in a piece commissioned by Australia's ABC. 'He wants to send three men to Venus and back in only ten years' time – a journey that will take a year. By 1978 he wants to send another three off to have a close look at Mars – that'll be a two-year flight.'

NASA scientists were almost as disappointed as their Soviet counterparts when, after international monitoring of the flight for nearly four expectant days, Luna 7 crashed into the Moon's Ocean of Storms. It was Russia's third failure to soft-land a robot craft on the Moon's surface – this time because the craft's retrorockets fired too soon. Despite all the photographs taken by both the US and Russia, it was still not known for certain whether much of the lunar surface was covered with a layer of dust so deep and soft that it could engulf a spacecraft attempting to make a landing. It was all the more disappointing for the Soviet scientists because by then they badly needed a success to counteract America's run of successful Gemini missions.

News of the Soviet failure reached me in London on the day that the Science Museum was able to put on public display Alan Shepard's Mercury capsule Freedom 7. It had a lifelike model of Shepard lying on his back inside, feet in the air ready for take-off, his left hand grasping the abort lever so that he could fire the escape tower if anything went wrong. Four years after launch it looked old-fashioned – already just a piece of space history.

But more setbacks also lay ahead for NASA, providing a series of exciting stories. First, however, after months of technical arguments and problems, their first successful rendezvous and station-keeping exercise was achieved a week before Christmas.

Gemini 6 and 7

The original plan, attempted in October 1965, was to launch an Agena upper stage on top of an Atlas rocket to act as a rendezvous target vehicle for Gemini 6. If that was a success, later Agenas would be fitted with a docking collar so that the first link-up of two vehicles in space could be conducted. There could be no flights to the Moon, it had been decided, until docking techniques had been mastered.

But NASA's first attempt at a double launch was not a success. Simultaneous countdowns were conducted at the Cape for the Atlas/Agena and the Titan/Gemini launches on nearby pads. Agena upper stages had been used 140 times since 1959, and no trouble was expected with this one, despite modifications which meant it was really an unmanned spacecraft, with propulsion systems that could be commanded either from Houston or by Gemini 6's commander. Walter Schirra, 42, and Tom Stafford, 35, climbed confidently into their Gemini couches 15 minutes before Atlas/Agena was due to lift off. They were to be launched as the Agena was completing its first circuit

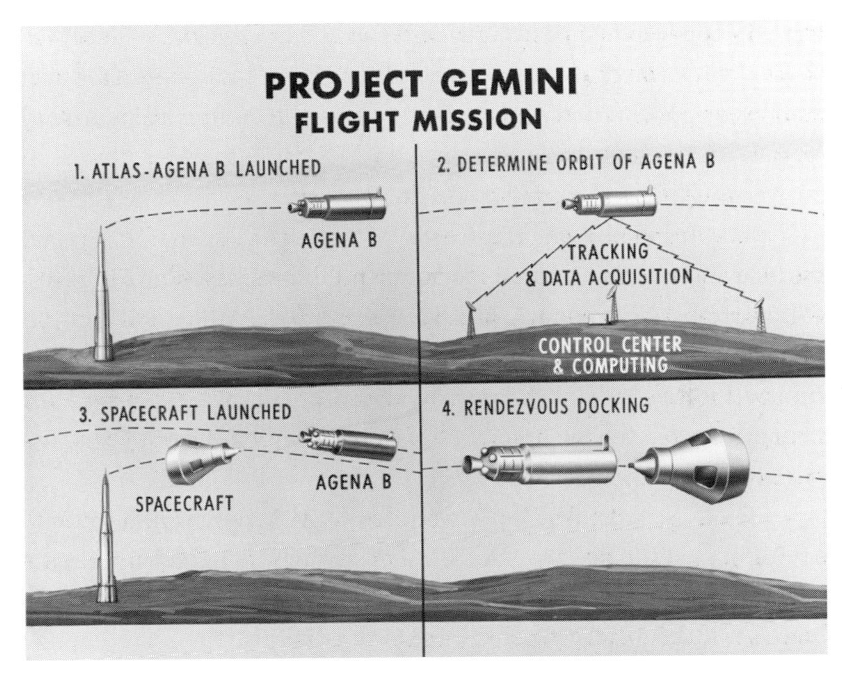

December 1965: NASA's plan for achieving the world's first 'rendezvous in space' with Gemini 6 and an Agena target rocket. When Agena was lost, the rendezvous was daringly achieved with two manned vehicles. Gemini 7 was sent up for Gemini 6 to use as the target. (NASA)

of the globe, and to be placed in a slightly lower orbit. Then with the help of radar and computers, they could catch up, lift their orbit, and take station alongside it on their fourth revolution. But to the astonishment of the Air Force men who had successfully launched so many military Agenas, this one exploded just after separating from the Atlas launcher. (It was ultimately discovered that the engine had 'backfired' and then exploded because fuel was injected into the firing chamber before the oxidiser.)

The Gemini countdown was abandoned, and Schirra and Stafford climbed disappointed out of Gemini 6, their flight postponed indefinitely. Weeks of argument followed. Two executives of McDonnell, makers of the Gemini spacecraft, offered a quick solution by suggesting that two manned Geminis should be launched in quick succession, the first to act as the rendezvous target instead of the Agena. They were turned down many times, but finally got their way because the doubters at the top could adduce no convincing arguments against it.

I decided to cover the start of the dual launch at Mission Control, Houston, rather than at the Cape. With Gemini 6 and its Titan 2 launcher stored under guard in a hangar, Gemini 7's countdown was trouble-free, and it was launched on Saturday afternoon, 3 December 1965, only five seconds behind the flight plan, into a 104 × 204 miles orbit – the apogee, or high point, only 6 miles lower than planned. TV coverage had been cut by three-quarters. The American public preferred to watch the Maryland versus Pennsylvania football match.

Frank Borman and Jim Lovell, both 37, were going ahead with their long-planned 14 days' long-duration flight, with completion of that as their prime task. But as soon as the launchpad had cooled down after lift-off, 300 technicians were swarming over it, repairing any damage caused by Titan's 200-tonnes thrust, and erecting the stored Titan and redesignated Gemini 6A craft in its place, racing for a second launch only nine days later. Borman and Lovell, who faced not being able to stand up or have a decent wash for two weeks, had sensors attached to shaved patches on their heads. That was because Houston scientists wanted to compare how they slept in space with similar tests they had already undergone on the ground – unpopular activities with the astronauts (and, I always suspected, of very doubtful value, like many of the other so-called 'life science' tests.) Because of concern about possible bone demineralisation, all their faecal matter was to be stowed in bags around their feet, for analysis when they returned. Jim Lovell's wife Marilyn, much favoured by the photographers for her good looks, was likely to have her fourth child during the mission, so we nicknamed Jim 'the first astrodad'.

After a quick visit to the Cape to see progress with erecting Gemini 6A, I went on to Washington, to cover the latest rows about Defence Secretary McNamara, who had decided to scrap 400 of America's 680 strategic bombers and to save a billion dollars on the

Astronauts Jim Lovell and Frank Borman, unshaven and unwashed, on the deck of the recovery ship after 14 days in December 1965 in the tiny Gemini 7 craft. They held the long duration record for 5 years until overtaken by Russia's 18-day Soyuz 9 mission in 1970. (NASA)

defence budget by replacing them with the 'all-purpose' swing-wing fighter, bomber and reconnaissance plane, the F1-11. That, as described elsewhere, was the aircraft that America was pressing Britain to buy after scrapping her own TSR2. Then it was back to England to cover the rest of the Gemini mission from there in between visits to the House of Commons to cover heated debates about whether Britain should or should not order the F1-11s.

Thus I missed the greatest drama of the double mission. Gemini-6A was ready for launch on the eighth day of the Gemini 7 flight, count-down went smoothly, and the engines roared into life. But at 1.2 seconds the malfunction system sensed something wrong and shut them down. All the rules said that Schirra should have pulled his D-ring to eject himself and Stafford from what seemed certain to be a rocket filled with 136 tonnes of propellants about to crash to Earth. But Schirra had sensed that the rocket had not begun to lift off, and was thus

The first space 'rendezvous'. Schirra and Stafford manoeuvred Gemini 7 (seen in foreground) to within 0.5 metre of Gemini 6, flown by Borman and Lovell, in December 1965. (NASA)

still securely anchored – despite the fact that the lift-off clock was ticking. When such crises came, it was the split-second decisions of the test-pilot astronauts who saved the missions – on this occasion stimulated by the knowledge that the 20g acceleration resulting from an off-the-pad ejection was certain to leave them with serious injuries.

An engineer found that a dust cover had been accidentally left in an engine during servicing, and three days later Gemini 6A soared safely into a 98.6 × 161 miles orbit. The Gemini 7 crew needed all their manoeuvring fuel if they were to complete their 14-day mission, so they were to be a passive target. Schirra in Gemini 6A, aided by a new computer considered remarkable because it was no bigger than a hatbox, did all the manoeuvring; he was able to squander all his propellant in a one-day flight. It still took him six patient hours gradually raising his orbit until the two vehicles were 40 metres apart, with no relative motion between them.

Then, for more than two revolutions the two craft stayed together at distances ranging from 0.30 to 90 metres, sometimes nose to nose, the crews eyeing one another through their windows. Schirra and Stafford commented on the 11 days' growth of beard worn by Borman and Lovell, and were so calm that Mission Control had to ask them to repeat how close they were: 'Ten feet between the two spacecraft,' said Schirra, ignoring, as the astronauts often did, instructions to use the metric system.

Gemini 6A returned triumphantly home, knowing that the techniques required for docking had been mastered; and Borman and Lovell stayed on for another three days, relieving boredom between experiments by reading Mark Twain's *Roughing It* and Walter Edmonds' *Drums Along the Mohawk*. Gordon Cooper had advised them to make room for a book after getting bored on his eight-day flight.

9 Overtaking the Russians

Gemini 8

The moment when NASA overtook Soviet space technology came during Gemini 8 in March 1966. It was clearly recognisable to those following the rival programmes. But at the time it made little impact. Attention was focused on the drama that followed within minutes, ending in an emergency re-entry and splashdown. Once again, having just returned from a US visit to cover the introduction there of Britain's new jet, the BAC One-Eleven, and having snatched time for briefings from NASA and space contractors, I covered the mission from BBC studios, wearing headphones linked to Mission Control.

I had had a dramatic demonstration of how the rendezvous and docking would be carried out during a visit to the McDonnell Douglas plant at St Louis, Missouri. There I watched it being done again and again on a simulator, while Bob Lindley, the Manchester-born executive in charge of Gemini spacecraft development, assured me: 'It's as easy as falling off a log.'

The demonstration was more subdued than usual, for only yards from the simulator there was a large hole in the hangar roof, and on the floor below it a jumble of wreckage. A few days before, on 28 February, the Gemini 9 prime and backup crews had flown from Texas to St Louis in two of the T38 jet trainers allocated for the astronauts' use as an effective way of keeping their pilot skills sharp. The four men had planned to spend several days practising on the simulators. Flying in formation, the aircraft, with prime crew Elliott See and Charles Bassett in one, and the backup crew, Tom Stafford and Gene Cernan in the other, found, as they approached in a mixture of snow flurries, rain and fog, that they were too far down the runway to land. Stafford followed the standard 'missed approach' procedure, and climbed straight ahead into the cloud, to make a second, successful instrument approach.

See, however, doing the flying as commander of the prime crew, decided to keep the runway in sight by circling beneath the cloud cover. There was always much competition between the astronauts, and he probably wished to demonstrate how his greater skill and daring would enable him to land first. But he was too low. Realising it, he cut in the afterburners and made a sharp right turn. Too late. The T38 struck the roof of the building only yards from the simulator below. To get to the simulator I had to walk around the aircraft wreckage on the floor, which was awaiting the arrival of Astronaut Alan Shepard and the other accident investigators. The major part of the aircraft crashed into a courtyard just outside, killing See and Bassett instantly. By the time Stafford and Cernan emerged from the fog to make their landing they had become the prime crew for Gemini 9.

Preparations for Gemini 8 continued without interruption. The explosion of the Agena target rocket on Gemini 6 had already cost NASA some anxious months ensuring not only that such an explosion would not recur, but that if a Gemini spacecraft docked with it the two vehicles would not be thrown out of control.

While they waited, the Gemini 8 crew rehearsed an elaborate spacewalk to accompany manoeuvres which would include docking and undocking with their Agena target at least four times. I commented that while it seemed strange that on this mission neither of the astronauts had made a spaceflight before, there was great confidence in the commander – Neil Armstrong. He was the first civilian to fly for NASA, but he was among the country's foremost test pilots. During seven years flight-testing experimental aircraft, he had travelled at over 4000 mph in the air-launched, rocket-powered X-15. But despite this, the pre-mission limelight fell upon his USAF companion, David Scott. He was the prototype, fresh-faced, all-American boy. He was also assigned the glamorous part of the mission. That involved spending 2.5 h outside Gemini 8, clambering outside and around it to the service module at its rear. There he had to strap himself to an astronaut manoeuvring unit, clipped like a bicycle to the rear because it was too big to be carried inside the spacecraft. He also needed to extend the 8-metre oxygen-hose tether with which he left the spacecraft by inserting a lightweight 23-metre addition. Finally, he had to pick up a much improved 'zip gun' which used Freon instead of the oxygen in White's EVA gun, together with a do-it-yourself power tool.

With all this equipment, Scott was to 'swim' to and fro between Gemini and Agena, practising bolting and unbolting a metal plate. The power tool had been designed with a contra-rotating barrel so that it would turn a nut and not the unanchored, weightless astronaut. Practising for these activities, Scott had done more than 300 zero-*g* parabolas in the aircraft known as the Vomit Comet.

On the day, all this was to prove a sad waste of time and effort. On 16 March the Agena target vehicle was successfully launched into a 298 km circular orbit, followed 41 minutes later by Armstrong and Scott in Gemini 8. From their 160 × 272 km orbit, it took them 5 h 43 min to station their craft within 46 metres of the Agena target; 36 minutes later Mission Control passed the instruction: 'Go ahead and dock!' Seconds after, Armstrong reported: 'Flight, we are docked . . . really a smoothie. No noticeable oscillations at all.'

Cheers and backslappings at Mission Control did not last long. The Agena was programmed to obey orders from either the spacecraft or ground control, and soon after Gemini had taken it over, Scott reported: 'Neil, we are in a bank!' Armstrong succeeded in stopping the motion, but not for long. At first they blamed the Agena's control system, for this was what had been feared. After Scott turned off the Agena's attitude control system the two craft steadied and straightened. But suddenly the roll began again, steadily increasing. In desperation they undocked – and learned that it was not Agena's controls that were at fault, but their own. 'We have serious problems here. We're tumbling end over end up here!' Scott reported. Spinning at one revolution per second they were becoming dizzy, with their vision blurred. Had they been over a ground station, Mission Control's telemetry would have enabled them to tell the crew immediately that the trouble was caused by No.8 thruster firing continuously, having stuck in the open position – and how to stop it. But communications at that time were far from continuous. Armstrong, with no way of knowing what was wrong, cut in the re-entry system, which enabled him to regain control, but started an emergency return to Earth, with David Scott's glamorous jetgun activities all abandoned in a flight that lasted only 10 hours.

There was little interest in my reports that the docking was something the Soviets were still far from achieving. There was more interest in routine Soviet complaints that the mission had had military objectives. This was vigorously denied by NASA; but I did point out in one broadcast that the US military were delighted at the proof that had been provided that it would be possible for their MOL astronauts to go alongside Soviet military satellites to inspect them, and, if they proved hostile, to disable them.

And despite their lack of manned flights during the Gemini missions, Soviet space scientists were far from idle. They had stolen some of the limelight on the eve of Gemini 8 by bringing safely back to Earth two dogs, named (when translated) Breeze and Little Lump of Coal. They had survived in space for three weeks, after being placed in an elliptical orbit which passed them repeatedly through the Van Allen radiation belts surrounding the Earth. This was to see if they, and the spacemen to follow, came to any harm. Strapped in corset-like suits, with little freedom of movement, the dogs had been fed with pellets of meat, potato flour, vitamins and water, placed automatically in their stomachs at regular intervals.

When being interviewed by the BBC *Today* programme I was thankful not to be asked whether this constituted grave cruelty to animals, and forecast that their survival would soon be followed by Soviet cosmonauts undertaking a flight around the Moon.

There had also been a delightful row involving Jodrell Bank the previous February, when at last the Soviets had achieved the first soft-landing of a spacecraft on the Moon's surface. Five days after it we were still awaiting Soviet announcements as to whether the craft, Luna 9, had survived the touchdown, sunk into the lunar dust, or was sending back TV pictures. The Soviets, however, had revealed in advance to the scientific community the frequency on which their TV pictures would be transmitted – probably because they were unsure of the quality of their own receiving equipment, and might need some backup help from the West.

They got much more help then they desired or wanted. It was discovered, I believe by a *Daily Express* technician in Manchester, that the Soviets were using standardised international equipment to receive their lunar pictures, similar to the 'wire-photo machines' being used by newspapers to transmit pictures from race meetings to newsdesks, or from one newspaper office to another – the origin of the now ubiquitous 'fax', or facsimile machines.

The *Express* arranged with Sir Bernard Lovell to rush one of their wire-photo machines to Jodrell Bank, together with engineers to link it up and make it work. The result was a rare scoop for the newspaper and for the BBC. What I described as 'the first man's-eye pictures of the Moon' – four views of what a man standing on the Moon would see as he looked across the surface to the horizon – were shown to millions in the West long before they were published in the Soviet Union.

Soviet scientists justified their anger by pointing out the poor quality of the *Daily Express*/Jodrell Bank pictures. They had been holding back publication of their own, they said, until the quality had been enhanced through their computers. Lovell's relations with the Soviets, which had been somewhat cozy before that because he had been able to give them considerable technical help, turned frosty for a while. But since the Soviets needed his help more than ever, they soon forgave him; and he in his turn apparently accepted that it was rather unfair to steal Soviet pictures before the Russians themselves could release them.

The incident was a landmark in the East–West space war for all of us. Soviet activities had been forced out into the open for the first time. One assumed, with something less than complete confidence – for it was always easy to over-estimate their competence and abilities – that

Western surveillance agencies had also been secretly acquiring the pictures and keeping quiet about it; and space correspondents could interpret with much more confidence what might be going on behind the scenes.

The Luna 9 incident was quickly forgotten when NASA's Surveyor 1 spacecraft made the world's first fully-controlled landing on the Moon's Ocean of Storms on 2 June 1966. It dispelled any lingering fears that such areas of the lunar surface would be so unstable that landers would be swallowed up in dustlike quicksands. In the six weeks after the landing it sent back 11 150 TV pictures, ranging from horizon views of mountains to close-ups of its own mirrors.

Surveyor was the third of three unmanned lunar exploration projects carried out in parallel with the three manned projects aimed at placing men on the Moon before 1970. Of the six Surveyors that followed, Nos 3, 5 and 6 landed at sites spaced out across the lunar equator surveying the areas for future Apollo landings, and Apollo 12 was to land within astronaut-walking distance of Surveyor 3 in 1969. After touching down, Surveyor 6 lifted off and landed again to demonstrate the ability of such a craft to return to Earth; and Surveyor 7 landed on the rim of Crater Tycho to conduct digging, trenching and load-bearing tests amid debris that had once been molten. That was the most exciting area within Apollo's reach, and the astronauts were disappointed when their pleas to land there were turned down.

Gemini 9–12

Media interest in the last four Gemini missions, all completed before the end of 1965, rapidly declined as it appeared that no matter how much trouble they got into, the astronauts proved they were able to survive. But the public perception did not reflect attitudes inside NASA. In mid-May there was reason for relief when only 200 newsmen got themselves accredited for the Tom Stafford–Gene Cernan mission, to cover a series of mishaps. Gemini 9 was intended to complete the different rendezvous, docking and spacewalking activities left undone by the previous flight. The men were preceded by the

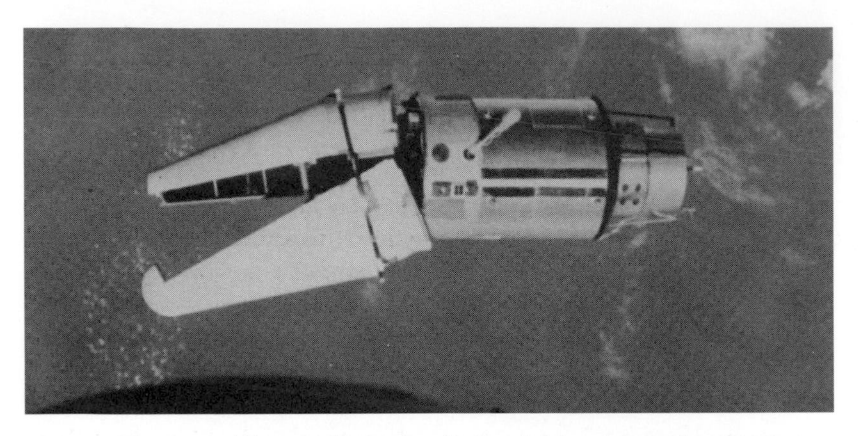

The Angry Alligator: this is what the Gemini 9 crew saw when they arrived in orbit to rendezvous and dock with the Agena rocket launched in advance as a target. The protective shrouds had failed to separate, making it dangerous to approach too closely, and giving the 'alligator' appearance.

Atlas rocket, carrying the Agena upper stage intended as the docking target; it took off on time, but two minutes into launch flipped over into a nosedive. To quote the official NASA record, it 'headed like a runaway torpedo back towards Cape Kennedy'. There was only partial relief when it plunged into the Atlantic Ocean 198 km from the launchpad.

Inquiries and backup plans led to a new Agena being launched on 1 June, but the Gemini countdown missed its 40-second launch window, and the crew once again descended in the launchpad elevator. Stafford complained that, while Borman and Lovell might have more flight time than he did, nobody could match the 'pad time' he had clocked up on the Gemini 6 and 9 missions.

Worse was to come. When they did get into orbit two days later they rendezvoused with what Stafford said looked like 'an angry alligator'. The clamshell-like shroud protecting the Agena during launch, and which should have been jettisoned by explosive bolts as it went into orbit, had not worked properly. Approaching within 30 metres, Stafford told Mission Control that the bolts had fired, but the half-open shrouds were still attached by neatly-taped lanyard wires. A later inquiry discovered that the Douglas engineer who knew what to do with the lanyards had gone off home to his pregnant wife, and on launch day a McDonnell crew followed the wrong Lockheed procedures. Not knowing what to do with two dangling lanyards, they had neatly taped them down!

The crew was willing to risk trying to knock off the shrouds with the nose of their Gemini, but Mission Control feared that it might cause them damage; and Gene Cernan's offer to cut the lanyards while spacewalking was also vetoed in case his EVA suit was punctured during the process. Stafford did carry out three different methods of rendezvousing with his target, but there was no docking. And then, when Cernan spent more than two hours outside Gemini, his facemask fogged up, and like Dave Scott before him he too was unable to try out the astronaut manoeuvring unit. While struggling blindly outside, he also broke off the 8 ft extendable antenna being used by the Department of Defense to measure the efficiency of communications through the ionosphere.

DoD was quite put out about the partial failures on the Gemini 8 and 9 missions. They had been hoping that NASA would not need their last two flights – Nos. 11 and 12 – and had made secret plans to take them over. Their own MOL astronauts were hoping to rehearse their ability to intercept hostile satellites passing over US territory, examine them for bombs or other weapons, and simulate their destruction. John Young, then 36, probably the most cautious but ultimately the longest-serving of the early astronauts, is on record as having commented that NASA chiefs 'must be out of their minds' when told that on Gemini 10, which he was to command in July, he must not only dock with his Agena target, but use its engines to fire his craft into the highest orbit yet to rendezvous with the 4-months-old Gemini 8 Agena with which Neil Armstrong had got into so much trouble.

Young's companion was Michael Collins, also 36. There was only a 35-second launch window if they were to have the best chance of achieving their double rendezvous. But this time the Agena was launched only two seconds late, and Gemini 10 followed exactly on time 100 minutes later. They duly docked with their Agena, and fired its powerful engine several times to boost their speed and take them into a 763×294 km (474×183 miles) orbit. After undocking from their own Agena, Young manoeuvred alongside Agena 8, and Collins scrambled out of the hatch to use his zip gun to propel himself between the two craft. Recovering an experiment package from Agena 8, he found it difficult to get a grip on Agena's smooth lip, and found like previous spacewalkers that EVA work was difficult, dangerous and deceptive. He also got tangled in the umbilical when trying to return to Gemini 10, and Young had to reach out to unwind him before he climbed back inside quite exhausted.

Neverthless the general impression inside and outside NASA was that Project Gemini was nearing a successful conclusion. With only two missions left, Gemini managers were having trouble holding their team together. Succeeding programmes, Apollo and Apollo Applications (which led to the Skylab missions), and most of all MOL, were all competing to persuade the uniquely qualified Gemini technicians to switch to their teams. Therefore NASA headquarters pressed Houston to speed up the Gemini programme and get it over.

This was not so easy as it might have been, for the Gemini 11 crew was led by Charles (Pete) Conrad, the most dynamic and adventurous of the astronauts. He campaigned hard for turning his flight into a trip around the Moon, or at least into a very high orbit. He and his co-pilot, Richard Gordon, had to settle for an ambitious plan to rendezvous and dock with their Agena target on the first orbit, and experiment with creating artificial gravity by undocking after linking the spacecraft to Agena with a 30-metre dacron tether. To succeed, Gemini 11 had to be launched within a 2-second window 97 minutes after the Agena. It needed three countdown attempts, but was achieved on 12 September. Conrad was the sort of man who made things happen, and the first-orbit docking – an important rehearsal for Apollo moonflights – was one of them. Gordon donned his spacewalking equipment, went outside and with a great effort pulled himself astride the nose of Gemini 11 to lean forward and attach the tether, housed in Agena's nose, to the Gemini's docking bar.

We heard Conrad's enthusastic shout: 'Ride 'im, cowboy!' – but as previous spacewalkers Scott and Collins had warned him, Gordon found it incredibly more difficult to do than rehearsals on earth had suggested. He succeeded after a 6-minute struggle, but was so exhausted and blinded with his own sweat that Conrad had to help him back inside – once more abandoning the rest of the spacewalk and other experiments, including the use of a power tool. The engineers had once again failed to provide the necessary restraints and handholds outside the vehicles needed by spacewalkers to control their movements. Discussing on TV news the astronauts' demands for improvements to the cumbersome, seven-layer spacewalking suits, I described the work going on at the RAF Institute of Aviation Medicine at Farnborough to develop a water-cooling system between the layers for the Apollo flights – at that time planned to start before the end of the Gemini missions.

Before undocking to test the tether, Conrad partially fulfilled his ambition to get into a much higher orbit. He used the Agena engine to increase the apogee to 1570 km, and with the hatch open and Gordon standing with his head outside, they got some wonderful views and pictures. After returning the apogee to 304 km, Conrad cautiously undocked from the Agena for a 3-hour test of the tether, which proved to be 'interesting and puzzling'. They found it difficult to keep the tether taut, but did succeed in inducing enough cartwheeling motion to generate sufficient 'artificial *g*' to cause the camera to slide to the back of the spacecraft, although the astronauts themselves noticed no difference. The tether was then released by jettisoning Gemini 11's docking bar.

Jim Lovell and Edwin (Buzz) Aldrin, the Gemini 12 crew, as well as DoD, were displeased when the Gemini Mission Review Board, considering what needed to be done on the last mission, decided to abandon the last chance to try out the US Air Force's AMU – the astronaut manoeuvring unit. Successive missions had failed to make use of it, either because it was difficult to reach and don, stowed as it was in the service module at the rear of Gemini, or because the astronaut was too exhausted by other activities before he was due to use it. The NASA chiefs were so concerned at the EVA problems encountered, that they decided to concentrate during the final flight on 'repetitive performance of basic, easily-monitored and calibrated tasks'.

The Air Force was particularly frustrated because they had just tested the hardware for their $500 million MOL space station by trying out a modified Gemini with the hatch built into the heatshield. The military astronauts would have to pass through that into the 40 ft-long laboratory in which they would spend 30 days or more – and an unmanned up-and-down lob had satisfied them, so far as we were allowed to discover, that re-entry heat would not penetrate the sides of the door. All that was then needed was a checkout in orbit of their new EVA suit – and once again NASA had let them down!

This, however, was a period when NASA was able to resist military pressure to include experiments and activities not strictly necessary for the moonlanding programme. By now the second of five Lunar Orbiter spacecraft was busy sending back photographs of possible landing sites for Apollo, and the prospects were bright at last that the Soviets could be beaten to the first manned lunar landing.

Lovell's recorded fear that 'essentially Gemini 12 didn't have a mission' proved to be unfounded. During a four-day flight starting on 11 November he and Aldrin demonstrated that NASA had the ability to dock and undock routinely with their targets. With the help of 44 handrails, handholds and foot restraints, plus rings to which he could hook his belt – compared with only nine on Gemini 9 – Aldrin completed three spacewalks totalling 5.5 hours without getting exhausted. He also tied the two vehicles together without the difficulties encountered by Gordon. Lovell spent four hours on tether exercises, proving that this would provide a useful form of station-keeping, although it was admitted later that the behaviour of the craft when tethered was not fully understood. [The use of such tethers was to wait another 26 years before more advanced experiments could be made with Space Shuttle Atlantis and Italy's tethered satellite – only to encounter more failures.]

Taking stock, George Mueller, NASA's Associate Administrator for Manned Spaceflight, said that seven different rendezvous 'modes or techniques' had been achieved: on the fourth, third and first orbits; optical rendezvous; from above, stable orbit and optical dual rendezvous. Alas, we got no guidance about these subtleties as we listened while they were being carried out!

The only goal Gemini had failed to achieve was that of setting down on land. Tests had been carried out with paragliders and Rogallo wings in the hope that Apollo craft could be brought back on land rather than splashing down in the sea – very expensive because of the heavy involvement of the US Navy. The paraglider tests, however, fell far behind the development of the Gemini spacecraft, and were never seriously considered for the much heavier Apollos returning from the Moon.

The many benefits which accrued from things that went wrong could best be illustrated by the explosion of the Agena target vehicle intended for Gemini 6. The lesson learned was that oxidiser should be injected into the firing chamber before the fuel. This modification was made to the ascent engine for the Lunar Module, and could well have avoided a situation in which astronauts successfully reached the Moon, but were killed by an exploding engine when they attempted to take off for the return flight.

As for costs, Project Gemini started at an estimated $531 million and actually cost $1147 million; but political and public relief at the way its achievements seemed to have put the US ahead of the Soviets in the moonrace was such that I heard no serious complaints – even from the Senator for Wisconsin.

By this time international lawyers had started to worry about ownership of the Moon. In the US there was much concern about the weak position they would face when it seemed the Soviets might get there first. Now that the odds had swung in favour of the Americans, it was the Russians' turn to worry. 'In the old days', I said on the *Today* programme, 'there was something called "the principle of national appropriation". When Captain Cook or some other explorer found a new continent or even a new island he raised his country's flag, and made a little speech to the astonished natives proclaiming that this terrirory belonged to his nation. And though there were arguments about who really got there first, the [civilised] world accepted the system.'

The East–West moonrace had started because each side had feared the other might invoke this time-honoured principle, and claim the Moon as a military base. Although Russia's Luna 9 and America's Surveyor 1 had had no men aboard, each had carried the rival national flag to the lunar surface. But by the time Project Gemini had been completed, the parallel development of long-range ballistic missiles had made military nonsense of such strategic theories. All the same, neither side was anxious to take any chances, and had apparently concluded that it would be best to apply to the Moon the Antarctic agreement of six years before: it would be declared international territory, with the Super Powers agreeing that no military bases would be permitted.

The long-term bonus for humankind was that unspoken plans in the minds of the military to use the Moon for missile target practice and nuclear testing were suppressed for ever.

10 Apollo's Bad Start

Apollo 7

On the last day of 1966, I found an opportunity to stake my claim to cover the Apollo moonlandings. NASA had scheduled the first three flights in 1967, so I did my reminder in the form of a 'curtain raiser' on *Radio Newsreel* – still the BBC's most important news programme – to update both our listeners and BBC News executives on the excitements that lay just ahead.

The first manned mission, Apollo 7, already several months late, was by then targeted for 21 February, commanded by Gus Grissom – and as a reminder of the dangers, I pointed out that he had had to swim for his life at the end of his first orbital flight aboard a Mercury craft five years earlier. The first three Apollos were to be 'open ended', lasting up to a fortnight, depending upon what they achieved. The second, then Apollo 8, was to be the exciting one. The Lunar Module was to be sent up the day after Apollo was orbited, and Jim McDivitt was scheduled to dock with it, go aboard with another astronaut, and make a test flight in the Lunar Excursion Module (LEM) before redocking, jettisoning the LEM and returning to Earth. Saturn 1B, the first stage of von Braun's moonrocket, was to be used for these low Earth orbit operations; and before the end of the year the complete 350 ft high Saturn 5 was to be man-rated for the first time, by placing three astronauts in a 4000-mile high orbit. If all that went according to plan, Apollo 10 would have the capability to reach the Moon in early 1968, and privately NASA chiefs expected the first moonlanding to take place by the end of that year.

At the end of January, after a tedious time trying to unravel the deliberate obscurities of Denis Healey's plans to reduce Britain's defence budget by bringing home thousands of British troops stationed overseas, I was having a rare weekend off. It was late on Friday

January 1967: the charred interior of the Apollo 1 spacecraft, in which Astronauts Grissom, White and Chaffee died in an inferno of fire. Locked inside for a ground rehearsal of their launch three weeks later, they tried in vain to open the hatch. Many warnings had been given, including one only three weeks earlier by Dr Charles Berry, the astronauts' medical director, of the dangers posed by Apollo's 100% oxygen atmosphere. Apollo's subsequent success was due to the complete redesign that followed. (NASA)

27 January 1967 that the agencies broke the news that Gus Grissom, Roger Chaffee and Edward White had died on Cape Kennedy's Launch Complex 34. There had been some sort of electrical 'short' inside the Apollo Command Module in which they had been locked for 6½ hours while going through a complete simulation – in effect, a full dress rehearsal – for their launch three weeks later. There had been warnings of fire hazards in a spacecraft with a 100% oxygen atmosphere; now it had happened. Once fire started, the very air they breathed burst into flame; and although Grissom and White had made desperate efforts to unlock the exit hatch from inside, they failed to complete it before they succumbed.

It was ironic that the ability to punch the hatch open with an explosive charge had been removed in favour of a manual system after Grissom's insistence that when he was bobbing on the sea awaiting the recovery team at the end of his Mercury mission the hatch cover jettison mechanism had exploded spontaneously, so that he had to swim for his life. NASA's official history records how Gary Propst, an RCA technician stationed in front of a television monitor trained on the command module's window, saw a bright glow inside, followed by flames flaring around the window: 'For about three minutes the flames increased steadily. Before the room housing the spacecraft filled with smoke, Propst watched with horror as silver-clad arms behind the window fumbled for the latch. "Blow the hatch, why don't they blow the hatch?" he cried. He did not know until later that the hatch could no longer be opened explosively.'

Despite all my background knowledge of the space programme, there was little I could contribute to the story over the weekend. But within a few days I was at Mission Control at Houston – probably to the annoyance of the BBC staff in Washington and New York. Lockheed and General Dynamics had jointly decided to fund a visit by aerospace and defence correspondents to their works at Los Angeles and Fort Worth – Lockheed to show us their supersonic aircraft mockup, and General Dynamics to pressurise Britain into ordering more of their F1-11 aircraft. I covered those stories in double-quick time, and hurried to Houston with my accompanying camera crew.

With their usual generosity in those days, NASA let me have the use of the Apollo simulator. I lay in the Command Pilot's couch in the left-hand seat, and talking over my head to the camera demonstrated how little chance Ed White had had, lying in the centre seat and reaching over his left shoulder, to operate the handle of the inner hatch. Grissom's job in such an emergency was to operate a pressure dump lever to blow out the oxygen, then roll around to help White lift the inner hatch down to the floor. Chaffee's job was to stay where he was and maintain communications. The minimum time needed for them to escape was 90 seconds; the evidence showed that they were overwhelmed by the fire in 15 seconds. Grissom and White were found with their bodies intertwined as they worked on the door; Chaffee was still strapped in his couch.

Soviet scientists, forgetful of the old adage about not throwing stones if you lived in a glasshouse, were quick to criticise the all-oxygen atmosphere in Apollo, pointing out that in their Voskhod spacecraft they had a normal two-gas system of oxygen and nitrogen. The NASA policy had been to pump in oxygen at slightly higher than sea-level pressure on the launchpad, to ensure that any leaks would be outward; as soon as it was in orbit, spacecraft pressure would be reduced to 5 psi (pounds per square inch) which was much safer and more practical when it had to be completely depressurised for space-walks. Space scientists at Britain's Royal Aircraft Establishment had suggested before the accident that it would be wise to carry a pressurised nitrogen cylinder, the contents of which could be released to smother the flames if fire broke out, and this was one of the remedies considered after the accident.

The report of the Board of Inquiry set up by James Webb, the NASA Administrator, was completed in eight weeks, and totalled 3000 pages, divided into 14 booklets and six appendices. But much was known about about the many weaknesses of the Apollo and LEM spacecraft long before the accident. A NASA friend working on them showed me with some relief – he was anxious to demonstrate that the technical staff were not so dumb as the media was suggesting – an internal report compiled in 1964 which spelt out the hazards posed by space cabins with all-oxygen environments. And three weeks before the fire our friend Chuck Berry, the Medical Director, had complained that it was harder to eliminate hazardous materials from the Apollo spacecraft than it had been in either Mercury or Gemini. In the same month, Douglas Broome, of NASA's Apollo office, had pointed out that the wiring used in the communications system was too flimsy and subject to damage.

The report accepted that what had happened in the Command Module could just as well happen to the Lunar Module; and to the relief of the surviving astronauts, who had been well aware of the spacecraft's deficiencies, there was an 18 months' delay while the systems were completely redesigned with no expense spared. During ground tests a normal two-gas Earth atmosphere was substituted, which could be gradually replaced with the desired 5 psi oxygen atmosphere after launch. All combustible material inside the two spacecraft was replaced, and the double hatch was replaced with a single hatch with a manual release which could be opened in three seconds from inside or ten seconds from outside. Looking back, it is astonishing how completely the spacecraft systems were redesigned during those 18 months.

I had already reported Grissom as saying of spaceflight risks: 'If we die we want people to accept it . . . we hope it will not delay the programme. The conquest of space is worth the risk of life.' There is little doubt that the tragic deaths of Grissom, Chaffee and White resulted in the release of funds for improvements that would not otherwise have been made; later accidents during flight which could easily have led to the abandonment of the whole project were thus avoided.

As always, the bad news made it possible to get some of the good news used as well. Amid my TV and radio explanations from Houston of what had gone wrong, I was able to insert some coverage of close-up pictures of the Moon's surface being sent back from a height of only 39 miles from the third Lunar Orbiter spacecraft. The search was underway for safe landing sites, and while pictures of the Ocean of Storms showed craters between 20 and 9 miles in diameter, there was plenty of room to land between – as indeed Apollo 12 demonstrated three years later. Other pictures we showed on TV included Central Bay and the Ocean of Vapours, pitted with small craters and a deep valley, and quite unsuitable for manned landings. Even more remarkable, the Lunar Orbiter got a picture of the Surveyor 1 spacecraft which had soft-landed a year earlier; it was pin-pointed by its 30 ft-long shadow.

Soviet politicians' smug criticisms of the Apollo accident had barely died away when, only three months after it, Russia's Vladimir Komarov became the first man to die in a spaceflight accident. He had been launched on his own in the first test flight of the three-person

Only three months after the Apollo 1 disaster, Vladimir Komarov became the first man to die in space. His spacecraft crashed when the re-entry parachutes failed. (Novosti)

Soyuz. It was the Soviet Union's first manned flight for over two years, and I had indulged in much speculation – based on hints from Moscow radio – that two spacecraft would be sent up, and that the Russians would, as usual, try to upstage NASA by having some cosmonauts spacewalk from one craft to the other. Komarov's problems throughout this ill-fated flight were fully described for the first time in David J. Shayler's *Disasters and Accidents in Manned Spaceflight* (Praxis Publishing Ltd, 2000). The second launch was cancelled and Komarov was ordered to re-enter on his eighteenth orbit.

As he came down, the main parachute-harness twisted, the spacecraft went into a spin, and from a height of 6 km Komarov crashed and was killed. George Brown, Britain's maverick Foreign Minister, sent a message of sympathy to his Soviet opposite number. The Americans, behind similar messages of sympathy, concealed their relief that the Soviets too had been delayed in the race to the Moon, and James Webb, head of NASA, expressed the rather hypocritical hope that the two accidents would lead to more effective co-operation between the two countries in future space programmes.

If the Moon missions had not been delayed by the need to redesign the spacecraft systems, they certainly would have been by problems with the three stages of the Saturn 5 launcher. When the S2 second stage was delivered to KSC in February 1967 the quality control teams counted 1407 'discrepancies' – mainly wiring errors. With much difficulty therefore the unmanned flight test of Apollo 5 was completed by the end of the year, and designated successful despite various malfunctions explained away as easily rectified.

NASA, charged with achieving the moonlanding 'before the end of the decade', then faced 1968 with a mixture of confidence and apprehension, while space correspondents looked forward to it with tense excitement – the tension in my case due less to anxiety for the astronauts than to fears that somehow I might be manoeuvred out of covering the biggest story of the twentieth century.

The delays meant that 29 major space launches were due in that year, starting with Surveyor 7 in January and culminating with the first two manned Apollo flights. Surveyor 7 duly added to the 65 000 lunar surface pictures sent back by its predecessors another 21 000 of Crater Tycho. This was a dramatic but more southerly area, which astronaut pleas to visit were to be denied on cost and safety grounds.

I was too busy covering Healey's defence cuts and Anglo-French rows over the delivery of Concorde engines to get involved in the same month in the Apollo 5 launch. This was the first unmanned test of the combined Apollo/Saturn 5 combination, and although the engine of the Saturn 1B (the topmost stage, intended to take Apollo to the Moon) misbehaved, Houston's mission controllers successfully tested the Lunar Module systems, including firings in Earth orbit of both the descent and ascent propulsion systems to be used for landing on and taking off from the Moon's surface. Despite the significance of this mission it got little media coverage.

What impressed me at the time, and even more when looking back, was that amid all these preoccupations NASA executives found time to think ahead about the future of its dedicated space scientists and technicians. Theoretically, if not actually, the moonlandings had been achieved – so what were all these people going to do in two or three years' time? I reflected these anxieties in a contribution to the *Today* programme, which I suspect was never used. NASA called this forward-look 'Apollo Applications Program', and a much-reduced version of it was to become the Skylab Space Station of 1973. The 1968 idea was to place two of the 58 ft-long 2nd-stage Saturns in near-polar orbits, with rotating crews of three, each staying for 56 days, in the first, followed by a nine-strong crew in the second, with one astronaut-volunteer staying for a year. Even though NASA's future budget was already being threatened by the cost of the Vietnam War, the announcement of AAP, as it was known, had an immediate effect in raising the morale of the thousands of space workers who were beginning to fear that their careers would end abruptly once the moonlandings were completed.

The second and last unmanned test flight of the Apollo/Saturn 5 combination in April was officially described as 'less than perfect', because von Braun's huge rocket 'ran into a sea of troubles'. Monitoring the mission from London, I found it hard to believe that five manned flights could be undertaken, let alone successfully completed, to culminate in a moonlanding within only 15 months. Two minutes after lift-off, stage one of Saturn 5 went through 30 seconds of severe longitudinal oscillations – the 'pogo effect' because it went up and down like a child's pogo stick as a result to sloshing fuel – producing g-loads in the spacecraft which no astronaut could be expected to withstand. Every major space launcher in the US, Europe, Russia and Japan had pogo problems in its early stages, caused by fluctuations in fuel burning and engine thrust. The cure, NASA found, was to 'tune' the vehicle, by filling a series of cavities with helium gas, thus getting rid of the interacting frequencies that caused the oscillations.

On the S2 second stage, two of the five J2 engines shut down too soon, and the other three had to fire longer to compensate for the loss of power. The S4B third stage, instructed by its computer brain, dutifully fired much longer to compensate for the S2's bad behaviour – but even so the Apollo spacecraft, with the S4B still attached, ended in an orbit of 178 × 367 kilometres, instead of the planned circular 160 km orbit. And finally the S4B failed to fire a second time – the all-important requirement on a lunar mission when it must be relit in Earth orbit to raise Apollo's speed to 24 400 mph (39 270 km/h) and thus place it on course for the Moon – technically 'injecting it into a translunar orbit.'

Far from being depressed, von Braun and his rocket team were quite pleased. All the things that went wrong they knew how to put right. And other things, like altitude control, navigation and guidance systems, and the Apollo Command Module's heatshield during recovery, had all worked perfectly. The US public and media were less than convinced, and were also constantly distracted by the Vietnam War and domestic sensations like the assassinations of Robert Kennedy and Martin Luther King. But NASA and its team of German rocket engineers knew how to swing interest back to the space programme by playing the Russian card.

In June, taking advantage of a visit to Boeing and Lockheed, I went on to the Cape to catch up on Apollo progress there, and talked to Dr Kurt Debus, Director of the Kennedy Space Center. He told me he still expected the Russians to make a bid to reach the Moon first, and revealed that he had evidence (no doubt provided by the increasingly efficient satellite surveillance) that the Soviets had developed a huge new rocket, even bigger than Saturn 5, providing a thrust of 10 million pounds. Western fears of being upstaged by the 'wicked Russians' were revived, for we could not know then how ineffective 'Webb's giant', as it became known, would prove to be. While few knowledgable scientists thought the Russians still had any chance of landing the first man on the Moon, it was acknowledged that they could still snatch a propaganda victory for the record books with the first manned circumlunar flight. Stories like that, emanating from Debus and his leader von Braun, were most effective in dissuading the politicians from carrying through their regular threats to cut NASA's budget.

For once on that visit I had time for a leisurely look around the Cape area, and noted the massive developments that had occurred since my first visits ten years earlier. It resulted in a piece entitled 'The Plight of the Roseate Spoonbill':

It was only hours before the last Saturn 5 rocket was due to be fired, 3000 tons of it, that the plight of the roseate spoonbill was revealed. A final check to make sure that the launching area was clear revealed her, calm and graceful, sitting in the scrub on a nice clutch of eggs, right alongside the pad. Just one of over 200 species of birds who've discovered that Cape Kennedy's moonbase makes a natural sanctuary.

Jack King, the public relations man with the world-famous voice, because he does the countdown and lift-off commentary, is responsible for the birds as well as the people. So, at the last solemn conference at which they make sure, for instance, that the right number of helium bubbles have been injected to stop the rocket from rattling, King reported on the roseate spoonbill. It never occurred to anybody to leave her to her fate. Two police cars were despatched just before lift-off, with sirens screaming. Mrs Spoonbill took off indignantly just before the rocket, and thus survived the seven million pound blast to return to her nest and complete the raising of her family.

Just as well. In numbers at any rate the birdlife in this area has been devastated in the ten years since I first came here. Then the marshes around Cocoa Beach looked like pictures of the world before man arrived: pelicans, egrets, cranes, storks and sandpipers were there in their thousands, and didn't even look up when you walked around. I've just had a drive around the same area, and saw one solitary sandpiper. The marshes have been filled in, and houses built for the spacemen's families in such a way that they can park their two or three cars at the front, and moor their boats at the back. Where the Banana and Indian Rivers meet nearby the map says there are a thousand islands. Every one is within reach of nearly 10 000 boats, looking for a barbecue spot away from it all. A golf course and the Minuteman High School, named after America's most important ballistic missile, were the ultimate developments that led to the bird population deciding that his rockets were preferable to man himself.

Of course, it's a wonderful place to live. The bathing's fabulous – though there is a problem on Cocoa Beach. So many cars now drive up and down the sands that, when you put down your towel, the ten-yard walk into the tepid sea is as hazardous as crossing Piccadilly.

These were the pieces that I most enjoyed doing, and the programmes that used them – in this case *Today* and *Outlook* and *Pick of the Week* – also welcomed them, especially as they cost them nothing. That was why they infuriated the new Foreign Editor, Arthur Hutchinson. He thought I should confine my contributions to news bulletins and the solemn and limited news programmes associated with them; and when I protested that transmitting such 'colour' pieces only cost a few pounds extra since I was already there, he exploded: 'As far as I am concerned, they are not worth £5!' The days of John Birt and 'multi-role, multi-media' correspondents were more than 20 years ahead, but happily I was not the only correspondent unable to resist doing what seemed to be a 'good story'.

By August I was able to report that NASA's target date for landing their first two men on the Moon was Friday 25 July 1969, which every one, including me, thought highly optimistic. But it was of course to be achieved five days before that.

I had mentioned several times in my broadcasts that NASA was unofficially considering a manned flight around the Moon during Christmas 1968 – a project which Apollo Program Director General Sam Phillips had done his best to keep secret under the code name 'Sam's Budget Exercise'. What I did not forecast, nor expect, was that NASA would place Apollo 8 in orbit around the Moon, and not confine the mission to 'a free-return trajectory' – merely passing behind it and then returning to Earth.

The decision to go to the Moon after only one manned flight in Earth orbit, and that using only the Saturn 1B and not the full Saturn 5 stack, was sparked by the discovery that there was so much wrong with the Lunar Module that it would not be ready for the Apollo 8 mission as originally planned. Jim McDivitt and his crew had been due on that mission to do a complete moonlanding rehearsal in Earth orbit, by switching from Apollo to the Lunar Module in Earth orbit, separating as if they were going off to the Moon, then returning and redocking with Apollo. That done, Apollo 9 would have carried out the circumlunar flight early in 1969. However, although the Lunar Module was late, Apollo was not, and several command modules were ready for flight. Wernher von Braun pointed out that, once it had been decided to put men on Apollo 8, it did not matter how far they were sent. So it was decided to reverse the Apollo 8 and 9 missions, which would keep the programme on schedule for the 1969 landing – and, perhaps even more important, beat the Soviets to the circumlunar flight.

All this had to be decided in advance of the launch of Apollo 7 in October, but was of course completely dependent upon its success. The Soviets gave the revised plan a fresh impetus by fulfilling Dr Debus's prophecy that the moonrace was not won yet with a mysterious launch on 22 September. With the Proton rocket – which the West had known for a long time was capable of sending at least one cosmonaut around the Moon without landing – Zond 5 was launched to continue the programme described by Soviet scientists as 'testing new systems in distant regions of circumterrestrial space'. It was several days before we could be certain that there was no astronaut on board – especially as a Russian voice was heard reading what proved to be recorded and simulated instrument readings. The recovery of Zond 5 in the Indian Ocean at the end of a seven-day flight was the first and almost the only time Russia has recovered a spacecraft at sea instead of on land; and, because winter recoveries were obviously much more difficult in the snow, plus the announcement that the cargo had included tortoises and insects, it did look as if Zond 5 was a dress rehearsal for an imminent manned circumlunar flight and ocean recovery.

Those of us anxious to build up the moonrace story were gleeful when Sir Bernard Lovell of Jodrell Bank went much further than we would have done, and declared that it was highly probable that a

Russian cosmonaut would get a close look at the Moon long before an American astronaut.

In spite of all this my unsympathetic new Foreign Editor remained unconvinced that he should authorise funds from his budget to allow me – nominally a 'Home' News correspondent! – to go to the Cape to cover Apollo 7. Fortuitously, and not for the first time, Boeing offered me free a first-class return ticket to go to Seattle to cover the roll-out of the first Boeing 747 – which at that time they were trying to discourage us from calling a 'jumbo jet'. By getting the return half of my ticket converted to economy, I could fly back via New York, Mission Control at Houston and then on to the Cape without any extra charges. Now nothing could prevent me from being there!

From Seattle I returned first to Grumman's works at Beth Page on Long Island. There they were making the Lunar Module – a sort of jet-powered, rotorless helicopter, with a crew compartment the size of a telephone-box in which astronauts only had room to stand up, and in which they were to land on the Moon and take off from it again. It was all so improbable that it was no wonder that they were behind in perfecting it. But that was an 'old' story, and to give me something new for the TV cameras, they let me try out their proposed Lunar Rover vehicle. It was a weird-looking vehicle, with splayed out wheels each resembling half a tennis ball on its edge, with aluminium teeth to grip the edge of craters. I was assssured that, trundling along the lunar surface at only 10 miles an hour, it would bounce 20 ft if it hit a 4-inch bump in the one-sixth gravity conditions. However, when I strapped myself into its single seat – it was more like an open farm tractor than a car – it carried me through the craters on a simulated lunar surface without complaint, although, with each wheel driven independently by a battery-powered motor, it did have an alarming tendency to run away with me.

Grumman's concept at that time was that, having carried up one mooncar, before leaving, the astronauts would hook on two more wheels with an 8 ft square solar panel. It would then have been possible for Mission Control to direct the vehicle as much as 700 miles so that the next crew could touchdown beside a fully-powered vehicle, waiting for them with motors running. But there were grave doubts as to whether such a solar-powered mooncar could survive the two-week long lunar night. Grumman made many changes, and from Apollo 15 each of the last three missions took its own Lunar Rover.

By the time I arrived at Cape Kennedy the five-day countdown for Apollo 7 had been under way for two days. My first story was an interview with Wernher von Braun, still playing the Russian card. He talked of the 'spectacular performance' of Zond 5 and rated Russia's chances of achieving a circumlunar flight before Christmas Day – Apollo 8 having slipped three weeks to that target date – as better than America's. As for getting a man on the Moon, it would be 'a photo finish'.

So, although Apollo 7 nowadays merits only a passing reference in the history of manned spaceflight, it set the style and pace for the whole moonlanding programme. As well as being a first trial run for the first team of Apollo astronauts, it was a rehearsal for NASA in the politically sensitive task of handling the media and the public. In accordance with its boasted 'open policy' NASA deluged the media with paper – very effective, and much copied by other organisations, for journalists had little time and less inclination to wade through it all, so all sorts of issues that officialdom did not want revealed were officially admitted but never noticed by the Press!

The privileged space correspondent had already been given elaborate loose leaf books containing enough technical details to enable a smart engineer to build his own Apollo spacecraft and Saturn rockets; now there was an 81-page NASA Apollo 7 Press Kit, which the US Navy, not to be outdone, matched with one of equal size detailing their recovery operations. This bore its own 'mission patch' around the title 'Manned Spacecraft Recovery Force Atlantic, TF-140'. The biographies of the astronauts in the NASA kit were outshone by the biographies of the two admirals and ten captains of the recovery ships whose marriages and up to seven children apiece were fully detailed. [That contrasts vividly with current procedures. In the years since then, the astronauts, and especially the women astronauts, have insisted, following many divorces, that their marital status is now their own affair.]

For Apollo 7 the US Navy task force was led by the aircraft carrier USS *Essex* 'the Fightin'est Ship' (unsayable for a broadcaster), which had been responsible for destroying 1564 aircraft and sinking 25 Japanese ships during World War II. The pararescue team, who would be doing the most arduous work when the spacecraft splashed down, got a page and half at the end, with no names or biographies.

Yet more handouts came from the US Air Force, jealous of both NASA and the Navy, with dozens more, together with photographs, film clips (and later videos) available at the JIPC (pronounced Gipsy, but standing for Joint Industry Press Center) set up in the early days in the press corps' most popular hotel. Funded by the contractors involved in the space programme, and fiercely independent of NASA (although the companies always worried about their future if they gave NASA any serious offence), the JIPC has always been of great importance to the media, both morally and materially. First because there was always hot coffee and doughnuts available to help one come to terms with frequent 3 am starts; and secondly because the nicely displayed handouts, with public relations officers usually more knowledgable and altogether more helpful than many of NASA's, provided the knowledge and understanding that gave depth to one's stories.

On Apollo 7 there were samples – I have one in my hand as I write – of the non-combustible Beta fabric, provided by Owens-Corning Fiberglas, for spacesuits that would not burn even if there were another fire in an all-oxygen atmosphere. A grave and elderly gentleman named Fisher, always formally dressed and perfectly shaved, would go around selecting journalists he respected, and present them with a Fisher Space Pen, specially developed for the astronauts because it would write at any angle – including upside down – and definitely would not leak at high altitude in aircraft. (In the early days of jetflight, it was common to arrive at journey's end with ink-stained pockets as a result of the pressure inside the pen being greater than that in the aircraft.)

Mercury astronauts first encountered the pen problem: no gravity equalled no ink, because the pen would not flow. Fisher made himself a millionaire with a space pen with a cartridge packed with nearly solid thixotropic ink, and pressurised with a tiny squirt of nitrogen. Friction from the revolving ballpoint liquefied just enough to produce smooth writing whether or not there was gravity. These were two more examples of 'space spin-off', usually derided by politicians with references to 'non-stick frying pans' – a gibe still in use!

Give-aways like this, plus mission patches and 'decals', attracted hordes of 'space groupies' to the JIPC, to the great annoyance and inconvenience of those of us fully occupied writing and broadcasting, and rushing in and out to refill our paper cups with coffee. Among the handouts, the kits provided by Rockwell (once famous as North American Aviation) for advanced aircraft and then successively makers of Apollo and the Space Shuttle, were always the most readable and understandable, thanks to a brilliant technical journalist named William Green.

This mass of paper, accumulated after visits to the NASA Press Desk and the JIPC, meant that the first thing most of us did after checking in at the accreditation desk was to find a large cardboard box in which to keep the paper all under control. It has always been a mixed blessing; TV and radio broadcasters for NBC, ABC, CBS and BBC spent every available minute and much sleeping time absorbing it all, and especially the elaborate 'abort' modes and rescue procedures if things went wrong. Unlike newspapermen, we would have to be knowledgable on the subject instantly, with no opportunity to look it up once accidents happened.

Most newspeople preferred to rely on the NASA and industry public affairs officers to know these things, and give them a quick briefing when it was needed. But the NASA people and almost all the industry public relations people never got through all their own handouts, and would merely refer inquirers to the Press Kits. 'It's all in there!' they would say, handing over yet another massive copy of the relevant document. Newspapermen who had spent their time enjoying the Port Canaveral oyster bars, cheap Californian wine and not-so-cheap local girls would then badger those of us who had done our homework for explanations – even on occasions desperately tapping me on the shoulder in the middle of a live broadcast.

NASA public affairs people covered their own reluctance to read all these millions of words by furtively giving the genuine correspondents (perhaps 20 or 30, as opposed to the 'space junkies', hundreds of whom got themselves accredited as media under the guise of representing semi-fictional space and school magazines) under-the-counter copies of the Flight Plan, etc. so that the handful among us (and often I was not included) clever enough to decipher them could identify what was supposed to be going on – and what would happen if it didn't – at any given minute during the 10- or 11-day missions.

While I continued to plod away on my own for TV and Radio news bulletins with my makeshift equipment, Richard Francis, a wildly enthusiastic producer who had recently acquired the title 'Projects Editor, Current Affairs TV' arrived at the Cape with a posse of assistants and secretaries, to mastermind the live TV coverage. He maddened the deskmen back in London by joyfully adopting 'space-speak': Yes and No became Negative and Affirmative; the possible and impossible were Go and No-Go, and on the rare occasions when planning seemed complete 'Everything is A-O.K.' When short of paper for my scripts I used the backs of some his cables. A typical page reads:

> Given big enough emergency spacecraft could be brought down anyplace anytime. All a question of how great the emergency. Next best is anytime near any recovery vessel. Next best daytime by any vessel. Next best daytime by USS Essex. Next best is anyday by Essex at approx 0800 EDT. Aim is 0800 by Essex on Day 11. Suggest we lose no sleep covering that range possibility. Only near Essex can there be live recovery pix.

Dick Francis's zeal for Europe's first space spectacular intended to promote colour TV was surpassed by America's big-three TV networks. They were planning to spend $3 million on live transmissions during the 10-day mission, and expected NASA to co-operate. It was all to lead to the first human revolt in orbit, and was to blight the future careers of the three-man crew – commander Wally Schirra, 45 (who had flown on both Mercury and Gemini missions); Walter Cunningham, 36, a research scientist; and Don Eisele, 38, a test pilot. The last two had not flown before, and none of them was to fly in space again.

Two days before Apollo 7 was launched on the two-stage Saturn 1B rocket we enjoyed a splendid visual story as we watched the first real moonrocket being rolled out of the world's largest building, the 500 ft-high VAB, or vehicle assembly building, so vast that four moonrockets could be put together in it at one time.

It was, I reported, rather like moving Nelson's Column vertically from Trafalgar Square to Tower Bridge – although Nelson's Column would actually have looked small beside it. What the Cape workers still call the 'incredible crawler' inched out carrying a 6000-tons load: a complete three-stage Saturn 5, with the three-man Apollo at 330 ft artistically surmounted by its escape tower to lift the spacecraft to safety if the millions of pounds of kerosene and hydrogen fuel below it should blow up in the first two minutes. With the 'umbilical tower' holding the stack with nine vice-like hands, it took six hours for the crawler-transporter to reach the launch pad, less than 4 miles away.

> Just as Apollo 7 is on time for Friday, Apollo 8 is so far on time for a December 20 launch. And the three men who expect to go round the Moon in it were with us watching the roll out. They're led by Commander Frank Borman, holder of the world record for spaceflight, with two weeks in orbit. Whether he and his companions, Astronauts Lovell and Anders, become legendary heroes by swooping within 60 miles of the Moon's surface, depends entirely on the success of their fellow astronauts in this week's launch.

It had taken ten years to reach this point in the space programme. Things had changed dramatically since Wally Schirra's previous flight as commander of the two-man Gemini 6, when he smuggled up a mouth organ for light relief. To stop any such nonsense happening again (though happily it did!) the astronauts had to confine anything personal they wanted to take to a tiny 'pilot's preference kit' measuring 8 × 4 × 2 inches; officialdom expected that that would enable them to take, for instance, a pocket Bible and photographs – on this mission of a total of three wives and seven children.

> That rigid discipline illustrates the difference here at the Cape now, compared with ten years ago when I started coming. Then everybody went about in the heat in shorts and open shirts, felt like a pioneer, and was happy. Now, everybody wears suits – ties, jackets, the lot – and of course worries. The climate is irrelevant. Every motel, coffee shop, space centre and Press site is air-conditioned. And you travel from one to another in an air-conditioned car.
>
> There's just one thing that hasn't changed. There's always been intense rivalry between the US Air Force and the National Aeronautics and Space Administration as to who really runs Cape Kennedy. For this shot, we visiting correspondents have for the first time been given passes, allowing us to enter and leave the Cape unescorted – and whichever way you go it's a 20-mile drive – by Gates Two and Three. Yesterday I got lost in the dark, and emerged at Gate One. With two other British correspondents who'd rashly asked for a lift, I was promptly arrested and made to drive all the way back again. That was at the end of a 12-hour day; but a grinning US Air Force sergeant said we'd be detained for four hours more before we were allowed to go. Happily, no one stopped me from using a telephone, and after ten years at the Cape I had quite a list of telephone numbers. We were released with apologies, ten minutes later. But I can't help feeling that it's no accident that you can't see the numbers on the gates until you actually reach them.

The last day of the countdown was so uneventful that there were opportunities to build up the story with graphic descriptions of escape procedures if anything went wrong – for Pad 34, from which Apollo 7 would be launched, was the one on which Astronauts Grissom, White and Chaffee had died when their spacecraft caught fire:

> The precautions against that happening again are fantastic. Here's just one example. In case the 225-ft high rocket, full of highly-inflammable liquid oxygen and hydrogen looks like blowing up, or there's another fire in the spacecraft, they've installed a special 'slide wire'. In these last critical hours, only the three astronauts and half a dozen technicians are allowed at the top. In an emergency the astronauts hook themselves to little trolleys at five-second intervals at a point over 200 ft up. At 50 mph they whizz down the slide wire, ending 1200 ft from danger. They've thought of everything. Wally Schirra, the astronauts' commander, is, at 45, heavier than the rest. Etiquette of course demands that he must escape last. So the slide wire has been designed so that his extra weight won't cause him to catch the others up in a sort of disastrous domino sequence.

Shortly before the flight most of us thought that Schirra, as one of the original Mercury Seven, and the only one at that time to fly on Mercury, Gemini and Apollo, was most likely among the 52 astronauts by then selected to win the fierce – though well-concealed – contest among them to be the first man on the Moon. (There were no women in the running.) So it was quite a surprise when, a week or two beforehand, he announced that Apollo 7 would be his last flight and that he would be retiring from NASA and the Navy at the end of it. That no doubt partly accounted for what had already happened as well as what was to come. We knew he had been involved in the many rows over whether a TV camera should be carried on the mission, and was probably the most vocal of the astronauts in opposing it. The official NASA history records that ever since 1963 the camera 'had been going in and out of the craft as though it were caught in a revolving door . . . When kilograms, and even grams, were being shaved from the command module, the camera was among the first items to go.'

Schirra had lost the long-running argument about TV shortly before the launch, when General Sam Phillips, Apollo's HQ Director, had ordered the Program Manager at Houston, George Low, to reinstal it. Three days would have been sufficient to fulfil the objectives of the flight, which were to demonstrate that the spacecraft, the crew and Mission Control could meet their objectives, and that the Command Module was capable of carrying out rendezvous manoeuvres. Despite that, the flight was to be continued for 11 days to evaluate 'long duration' capabilities.

Heavy rainstorms cleared the Cape just in time, and their presence made it one of the most beautiful launches seen before or since. The brilliant white Saturn 1B lifted off, hung for a moment parallel with the orange launch tower, and then rose steadily into an area of clear blue sky. We saw it roll majestically on to its flight path; brilliant white smoke ceased abruptly as the first stage fell away after two minutes, dwindling slowly as what had been reduced to a 34-ton package moved out of sight.

'She's riding like a dream,' said Schirra, and a classically educated Mission Controller said that the first manned Saturn proved to be 'a very gentleman'. Europe got brilliant colour pictures thanks to the US Air Force having lent the use of their new $100 000 telescopes to the TV networks. We promised viewers that next day the crew would unpack their onboard equipment, and send equally good pictures of what things looked like both inside and outside the spacecraft.

Next day was a Sunday, and an expectant world was standing by its expensive new colour TVs. But, instead of the pictures, we gave them voice recordings of a contest between the spacecraft and Mission Control for most of the 15th orbit. We first sensed the coming row in a brief exchange between Walter Cunningham in the spacecraft and 'Capcom' – short for 'Capsule Communicator'. That was usually an astronaut taking a turn at Mission Control, and whose primary role in those days was to read out interminable lists of numbers for the astronauts to write down (hence the need for reliable pens), and then read back before feeding them into the onboard computer to control the spacecraft's manoeuvres and systems. The spacecraft's safety as well as the success of the mission depended entirely upon the accuracy of these readouts and readbacks. It is not difficult to imagine the tension they could cause with astronauts who were both tired and suffering from bad head colds (to say nothing of space sickness) especially if the procedures fell behind time:

> CAPCOM: . . . Now concerning the matter of the television, there's been considerable discussion here in the Center. The Flight Director wants you to turn on the television at the appropriate time.

This was the nearest that Mission Control ever got to issuing a direct order. But all they got was Cunningham's reply: 'Wally will be on the air shortly . . . Go on with your maneuver pad.'

> CAPCOM: Let's wait first and get Wally's comments on the television.

Then came LOS – loss of signal – as Apollo passed around Australia and out of range of a ground tracking station. When communications were resumed, Wally was there, but busy:

> SCHIRRA: Roger, Jack. I got 25 hours 41 minutes of the NAV check. I didn't get the seconds. Continue after that.
> CAPCOM: Okay, starting at the seconds: 5500 + 2766. 05376, 1125359284359. You have the align of 23 + 24 + 0800.
> SCHIRRA: Roger. The align was 23 + 24 + 0800, NCC 1, 26245510 + 00617 – 00010 + 019851960 + 12430197832398 2 090 2 03001035198115102541 15500 – 2766 – 0537612263592484539. Over.
> CAPCOM: It is correct except in noun 43, the latitude. The sign should be +2766.
> SCHIRRA: Roger, I have the plots here.
> CAPCOM: OK, you got it. Go ahead Wally.
> SCHIRRA: Roger. You've added two burns to this flight schedule; you've added a water urine dump, we have a new vehicle up here, and I tell you this flight TV will be delayed without further discussion until after rendezvous!

By now Deke Slayton, the astronauts' much respected boss, had taken over from Capcom, and he did go on discussing it – with some urgency, because they were now within 15 minutes of the promised 10-minute transmission into the expectant TV networks:

> SLAYTON: All we have agreed to do is flip it. Apollo 7, all we have agreed to do on this particular pass is to flip, flip the switch on. No other activity associated with TV. I think we are still obligated to do that.
>
> SCHIRRA: We do not have the equipment out, we have not had an opportunity to follow setting, we have not eaten at this point, I still have a cold. I refuse to foul up our time lines this way.

A brief silence, and Capcom was back asking the spacecraft for 'the opposite OMNI please, and your PMP power to OX.' The technicians at their consoles wanted to get on with things they considered more important, such as oxygen percentages remaining, and sorting out a problem with the primary evaporator.

Mission Control told us that they 'had decided to accept the crew commander's judgment' – the only time such a thing has happened, for the simple reason that one of the NASA bosses listening to it announced explosively that these astronauts would never fly again – bad luck for Cunningham and Eisele, who had little choice but to back their commander.

Schirra could reasonably argue that his attitude was justified by the results. Just before and after the altercation, his crew successfully completed the two most important tasks of the mission. Firstly they turned the spacecraft around, manoeuvred accurately into the flight path of the second stage of the Saturn rocket as it followed them in orbit, and demonstrated that if it had been carrying a Lunar Module they could have docked, drawn it out, and turned with it on course for the Moon. Then, after the row, the crew demonstrated the spacecraft's ability to carry out a rescue in lunar orbit if the Lunar Module got into difficulties on its return from the Moon. Making use of their computer, a telescope and an ordinary ship's sextant, they succeeded in

The Apollo 7 astronauts, Schirra, Eisele and Cunningham, at the end of that project's first manned mission. Though it was rated a technical success they were written off as an 'awkward squad' and none of them flew again. (NASA)

manoeuvring within 70 yards of the rocket stage after it had been allowed to drift 12 miles ahead and 7 miles below them.

There was noticeable tension at Mission Control on the third day. With the major targets achieved, the only reason why the crew should not do their promised TV transmission was the colds with which they had infected one another. No one was confident that Schirra would not repeat his refusal to carry it out, and this seemed possible as we watched a blank screen for a whole minute after it was due to start. But suddenly all three astronauts appeared, in a sitting position.

Schirra, centre-stage, was holding a card up to the lens, just readable: 'Hullo from the lovely Apollo Room high atop everything' – a take-off of American radio and TV techniques. Then Walt Cunningham dismounted the camera and took it to the window. We saw the Earth swinging madly. After pleas from Houston to 'hold it steady', areas below the spacecraft came into focus. New Orleans, the State of Louisiana and then the Florida Peninsular were vaguely recognisable. Then back inside for more mickey-taking; Schirra held up another card: 'Keep those cards and letters coming, folks' – the anxious phrase used on sponsored TV shows when the audience began to fall. It was a 7-minute TV show instead of the promised 12 minutes, and Donn Eisele reinforced the crew's antipathy for show business by explaining how much the TV camera was in the way when they were moving around inside the spacecraft. But for NASA it provided a much-needed boost for the space programme, and another show was promised for the same time the following day.

I stayed at the Cape just long enough for the second Wally Schirra show, and of course by then he was enjoying it. The public saw the first demonstration of the weightlessness with which we were to become all-too-familiar as the years rolled on. Floating camera lenses, pencils and water globules, were followed by much clearer views of Earth through the windows because Apollo's altitude had been reduced by 40 miles to only 90 miles just in case an emergency re-entry was required.

Schirra had his little joke again, revealing the basic, earthy astronauts' humour which I have always found jarring. We were mildly surprised by the roars of laughter in Mission Control when the card he held up this time read: 'Are you a turtle?' It took a little while to establish that this referred to the exclusive 'International Turtles Club' formed by the astronauts. Its members had devised a series of embarrassing questions which could be asked in public, with those addressed being required to buy a round of drinks if they did not have the courage to give the correct reply. Schirra's question was directed to his TV opponents Deke Slayton and Paul Haney, head of public relations, and the correct reply was: 'You betcha sweet arse I am!' With millions listening, they preferred to buy the drinks when the crew returned.

Back in London I found that the mission had, as expected, caught the public imagination, and there were slightly shamed-faced congratulations from the Foreign Editor on the coverage given to my stories. One reason for their success was that they provided a welcome contrast to the dreary lead stories about Britain's unsuccessful efforts to dissuade the Rhodesian Government from declaring unilateral independence – 'UDI'.

Since the mission was continuing I was required to forget my jetlag and thoughts of spending time with my wife and children in order to provide frequent updates and explanations for TV and Radio News. The crew duly provided material with grumbles about the body-sensors insisted upon by the doctors at Mission Control to enable them to monitor heart and other bodily functions. The doctors continued, as doctors always have, it seems, to take a morbid interest in their faeces, all of which had to be brought back to Earth for analysis. And the astronauts also complained about the food. Schirra, we found, having declared at the end of his Gemini flight that next time he would insist on taking some coffee, had in fact done so. He was able to brew it with hot water from 'the British-designed hydrogen fuel cells'.

Re-entry after 10.7 days and 164 revolutions was mid-morning in Britain, and as Dick Francis's team covered it live on TV, I was invited to do it live for Radios 1 and 2. In addition my TV News chiefs, gleeful at having me back in London and under their control and supervision, encouraged me (though I needed little encouragement) to give the studio pundits, whose knowledge was theoretical rather than first-hand, vigorous internal competition with clearer straight reports and explanations in the TV news bulletins of what it all meant.

The astronauts' nerves were noticeably a little frayed, and the splashdown untidy, I reported: 'What we've been seeing is that while spacemen need to be supermen, they're still human. But I think in future flights America will try to ensure that the fact that they're human doesn't show quite so much!'

Splashdown was in fact 18 miles from the recovery ship, and Apollo 7 rolled upside down – 'Stable Two' as Mission Control insisted on calling it, to emphasise that even if uncomfortable it was not dangerous. There was an anxious wait of 15 minutes before contact was established, and it was 50 minutes before the frogmen's helicopter reached it. As soon as the crew was safely aboard the recovery ship, Mission Control swallowed hard, announced that the Apollo 7 flight had been 'perfect', and that they were on course for Apollo 8 to fly around the Moon during Christmas two months later.

That night I began all my radio and TV pieces: 'It's no longer IF America can land a man on the Moon, but WHEN.' But despite the alleged perfection of their mission, Messrs Schirra, Eisele and Cunningham were never allowed into space again

As for NASA's Administrators in Washington, from the deep depression that followed the Apollo 1 fire at the start of 1967, less than two years later – three days before the Apollo 7 launch – their optimism had risen to heights that can only be described as ridiculous. Dr George Mueller, in overall charge of manned spaceflight, and one of the most respected of NASA's leaders, forecast in a speech to editors and publishers the likely benefits when the Space Shuttle arrived. It would evolve into a system able to deliver usable payloads into space at a cost of about $5 per pound compared with the first Explorer flights costing a third of a million dollars per pound, he said. Estimates 'which already existed' – no doubt emanating from von Braun's office – anticipated that a Shuttle would be able to fly passengers from New York to Tokyo in 46 minutes for 'a little above ten cents per passenger mile'. Alas, in 2002, Space Shuttle flights were still costing $400 000 each, and the occasional billionaire passenger was having to pay $20 millions for a 7-day trip!

11 Lassoing the Moon

Apollo 8

Typically in an ungrateful world, when the mission had ended, and NASA managers were satisfied that everything possible had been learned during the 'debriefings' – long sessions when the crew members discussed everything that had gone right as well as what had gone wrong – the Apollo 7 men quickly disappeared from public view.

After leaving NASA Schirra held various chairmanships and directorships, and ended his career as a consultant and public speaker. Cunningham joined engineering companies and wrote one of the best of the astronauts' books – *The All-American Boys.* Eisele was briefly given the appointment of Technical Assistant (Manned Flight) at NASA's Langley Research Center, then became Peace Corps Director in Thailand. He died of a heart attack in 1987 aged 57.

We of the media of course were the worst offenders of all in discarding the past in order to concentrate on what lay ahead. Only a week after Apollo 7's return the Soviets gave me a chance to build up what I called 'The Space Olympics'. NATO intelligence had as usual observed that the Soviets, who lacked NASA's round-the-world communications network, had sent out their usual tracking ships to provide global communications for a manned flight. It seemed logical that, following the successful recovery of Zond 5 in the Indian Ocean, a cosmonaut recovery at sea would now follow as a final rehearsal for a circumlunar flight – the sea recovery being safer when dealing with the 25 000 mph return from the Moon compared with the 17 500 mph re-entry from earth orbit.

Colonel Giorgi Beregovoi was sent up in Soyuz 3 a day after an empty Soyuz 2, but although the two craft were brought close together both automatically and manually, they were hopelessly misaligned, and no docking was possible. [David Shayler's *Disasters and Accidents in Manned Spaceflight* has full details of the problems that plagued this mission.] It all ended in anti-climax: Soyuz 2 was brought back first, obviously checking that the parachute system, which had failed with fatal results for Komarov 18 months earlier, was now working correctly. Beregovoi's re-entry followed, both craft landing as usual in the snows of Kazakhstan. Preparations for recovering Beregovoi in the Indian Ocean were apparently only a precaution in case the Soyuz 2 return showed that the parachute and retrorocket system was still unreliable.

But the 1968 'Space Olympics' were not quite over. Two weeks later what the Russians described as 'an automatic space station', Zond 6, made the sort of flight around the Moon that had long been expected. But it was unmanned, and though ships were still available to recover it in the Indian Ocean, it too was recovered on land.

A new 'double immersion' technique had been used, explained Soviet scientists, to reduce the re-entry speed of 25 000 mph. The spacecraft's aerodynamic shape had been used to cause it to bounce off the Earth's atmosphere, thus slowing it down, and then to lose more speed as it entered the atmosphere a second time. It was the 'skip re-entry system' originally proposed for Apollo's return from the Moon, but abandoned because of the danger that the craft would actually bounce right past the Earth and be lost irretrievably in solar orbit. Apollo rehearsals had shown that it was better to use that craft's aerodynamic shape to make it dip and lift twice within the atmosphere before the final descent.

By now the BBC's consistent coverage of the space programme had won me many friends in high places in NASA, and it was early in 1968 they told me, as explained earlier, that because it was unlikely that the Lunar Module would be ready for flight testing on what had been planned as Apollo 8, that misson would be switched with Apollo 9 – originally intended to be a rehearsal in Earth orbit by the Command Module of the manoeuvres it would have to make around the Moon. von Braun had won his point that once it was launched it

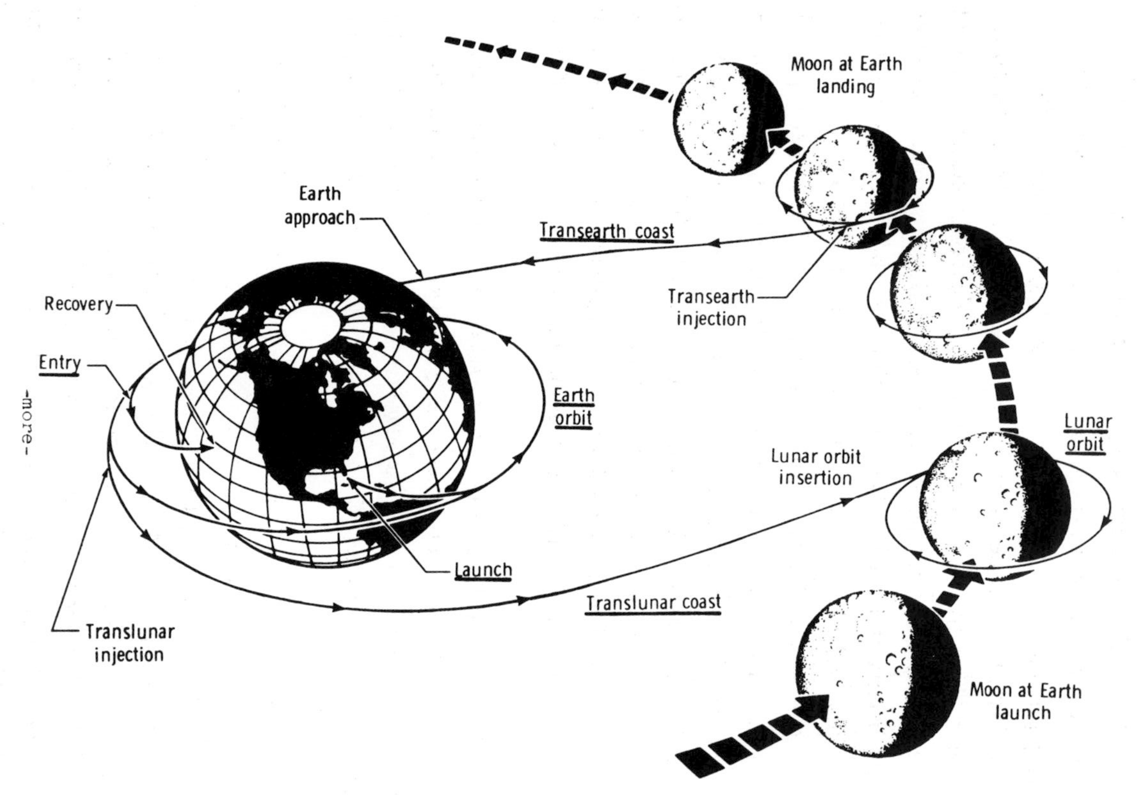

Apollo 8 mission profile. (NASA)

made no difference whether the Command Module did its manoeuvres in Earth orbit or was sent around the Moon to do them. And that, he added, would really shake the Russians!

Since only one manned Apollo flight had then been made, and the giant Saturn 5 had never been used to launch men, it was a breathtaking proposal. But it was clear it was getting serious consideration, and in June I broke the story that Apollo 8 could be sent around the Moon in December. It was met with a good deal of disbelief – and I got some rebuking letters from space buffs, pointing out that nothing like it had been announced by NASA! As for the BBC's news chiefs, they either did not believe the story – or more likely did not listen!

But in mid-November, at a meeting of the 17 executives providing Apollo with its systems and equipment, all except one supported the more ambitious mission which involved not only sending Apollo 8 to the Moon, but placing it in lunar orbit; only the McDonnell Douglas representative preferred the circumlunar option – passing once around the back of the Moon. Several others, while supporting the lunar orbit option, said that the public should be warned that it involved more risk.

When the official announcement came I took great satisfaction in the general panic it created on our newsdesks as they faced the need to plan coverage of the biggest story of the twentieth century. Anguished cries arose: Will there be any TV; and if so how much, and how shall we get it?

The Apollo 8 crew, although less vocal on the subject, shared the objections of their predecessors to carrying a TV camera. But this time NASA made it very clear to the public that there would be no reservations about television; America, Europe and Japan would all get live TV pictures transmitted by the Apollo 8 astronauts as they circled the Moon on Christmas Eve only 70 miles above its surface. I got to the Cape on 15 December, six days before the planned launch, just in time to cover that and the fact that the Soviets, via Moscow radio, had admitted that they were not yet ready to send a cosmonaut around the Moon. 'For the first time since the space race began ten years ago,

America is poised to demonstrate to the world that she's overtaken Russia in space techniques,' I reported.

The five-day countdown for the first launch of the complete Saturn 5 stack with men on top flowed on smoothly, and the main 'worry' – or more accurately the subject of my pre-flight stories – was whether Frank Borman, James Lovell and William Anders would catch the Hong Kong influenza which was sweeping through the US at that time. NASA dreaded a repetition of the bad temper that had accompanied the last crew's colds, and it was an indication of national priorities when the first batch of a new vaccine against this particular strain of 'flu' was diverted to the Cape, so that not only the astronauts but 1200 flight directors, officials and technicians could be inoculated against it.

As an additional precaution the astronauts, all married with a total of 11 children between the ages of two and 17, were kept in strict isolation for 14 days before lift-off – although Lovell and Anders, whose wives had come to the Cape to watch the launch, did insist upon an evening off with them two days before. We got to know Lovell's attractive and possessive wife Marilyn quite well as Project Apollo progressed. Borman, a man of iron will and unlimited ambition, was the exception. His wife had remained at home at Houston, content, so we were told, to watch the launch on television.

There was always something to keep the British correspondents out of bed at the Cape, starting with the 5-hour time difference, which meant a very early start (sometimes accompanied by frantic wake-up calls to the camera crew sent to work with me) to get my film reports despatched in time for use the following day. BBC TV News was still limited to ten minutes for its main bulletins at 5.50 and 8.50 pm, although the BBC2 channel had a 20-minute programme called *Newsnight*, starting then at 10.30 pm, and all these 'outlets' wanted reports from me, in addition to all the radio news programmes.

Two days before Apollo 8 went off, we were up most of the night watching a Thor Delta rocket launch Intelsat 3. The International Telecommunications Satellite Consortium, INTELSAT, already boasting 63 member-nations, was just beginning to establish its now vast system of geostationary satellites upon which much of the world's telecommunications depend. The previous attempt to launch Intelsat 2 had provided Florida with one of its fairly regular spectacular bonfires in the sky, so we could not afford to neglect what might be a bad omen for Apollo. This launch duly turned night to day, but nevertheless went perfectly – although the satellite would not be operational in time to add to the television channels available to cover the moonflight.

Sometimes, with a 4-minute 'think piece' commissioned for the weekend edition of *From Our Own Correspondent*, I was so tired and jaded I would take two hours in bed and leave only one hour to write it before transmission. These reports, as most journalists will confirm, usually turned out to be much better than scripts over which one had laboured for hours. The subconscious did its work much better when left alone – and this particular programme was always a good opportunity to pass on interesting titbits of information for which there was no room in short news reports:

> Florida's orange and grapefruit groves are heavy with fruit this year – not least the groves surrounding the Moonport area on Merritt Island, lying just behind Cape Kennedy. One little-publicised recovery operation following Saturn 5 launches includes the fruit shaken from these trees by the biggest noise, and probably the worst vibrations, ever created by man. Luckily, it's not often that a launch comes when the fruit is at its ripest and heaviest. And this year, anyway, there's a glut of Florida fruit – 42 million boxes of it.
>
> It's one of the few statistics that don't interest Cape Kennedy's 20 000 workers. They talk of 'birds' all the time, but are referring to their beloved rockets, not to the still-abundant wild-life in the groves and surrounding swamps, nor to their private lives. The pace of America's moon programme is so hard and fast, they've little time or inclination for either. The space technicians have to be so dedicated to their work, and put in such long hours, that there's a high divorce rate. The bitter conferences and extra work that followed the disaster two years ago, in which three astronauts

were burned to death, led to a new peak in Cape Kennedy's broken marriage rate.

Inevitably the newspapers are full of stories describing how the wives and children will be spending Christmas on their own. But the 40 000 men who will be involved in Apollo 8 are far fewer than in the early days, and one veteran actually told me that in his view the recovery forces were spread 'very thinly'. When John Glenn made America's first orbital flight nearly nine years ago, the Defense Department stationed 126 aircraft, 24 ships and 20 000 men around the world to ensure his recovery no matter where he came down. This time only half as many servicemen are being used to operate 68 aircraft and 16 ships.

The rest are split between Cape Kennedy and the Manned Spacecraft Center, a thousand miles away in Houston, Texas. To me, more astonishing even than the mission itself is the fact that the elaborate Launch Control Center at the Cape, with its 400 technical experts manning rows of consoles, goes out of business ten seconds after the rocket leaves the pad. At that point a similar control center at Houston takes over, and the rest of the flight is controlled from there. The duplication in communications must be immense, but the politicians have always insisted that the employment provided by the Moon programme, which peaked at over 300 000, must be spread around the country.

The variety of employment it's created is astonishing. It ranges from eight extraordinary-looking Boeing [Aria] jets with 7-ft long radomes on their noses, which alone must have cost over £30 million, to specially-designed underwater cameras, for the use of the frogmen in case the spacecraft sinks when it comes down. We are always being shown examples of new materials created for space-walking suits, or being given samples of ball-point pens which will still write in weightless conditions.

Safety devices have been one of the richest sources of work for the contractors. A special slide-wire running to the top of Saturn 5 cost over one million pounds to develop and fit. The astronauts themselves have always fought a losing battle against excessive safety devices, because they build in lots of extra things to go wrong, and reduce the payload. For instance, the launch escape tower, for use if there's an emergency on the launchpad or just after lift-off, has never been used. If that was left off, an extra five tons could be sent to the Moon – an immensely useful addition to the payload. Apart from extra food, water and oxygen, it could include a vehicle to enable the spacemen to move around once they get there, instead of being confined to within a few hundred yards of that landing point.

There's a lot of impatience among the American public with the immense cost and sophistication of everything associated with the Moon programme. And lately, against the background of the Vietnam war, where no safety devices are available to prevent men being killed at the rate of over 100 a week, the Space Administration has suffered some vicious cuts in its long-term budget.

At the top, they're very sore about that, for they can at last produce impressive evidence of industrial and social fall-out from the space programme. A list of 2750 such benefits shows that many of them are in the medical field. For instance, a device originally invented to count meteorite hits on a spacecraft is now being used to measure muscle tremors, and thus detect Parkinson's disease. And space helmets aren't just being copied as toys for children. A development from them is being used in hospitals to check the oxygen consumption of sick children.

The spinoff from the space programme into improvements in the quality of life for ordinary people was rapidly becoming a major political weapon in NASA's favour, with the result than an annual *Spinoff* book has been produced ever since. It was the brainchild of the late James J. Haggerty, a 20-stone journalist, who divided most of his postwar career between supporting the Washington Redskins and presenting in layman's language the technical benefits of America's current space budget of $15 billion per year. One of hundreds of examples of the fallout benefits is the use of spacecraft wind tunnels for training members of the US Ski Team, thus ensuring that they achieve national glory in international events.

On Friday 20 December, the day before the launch, I earned the displeasure of Dick Francis, planning to mount his biggest-ever 'TV outside broadcast', backed in the studio in London with a team of instant experts. All the radio bulletins carried my report that although there was a good Florida weather forecast for Saturday morning's launch – which would be at lunchtime in Britain – a good deal of cloud was expected, and 'hopes that the whole world can watch Apollo 8 soaring off to the Moon on the first hundred miles of its journey don't look so good'. Worse still I had to report that hopes for wonderful pictures of the Moon as the approaching spacecraft got closer had also been dashed, since the astronauts were unlikely to see it at all until they went into orbit. The Foreign Editor passed on to me Francis's sad comment at the daily news meeting that I had cost his programmes a million viewers; no one dared to say it, but there was no doubt that back at base they would have been much happier if I had found something else to talk about. I had known beforehand that the report would discourage people from staying at home to watch television, but decided that the public was entitled to know about NASA's warning. I would have felt terrible after all the buildup if there had been little to see, so I had no guilty conscience even when the launch weather and the view of lift-off were spectacularly perfect!

The other part of the warning – that the astronauts would probably see nothing until they got into lunar orbit – turned out to be more accurate.

Most people thought that the Christmas flight to the Moon, even though very costly in terms of overtime, was a clever public relations ploy by NASA, but that was untrue. In fact there was some influential pressure on NASA to postpone the December flight on the grounds that if it went wrong and the astronauts were lost, it would mar Christmas celebrations on Earth. But to postpone the mission would have seriously jeopardised NASA's prospects of achieving the moonlanding in 1969.

As we were all to learn, a moonflight had to start when the Moon was new – as delicate against the evening sky as a finger-nail paring. That occurred only once a month, and if launch was not achieved between 21 and 27 December the whole mission would have to be postponed until the next six-day launch 'window' opened on 18 January. The windows hinged upon the Moon's position and the sun angle on the surface at the time the spacecraft arrived, and even then of course launch could take place only if the weather on Earth was good both at the Cape and in the chosen 'abort recovery' or emergency landing areas if things went wrong during or just after lift-off. The mind boggled at the complicated calculations required to take into account all the factors, including the relative speeds of Earth and Moon, needed to produce an accurate trajectory to carry Apollo to the Moon, into orbit around it, and then back to the chosen splashdown point on an Earth which had moved hundreds of thousands of miles since the spacecraft took off.

The drive to the Press site at 2 am on the morning of the launch was easily the most nerve-wracking part of the whole mission. Because it was a Saturday with few people required to go to work, it seemed that millions – although the official count later was a quarter of a million – had scrambled into their cars all over Florida and in many States beyond, and driven through the night, hoping to reach well-publicised viewing areas. And as the police had rather despairingly warned, the Saturday sailors, notoriously indifferent to road problems, insisted on their right to have road bridges raised to enable them to chug through the Indian and Banana Rivers around Merritt Island. Every approach road to the Cape – and especially the two from Cocoa Beach – was solidly blocked and that was something that even NASA had not foreseen. The coaches laid on for second-grade visiting VIPs were hopelessly trapped. (First-grade VIPs had no problems, because they were helicoptered in!) We newsmen were slightly better off in our hired cars; we plastered our windscreens with Press labels, and forced a way through on the wrong side of the road – and on the verges and even in dry ditches when anything came in the opposite direction. The consoling thing was that the Americans stuck in this monumental traffic jam, unlike typical British motorists, did not resent our claiming priority; many did their best to help us. Happily the normally implacable Florida police made no objections. Every American wanted the media to be there to report to the world that this was America's day.

I arrived an hour late at my reserved seat, No. A13 in the Press Stand, and lost more time ejecting some resentful 'groupies' who had moved in. When I plugged in my telephone – one had to take the instrument away, because there was always a freelance journalist around, looking for a chance to dictate his copy to Europe or Japan at someone else's expense – it started ringing instantly, for it was already 10 am in London.

Those dawn vigils at the Cape were unforgettable. Soon, although the noise made it almost impossible for me to hear London, I was describing the scene down the phone and praying that they were recording my voice in good enough quality to broadcast. At the other end, they always complained about the quality, so I was never confident of the result.

> It's like a football match here . . . Saturn 5 is directly in front, like a scene-set. Even though it's three and a half miles away it looks enormous, for it's bigger than the Hilton Hotel in Park Lane. The rocket is dazzlingly white, floodlit with criss-crossing search-lights as we face the Atlantic. And on the right a lovely pinkish-green dawn is just breaking . . . Every American here now seems confident that America will reach the Moon before the Russians. But there are lots of hazards first. One estimate is that this mission is five times more dangerous than any previous flight, beginning of course about three hours after lift-off when the top stage of the Saturn rocket must be relit to push the speed up to the 25 000 miles an hour needed to aim for the Moon . . .

The astronauts, I explained, were not going to get rich by attempting this historic flight. Borman and Lovell, as the two senior men, were getting the equivalent of £22.7s. a day in basic pay and allowances, and Anders, as the new boy, only £17 a day. The one fringe benefit was a £28 000 insurance policy, paid by the publishers of stories about the astronauts' families. Like the rest of the media, I would not mention the *Life* contract by name; it was considered that NASA and the military had breached their obligation to conduct an 'open' space programme by agreeing to such an exclusive arrangement between the publishers and the astronauts.

The idea behind the agreement was to compensate the astronauts for the small financial reward they were getting in return for being turned into American 'role models'. NASA managers apparently believed that stories about their families would make them rich – but nothing could be further from the truth. By the time the TV networks and the newspaper reporters had done their worst, there was little material left for *Life*, who got a very poor return on the contract. For the astronauts it was even worse, for the money had to be shared equally among them all – by then a total of about 50 – whether they were involved in the actual Moon missions or not. So all they got at the end was a few thousand dollars apiece!

It was still quite cold at the Cape when, at 7.51 am, the countdown clock reached zero at last. Lift-off, a mere six-tenths of a second late, was surprisingly emotional. I was so busy I always found myself mercifully insulated against it. But the drama of the roaring flames struggling to lift more than 3000 tons, and turning the early morning sky, by now deep blue, into gold with a scarlet centre, seemed to release all the pent-up tensions in the watching crowds.

Even the British newsmen were on their feet shouting 'Go – Go'; most of the Americans were leaping off the ground, punching their fists into the sky. Others were clapping or crying.

It must have been a good five seconds before the noise reached us, rattling the corrugated roof of the Press Stand so that I half expected it to crash on our heads. At 20 000 ft a burst of pure white cloud, fluffy as a giant ball of cotton wool, marked Saturn 5's progress. Only then did I

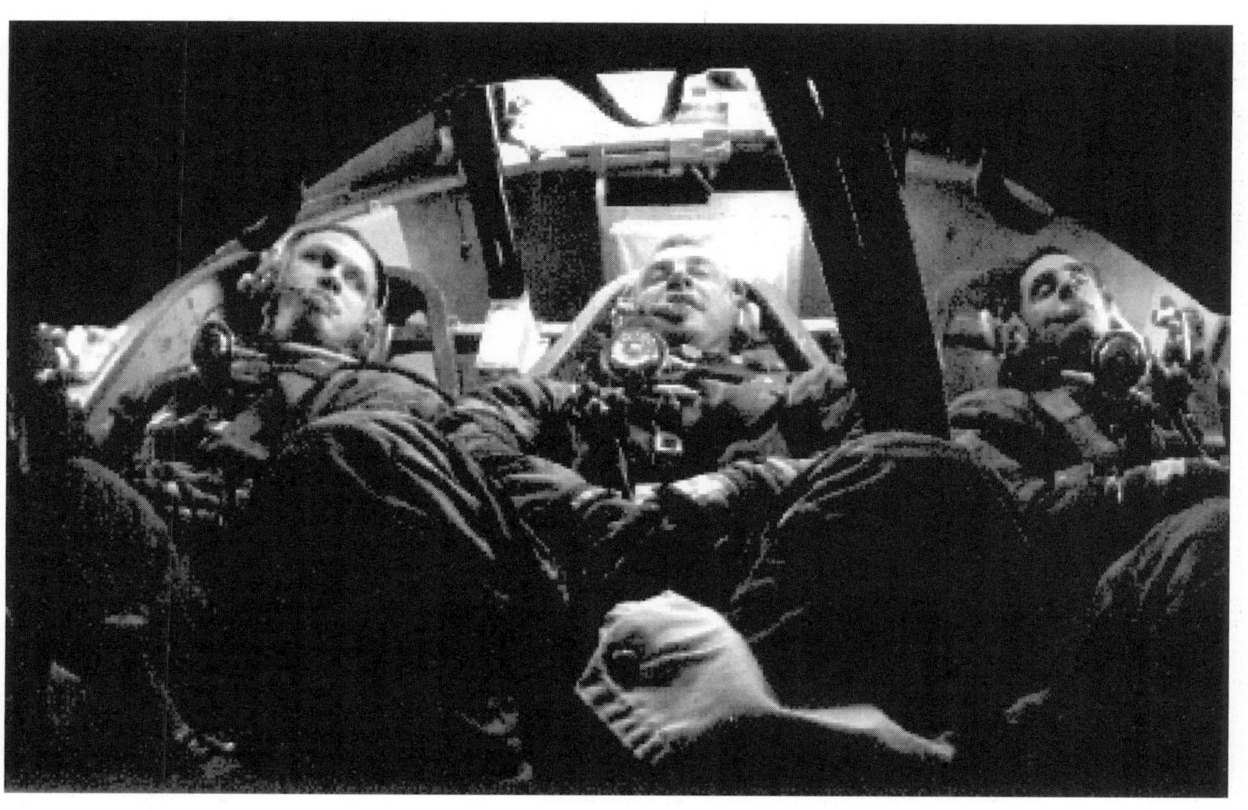

The Apollo 8 crew ready to leave for the first flight around the Moon. (NASA)

look back at the launchpad, and notice that a darker, dirty white cloud, mostly steam, was rising from it to join up with the cotton wool. It also contained acid rain, and we had been warned to cover our cars (not that we did, since they were hired!) in case the wind carried it over the car park and the descending acid pitted the paintwork.

I recorded the fact that we could see the start of the Moon journey for exactly three minutes: a searing line of flame cutting through the tranquil Florida sky as 2000 tons of fuel was burned. The first stage, Saturn 1, burned 531 000 gallons of kerosene and liquid oxygen in 2 min 34 s; then an expanding ball of white smoke as Saturn 2's five J-2 engines banged into life and started to consume 359 000 gallons of liquid hydrogen and liquid oxygen, taking the vehicle out of sight. After 6 min 10 s and an altitude of 108 miles Saturn 2 fell away in its turn, and the third stage, the Saturn 4B (no Saturn 3 was ever built) took over for 2 min 40 s to 'park' itself safely with the CSM – the Command and Service Module – still attached in an earth orbit of 119 × 103 miles. Borman reported that there was still slight 'pogo' during the second stage burn, but nothing to worry about. Anders, making his first spaceflight, commented that the 11 min 24 s journey from standing start to over 17 000 mph was 'like travelling an old freight train going down a bad track'.

For just over two hours, the astronauts and Mission Control cross-checked the systems on the spacecraft and the Saturn 4B third stage. Then Astronaut Michael Collins, just back from a neck operation which had lost him the Command Module pilot's centre seat in Apollo 8, and now acting as Capcom, told the crew: 'All right, you are Go for TLI.' That meant translunar injection, from which there was no turning back. Once injected, the spacecraft had to loop around the Moon before it could return to Earth.

We listened to yet another countdown as the spacecraft was perfectly positioned so that the S4B could be relit for a 5 minute 12 second burn, to raise the speed to 24 200 mph, the highest man had ever achieved:

'Another astonishing five minutes,' I reported. 'Within seconds the astronauts were 800 miles from earth. By now it's thousands. There were long, calm periods. Astronauts and ground controllers had little to say, everything was going so perfectly. The speed built up to seven miles a second. Over Hawaii the rocket was burning so fiercely, people could see it . . . So Frank Borman, James Lovell and Bill Anders really are on their way to the Moon. Whether they come back or not, they will always be immortals!'

But Frank Borman did not see the moment in the same dramatic light as we did on the ground. On the contrary, he said after his return, that was the moment when he felt much more relaxed: 'When you are in Earth orbit you are always aware that if something happens you have to react quickly to get down. Once you burn TLI . . . you really are not concerned with reacting quickly because it is going to take you two or three days to get home anyway.'

Having separated from the S4B, the crew turned Apollo around and went through the manoeuvres that later flights would need in order to dock with what had been the far end of the S4B in order to draw out the Lunar Module, which Apollo would have to carry on its nose and use for the moonlanding. Since there was no Lunar Module on this mission, Borman chose not to get too close, but reported that the manoeuvre would pose no problems. The crew then fired the small reaction control engines to ensure that, since both vehicles were on course for the Moon, they would gradually move apart, to ensure there would be no danger that they would collide. Both the crew and Mission Control became alarmed when, instead of separating, the two did in fact steadily get closer. It took the mission analysis experts some minutes to work out that Apollo had not been correctly positioned when the reaction control engines had been fired; and a new firing, of 2 metres per second, was ordered to counteract the earlier 1 metre per second manoeuvre.

As the astronauts settled down for their first sleep period it became necessary to put the spacecraft into 'the barbecue mode' to equalise the outside temperature. In outer space the side of Apollo exposed to the Sun grew enormously hot, while the other side, shielded from the Sun, was intensely cold. This problem was solved by putting the craft into a slow, once-an-hour roll as it travelled, thus equalising the temperature, rather like a chicken on a spit.

All this had put the spacecraft slightly off course, so 11 hours after lift-off and 61 000 miles from Earth the main Service Propulsion engine, giving a thrust of 5600 lb or 9300 kg, was fired for 2.4 seconds. It was a useful test of the engine upon which the lives of the crew depended; it had to operate perfectly and repeatedly if Apollo was to come safely home. That first firing added 16 mph to the spacecraft's speed and course, intended to take it within 69 miles of the Moon's surface, so that lunar gravity would then pull it round on to a 'free return trajectory'. That meant that a final decision whether to fire the engine again and place Apollo in a tight orbit around the Moon need not be taken until Christmas Eve.

There were still 55 hours to go before Apollo could go into lunar orbit, so I used the first 24 to fly back to London. There I could keep my TV bosses happy by being available in the studio for the rest of the mission, and try to convince my family that I was home for Christmas at the same time. Charles Wheeler, one of the two BBC staff based in Washington, had volunteered to spend his Christmas at Mission Control, near Houston, to file stories from there.

Charles covered the fact that when at last the astronauts were free to take off their spacesuits, they found that rapid movements made them spacesick. Borman retched and vomited twice, and had an attack of diarrhoea – dreaded by all astronauts in their closely-confined quarters, lacking both privacy and a proper toilet. Not surprisingly, Anders and Lovell did not feel so good either. Mission Control and all the other astronauts were alarmed when Borman asked for a private conversation with the physicians. 'Chuck Berry was in hog heaven' wrote Michael Collins scathingly six years later in *Carrying the Fire*. 'Here he had been waiting nearly a decade for someone in flight to solicit his advice, and by gum the first humans to leave the cradle had called for their pediatrician.' Collins was one of half a dozen privileged astronauts present with Berry and a few others during the private talk with the crew – the only transmission during the whole mission which the media and the rest of the world were not allowed to share.

In fact, not much was said. Borman reported that he felt miserable; the doctors, not knowing whether it was the motion sickness often reported by Soviet cosmonauts or the start of a serious illness, prescribed rest and fluids. It was agreed that there was no choice but to wait and see what happened. So far as the media was concerned the whole thing was played down; but everyone was relieved when Borman began to feel better the following day, and Lovell and Anders got no worse. Collins wrote later that it was probably NASA's first case of motion sickness; the Mercury and Gemini spacecraft had been too small for the astronauts to move about enough to get motion sickness as the Russians had done in their much bigger vehicles, and the NASA astronauts always – then and later – tended to dismiss the much greater cosmonaut experience as not really relevant to their activities.

Charles Wheeler also had to cover the first two TV transmissions, made possible by a special antenna swung out from the side of the Service Module (SM), giving the world the first-ever long-distance views of Earth. The first was not a success; Jim Lovell was using the wrong lens to point through a hazy window, and Borman apologised because they could not show their viewers the 'beautiful, beautiful view, with a blue background and just huge covers of white clouds'. Twenty-four hours later, and 60 000 miles out, the telephoto lens worked much better and Jim Lovell speculated as to whether, if he were some lonely traveller looking at Earth from that height, he would think it was inhabited – and whether he would land on the blue or the brown part.

Soon after that a new stage in manned spaceflight was reached. Like a ball thrown upwards, the spacecraft had been gradually slowing down, until its velocity was 2724 mph and its position was 202 825 miles or 326 400 km from Earth, and 38 900 miles or 62 600 km from the Moon. For the first time, men had reached a point where the pull of Earth's gravity was less than that of another body. Now the pull of lunar gravity was greater and the craft's speed began to increase again as it fell towards the Moon.

By the time I took my seat in the TV News studio on 23 December we were able to discuss some pictures of the crew which had been put through an 'enhancing' process by NASA. Chuck Berry and the other doctors had studied them with satisfaction. They showed a crew recovered from their brief space sickness, and with a much higher morale than their predecessors in Apollo 7. It was just as well. In less than 12 hours the choice had to be made: instruct them to 'free-wheel' around the Moon, only 70 miles from the farside, continuing on their present figure-of-eight course back to earth; or give them the Go to fire the SM main engine to enable them to make 10 orbits of the Moon, chancing the reliability of the engine to fire again 20 hours later to get them back into a trans-Earth trajectory. (During tests the SM engine they were carrying had never failed; but earlier versions had failed four times in 3200 attempts.)

The tension, I said in my pay-off, would fall most heavily upon the wives. NASA had provided filmed interviews with all three, but my filed script only gives 'in-and-out' cues, so I have no record of what they said. But one can be sure that they all made the dutiful, brave and supportive comments required by NASA of the spouses of the elite cadre of astronauts selected for the moonflights.

Despite the fact that Richard Francis now had his 'Apollo Studio' in full cry and crammed with experts brought on at peak moments, we devoted most of our main TV News that night – a mere 10 minutes at 8.50 pm – to explaining the LOI, or Lunar Orbit Insertion manoeuvre, timed to take place during the early hours of Christmas Eve. Soviet scientists helped me to build up the tension. *Pravda* quoted one as praising the courage of America's astronauts, but adding that too much was being left to chance. A Soviet manned shot, he said, would be backed up by fully automatic control systems, whereas the astronauts had to fire their own main engines while behind the Moon and out of communication with Mission Control. Ground signals could not be transmitted through the Moon's 2000 miles diameter. But it hurt when I had to conclude my report: 'We're now handing you over for further commentary to James Burke in the BBC Space Studio!'

Just before communications were lost as they passed behind the edge of the Moon Carr wished them luck and Lovell replied: 'We'll see you on the other side.' Eleven minutes later the crew had to fire the SM engine for four minutes, reducing their speed by 2000 mph. ('The longest four minutes I ever spent', confessed Lovell after his return.) If the braking was too much, they would crash into the Moon; if Apollo's attitude was incorrect and its speed was increased, it would arc away irrecoverably into solar orbit. Millions on Earth waited tensely, hoping that Apollo would NOT reappear too soon – for that would mean the attempt to get into lunar orbit had failed. They did in fact emerge from the back of the Moon 1 min 25 s *later* than expected – one of the rare occasions when advance calculations were even a few seconds astray; and when communications did crackle and clear Houston had more urgent matters than congratulations to deal with – a warning to the crew that a water evaporator was overheating, followed by a business-like exchange of technical data. Only then did Lovell reveal that at last they could see the target – and I can always hear him saying it:

> OK, Houston. The Moon is essentially grey, no colour. Looks like plaster of Paris, or sort of a greyish deep sand. We can see quite a bit of detail. The Sea of Fertility doesn't stand out as well here as it does back on Earth . . . The craters are all rounded off. There's quite a few of them; some are newer . . . Many look like hits by meteorites or projectiles of some sort.

We all wanted to know about the Moon's farside, never before seen by man. Anders' report on that was disappointing: 'The backside looks like a sand pile my kids have been playing in for a long time. It's all beat up, no definition. Just a lot of bumps and holes!' The others seemed to share his view of the Moon generally that it was 'a dark and unappetising place'.

After that first burn their lunar orbit was elliptical – 69 × 194 miles – and a planned second burn – LOI-2 – was made four hours later to tighten the orbit to a circular 69.5 miles. I commented in my reports on the calm, test pilot approach of the crew on this exciting day for the human race; the first view of 'Earthrise' sent back by Anders soon after they had gone into lunar orbit was perhaps the most memorable picture of my whole life.

But an excited world had to wait before seeing and hearing in detail the first human reactions to a close-up study of another astral body. First Mission Control and Apollo 8 wanted to check that they and the spacecraft computer were ready for a return to Earth if things went wrong and the crew had to fend for themselves. Throughout the Apollo missions we became accustomed to regular 'PAD updates':

> MISSION CONTROL: Roger. TEI [Trans Earth Injection] 10. SPS G&N [Service Propulsion System. Guidance & Navigation] 45597 minus 040 plus 157089191564 plus 35189 minus 01513 minus 00346 180007000 November Alpha plus 0018635223318350194092825 3 boresight star Scorpi Delta, another name for it is Deshuba, down 071 left 45 plus 0748 minus 1650012995363001465005 primary star Sirius. Secondary Rigel 129155010 four quads 15 seconds ullage. Horizon on the 2.9 window line at T minus 3. Use high-speed procedure with minus Mike Alpha. Over.

Despite all sorts of static noises, these figures would be read back with incredible speed and accuracy.

A major task for the crew, paving the way for a landing a few months later, was of course to check whether the human eye could identify the craters in the earlier Lunar Orbiter photographs, and they found that that was relatively easy. But even more detail was needed to ensure the safety of the landing, so it had been agreed with Houston that, as the television pictures of the surface flowed back and were recorded on video, the crew would identify previously unnamed craters with temporary code-names. Soon there were craters named after Borman, Lovell and Anders, and another for Collins, whose surgery had lost him his place on that flight; NASA Administrators and top brass were given craters – though poor Dr Berry was omitted. The astronauts who had lost their lives duly got craters – Grissom, White,

Chaffee, Bassett, and See; and when they were all so busy crater-gazing that they had to be reminded by Flight Controller John Aaron that their environmental system needed attention, Crater Aaaron was instantly created.

In addition to the TV pictures, it was Anders's task to devote much of the 16 hours spent in the close orbit taking colour film and photographs for study when they returned – especially of the five landing sites already chosen, the favourite among them being in the Sea of Tranquillity. He had an exhausting time, hurriedly fitting filters over lenses when the dipping sun angle turned the lunar surface into a dazzling white glare – and periodically moaning about the fogged-up state of the windows.

In my last TV piece that Christmas Eve I reported: 'One point worth noting is Lovell's insistence that many of the Moon craters appear to be fresh. With no atmosphere to burn up the meteorites, the Moon is under constant bombardment. But the chances that a spacecraft like Apollo 8 would get hit are only once in a hundred and fifty years. And so today has largely removed the terror of the unknown from the final landing manoevre.'

It was still Christmas Eve in the US but the early hours of Christmas Day in Europe when, during their ninth revolution around the Moon, the Apollo 8 crew provided the world with one of the most remarkable and humbling TV transmissions of all time. Making the best of their two unfogged windows they first showed a half-Earth view, blue and white against the bleak monochrome lunar landscape rolling in slow motion beneath them. And as the Moon revolved below them and seemingly below us as well, we listened to their commentary:

> FRANK BORMAN: This is Apollo 8 coming to you live from the Moon. We've had to switch the TV cameras now. We showed you first a view of Earth as we've been watching it for the past 16 hours. Now we're switching so that we can show you the Moon that we've been flying over at 60 miles altitude for the last 16 hours. Bill

Anders, Jim Lovell and myself have spent the day before Christmas up here doing experiments, taking pictures and firing our spacecraft engines to maneuver around. What we will do now is to follow the trail that we've been following all day and take you on through to the lunar sunset. The Moon is a different thing to each one of us . . . My own impression is that it's a vast, lonely forbidding type existence, great expanse of nothing, that looks rather like clouds and clouds of pumice stone. It certainly would not appear to be a very inviting place to live or work. Jim, what have you thought most about?
LOVELL: Well, Frank, my thoughts are very similar. The vast loneliness up here of the Moon is awe-inspring, and it makes you realise just what you have back there on Earth. The Earth from here is a green oasis in the big vastness of space.
BORMAN: Bill, what do you think?
ANDERS: I think the thing that impressed me the most was the lunar sunrises and sunsets. These in particular bring out the stark nature of the terrain, and the long shadows really bring out the relief that is here and hard to see, and is very bright –
BORMAN: It is here, and hard to see, at this very bright surface that we're going over right now. Now describe, Bill, some of the physical features of what you're showing.
MISSION CONTROL: Apollo 8, Houston, we're not receiving a picture now, over . . .
ANDERS (When the TV picture was restored): . . . The contrast between the sky and the Moon is a vivid dark line. Coming into the view of the camera now are some interesting old double-ring craters. They are quite common in the mare region and have been filled by some material. The same consistency of the other maria and the same colour. There are three or four of these interesting features. Further on the horizon you see the Messier. The mountains coming up now are heavily impacted with numerous craters whose central peaks you can see, and many of the larger ones. Actually I think the best way to describe this area is a vastness of black and white – no color. The sky up here is also rather forbidding, foreboding extents

of blackness with no stars visible when we're flying over the Moon in daylight. You can see by the numerous craters that this planet has been bombarded through the eons with numerous small asteroids and meteoroids pock-marking the surface every square inch. And one of the amazing features is the roundness that most of the craters – seems that most of them have a round mound-type of appearance instead of sharp, jagged rocks. All, only the newest of features have any sharp definitions to them, and eventually they get eroded down by the constant bombardment of small meteroids. How is the picture now, Houston? Houston are you reading us?
MISSION CONTROL: Loud and clear, and the picture looks real fine . . .
SPACECRAFT (Probably Borman): There is an interesting rill directly in front of the spacecraft now running along the edge of a small mountain. Rather sinuous shape with right angle turns. This area just to the west of the Sea of Crises is called the Marsh of Sleep and to the west of that the Sea of Tranquillity. Can you see the fracture pattern going across the mare in front of us now, Houston?
MISSION CONTROL: That doesn't quite stand out.
ANDERS: Roger. The series of cracks or faults across the middle of the mare, they drop down in about three steps to the south. The parallel faults pattern to the north and drop down in the center. I hope all of you back down on Earth can see what we mean when we say it is a very foreboding horizon, a very rather dark and unappetising looking place. We are now going over, approaching, one of our future landing sites selected in this Moon region called the Sea of Tranquillity – smooth in order to make it easy for the initial landing attempts, in order to preclude having to dodge mountains. Now you can see the long shadows of lunar sunrise. We are now approaching the lunar sunrise.

For a few silent moments the camera panned over the cratered Moon. What came next struck one instantly as a stroke of genius. Bill Anders continued: 'For all the people on Earth, the crew of Apollo 8 has a message we would like to send you,' paused, and began reading Genesis:

> In the beginning, God created Heaven and Earth, and the Earth was without form and void, and darkness was upon the face of the deep. And the spirit of God moved upon the face of the waters and God said 'Let there be Light', and God saw the light and that it was good. And God divided the light from the darkness.

Jim Lovell took up the reading at the fifth verse:

> And God called the light day, and the darkness he called night. And the evening and the morning were the first day. And God said 'Let there be a firmament in the midst of the waters. And let it divide the waters from the waters.' And God made firmament, and divided the waters which were above the firmament. And it was so. And God called the firmament Heaven. And evening and morning were the second day.

Then Borman took the last verse:

> And God said 'Let the waters under the Heavens be gathered together in one place, and the dry land appear.' And it was so. And God called the dry land Earth; and the gathering together of the water he called seas; and God saw that it was good.

With an actor's perfect timing Borman paused again and concluded:

> And from the crew of Apollo 8 we close with good night, good luck, a Merry Christmas, and God bless all of you – all of you on the good Earth.

It was all that was needed for the brief Christmas Day television news bulletins. After that for most of the public the seasonal festivities took precedence over what was for the crew, Mission Control and those of us covering the flight, the tensest moment of all: TEI, trans Earth injection, the computer programme which had so often been updated.

Now it was a 5 min 3 s firing of the Service Module's main engine to give Apollo 8 the additional speed needed to break out of lunar orbit and head for Earth and home. 'All the holy words in the universe would not budge Apollo 8 from its orbit,' wrote Michael Collins in *Carrying the Fire.* 'Its only salvation lay in the chemical energy locked within the SPS system.'

But in their 10th lunar orbit, out of sight and sound behind the Moon, Command Module Pilot Jim Lovell, with Apollo this time pointing forwards instead of backwards, punched the right buttons, the rocket fired, and added 2394 mph to their speed. As they emerged from the back of the Moon on course for Earth, Lovell told Mission Control: 'Please be informed there is a Santa Claus.'

Duly relaxed, the crew began their 57-hour cruise back to Earth by eating their own Christmas Dinner. They had to reconstitute it first, and suck it out of plastic bags, but on the Flight Plan it looked attractive enough: turkey chunks with cranberry and apple sauce, washed down with grape punch, and followed by coffee. Radio was much more conscientious than TV in covering all these activities, and I had a busy few hours in Broadcasting House describing them in Radios 1, 2 and 4. ('The Third' as it was then known, remained aloof, devoting itself exclusively to religious services around the world.) But I too was finished in good time to drive quietly home through London's deserted streets to my own family Christmas dinner at Sydenham.

Twenty minutes before splashdown was due, Borman punched the buttons to fire the explosive bolts to jettison the 20-ton Service Module and its faithful engine. Of the 3000 ton vehicle launched from the Cape six days before, only the 6-ton Apollo remained. However, its 2-inch thick heatshield proved adequate to protect the crew against the fiery, 25 000 mph re-entry – 8000 mph faster than from Earth orbit – as it blazed through the atmosphere.

It was hard to believe, I reported, that the first men to go to the Moon could return to Earth on the very minute predicted before they left – but that fact summed up the magnitude of America's technical achievement.

For the United States, 1968 had been a year of riots, assassinations, and a death toll of 15 000 in the Vietnam War. But as NASA's history of Apollo records, it was Apollo 8's voyage in its last days that newspaper editors voted to be the story of the year; and a friend telegraphed Frank Borman: 'You have bailed out 1968.'

12 What Makes an Astronaut?

Two weeks into the New Year of 1969, and on the day when Borman, Lovell and Anders were being decorated by President Johnson in Washington prior to a 'ticker-tape' parade in New York, the names were announced of the men assigned to the first moonlanding mission: Neil Armstrong, Edwin 'Buzz' Aldrin and Michael Collins.

NASA had tried in vain to stem 'the cult of personality', as it was called by the Russians, so far as their own spacemen were concerned. Project Mercury, NASA's first manned spaceflight programme in the early 1960s, had originally been called Project Astronaut, but that had been changed because it was thought it would increase the excessive publicity the spacemen were bound to get. NASA's *The New Ocean* records that Abe Silverstein, then Director of Space Flight Development, advocated 'a systemic name with allegorical overtones and neutral underpinning. The Olympian messenger Mercury, denatured by chemistry, advertising an automobile, and Christianity, was the most familiar of the gods in the Greek pantheon to Americans.' Another consideration was that there was neither wish nor intention to give the astronauts supermen's salaries; they were to be paid the service rates of their ranks, and told to regard their selection as astronauts as a privilege and a compliment. Complaints about the policy came later; initially there was no fall in the number of astronaut-applicants.

NASA's Public Affairs Officers, often recruited from the ranks of newspapermen, understood media attitudes to personalities and 'human interest' stories rather better than their Administrators. It was inevitable that TV, radio and newspapers would want to know everything about the personalities and families of these descendents of history's pioneers. The first humans to venture from Earth to visit another world were, after all, the Ultimate Explorers.

I was personally very relieved when the names were announced; otherwise I faced trying to get them unofficially or relying upon speculation, since I was busy using my limited spare time to write a short book which I had called *Moonslaught*. It had been commissioned by a young publisher called John Selwyn Gummer, famous 25 years later as Britain's Minister of Agriculture, Fisheries and Food, and it was to make me what seemed a lot of money in those days – a total of over £3000! Having gone cap-in-hand to my BBC boss for permission to do it, I urgently needed to finish a chapter called 'The Men with the Moon Tickets', because it was to be in the bookshops with explanations of the whole mission well before Apollo 11 was launched, and it would be out of date when it was over.

The more I had studied the motives and methods involved in the selection of the astronauts and cosmonauts, the more fascinating the subject became. In both East and West, management had clearly started with the determination that these 'puppets' they were creating should respond to the strings that they intended to pull, and not become monsters, getting the credit for their masters' foresight and taking over the whole propaganda machine. Yet every time a big step forward in space techniques was achieved, the only way to claim the national prestige it merited was by elaborate celebrations built around the spacemen involved. It happened first of course with Yuri Gagarin, whose name was not allowed to be published until he became the first man in orbit, and then had to be turned into a celebrity symbolising Soviet technological superiority.

Now, ten years after manned spaceflight began, it was just becoming possible to compare the very different East/West personalities involved. We knew a great deal about Dr Wernher von Braun, the German behind the American moon programme and the subject of chapter 1; and after Sergei Korolov's death three years earlier, in 1966, the Russians had at last acknowledged that he had been their anonymous 'Chief Designer' and the driving force behind their manned spaceflight programme. They were, however, still years away from acknowledging that he had done most of his spaceflight planning and designing while imprisoned for six years under the Stalin regime.

Originally NASA had intended to appoint 12 astronauts for the six tentative one-man Mercury missions. But as the selection process continued, with endless physical and psychological tests, it became evident that it would be unfair to appoint so many, when only half could expect to fly. The number was cut to six; but the final stages of selection were so difficult that the judges weakened and finally appointed 'The Mercury Seven'. Since then five more groups of astronauts had been appointed, and the qualifications had dramatically changed. Originally they had had to be military test pilots, under 40 in age, with at least 1500 hours of flying on their log books, and a university degree behind them so that they could be turned into scientists as well. That process had been reversed for the last selection. Eleven scientists had been chosen who could be trained as jet-pilots.

What caused most surprise was the age, not the youth, of the Americans selected. The initial qualifications demanded were so high, and took so long to obtain, that the average age of selection was 32 – and it was usually six years or more after appointment before they actually got a spaceflight. (Later, because of the long gap between the end of Apollo and the start of Shuttle flights, astronauts like Robert Crippen had to work and wait for 16 years before getting a flight.)

On the personal side I noted that because astronauts were usually over 30 they were nearly always family men. Of 52 'active' astronauts out of a total of 65 appointments in early 1969, only three were unmarried. The well-balanced Superman, with the required combination of temperament, physical fitness and technicals skills, emerged as being usually married, with an average of 2.5 children. The bachelor possessing the required Superman qualities was, and still is, very much the exception.

Gathering enough information about Soviet cosmonauts to make a reasonable comparison of their qualities was in those days a challenge for the most industrious investigative journalist. But over many years it was possible to put together facts gleaned from sources as various as Soviet brochures issued for schoolchildren to postage stamps containing details which somehow evaded the censors.

During the cold war, deducing how much the Soviets had achieved in space was an intriguing activity. The Russians loved their stamps and it was soon noticed that they were escaping the rigorous censorship applied elsewhere. From this stamp, I learned for the first time how Cosmonauts Lyakhov and Ryumin, during nearly 6 months aboard the Salyut 6 space station, had made a daring spacewalk and used a pair of pliers to cut away the jammed umbrella-shaped antenna so that their second docking port could be re-used. (Graham Turnill)

It was clear that Soviet scientists, like their opposite numbers in America, looked first among their top-quality military test pilots when they began selecting their cosmonauts. After that, divergent qualities became apparent. It seemed to be more than coincidence that Cosmonauts 1 and 2, Yuri Gagarin and Gherman Titov, had proved to be superb at the postflight propaganda game; but basically it was very clear that Soviet cosmonauts were first and foremost military men detailed for duty, and expected to do as they were told. That is not to say that they were any less eager for such an appointment than most, but not all, of America's test pilots. But it was clear that the Soviet space programme always nursed many secrets which even the cosmonauts were not allowed to share.

Most striking of all about the rival breeds was the difference in the selectors' attitudes to their physical condition. While every American chosen as an astronaut had to be a nearly-perfect specimen of the human race, the cosmonauts were often people who had suffered and recovered from serious accidents or injuries. The German invasion was reflected in most of their personal histories, starting with Gagarin himself, whose schooling was interrupted when he and his parents joined the refugee columns from the Smolensk area. But that had not affected the engaging personality, ready smile and sense of humour that had made him the perfect travelling ambassador after he became the first man in orbit. His death at 34 on what was claimed to be a routine training flight in a jet fighter remained something of a mystery for many years; why was the great Yuri Gagarin involved in such a low-level activity? Later there was a suggestion that he had been assigned as Komarov's backup for the Soyuz 1 mission. But there can be no doubt that Gagarin, dead after smiling his way around the world, quickly grew into a legendary ageless figure much more useful to the Soviet Union than he would have been had he lived into querulous and critical old age.

The third Soviet cosmonaut, Andrian Nikolayev, lived both to be old and a General (as did 11 other military cosmonauts) but his beginnings were modest. The son of a farmer, he changed his mind in the middle of medical training in favour of timber-felling, and then was conscripted into the Soviet Army and proved to be a brilliant fighter pilot. His marriage to Valentina Tereshkova, the first and only spacewoman for the next 19 years, did Nikolayev's career no harm at all. They married five months after she became Cosmonaut No.6, and their healthy daughter, born within a year, reassured Soviet scientists that there were no harmful radiation after-effects as a result of their spaceflights. Tereshkova's hurried selection as a cosmonaut was clearly a rather cynical political ploy by the Soviet leaders to ensure that the first spacewoman was a Russian. She herself confessed that when Gagarin and Titov went into orbit she 'was not even dreaming of becoming a cosmonaut'. Yet somehow she graduated from a technical textile school to a brief three-day stay in orbit, the sole occupant of Vostok 6, only 21 months after Titov's flight.

Vladimir Komarov, who (as described in 'Apollo's Bad Start') became the first human to die during a spaceflight, was much the most professional of the early cosmonauts, but despite that was dogged by ill luck. He began adult life by earning a living on a collective farm during the grim war years. Then he won his way into schools of aviation, finally achieving his ambition to become a test pilot. He had 20 years of aviation experience behind him when he was appointed a cosmonaut, and would probably have been preferred to Gagarin for the first space-flight until ill-health struck. An operation was necessary, after which the doctors ordered: 'No parachute jumps and no overloads for six months.' He had no sooner recovered and been appointed backup to Cosmonaut No.3, Pavel Popovich, than the doctors ordered his discharge from the programme. Strenuous sessions on the centrifuge showed that his heart had developed an 'extra-systole' or slight irregularity of the beat.

It was a similar condition that had wrecked the chances of Deke Slayton, the only one of the original Mercury Seven not to get one of those flights. After being nominated for NASA's second orbital mission, he was replaced. Like Slayton, Komarov appealed again and again. Slayton lost; Komarov won. After proving, just as Slayton did, that in training he could withstand greater physical stresses than he would ever encounter during an actual flight, the doctors were over-ruled, and Komarov was restored to flight status. Perhaps the Soviet space doctors thought there would be positive medical benefits, because Voskhod 1, Russia's first three-man flight, which lasted only one day, was intended to test the men as much as the spacecraft system.

Komarov's companions were a physician and a scientist. Boris

Yegorov, the doctor, was the son of a surgeon and had got to know Gagarin at medical school. He was seconded to the cosmonaut corps only a few months before the flight, and within minutes after launch was checking on the health of his patients, and no doubt paying particular attention to Komarov's heart – which stood up to the stresses so well that he became the first cosmonaut to be given a second flight. Yegorov returned to his medical career shortly afterwards.

Yegorov's other 'patient' – and with good reason – on Voskhod 1, was Konstantin Feoktistov, who had decided with his elder brother when they were small boys that they would be spacemen and go to the Moon. As boys they had even estimated fairly accurately when it should become possible – 1964. But their plans received a setback when the elder boy was killed in action in 1941; and soon Kostya, as Konstantin was known, then aged 15, had to join the stream of refugees, fleeing with his mother from the advancing Germans.

But being young and innocent-looking as well as highly intelligent, Kostya was recruited as a scout, and sent off into German-occupied areas to see what he could find out. He returned safely from many such reconnaissances, and won high praise for the information he brought back. But in July 1942 his luck ran out. He was put across a river at night, with another boy two years younger, and the following day they walked around the streets of Nazi-occupied Voronezh. A patrol stopped and questioned them in a restricted district, and after questions marched them off. Beside a pit they were ordered to stop. The Germans opened fire, and the boys fell into the pit, Kostya feeling a sharp pain in his chin as he fell.

The Germans left them for dead, but when night came Kostya managed to crawl out and escape. After a long spell in hospital he had a scar on his neck and a medal: 'For Victory over Nazi Germany'. At 19 his enthusiasm was undiminished. He got himself into the Moscow Higher Technical School, graduated, and kept presenting himself as a candidate-cosmonaut until his persistence won him a place.

He was 43 when at last he went into space, still – we were told – determined to be the first Russian on the Moon. Realistically, his age being against him, he relinquished the ambition two years later and left

Cosmonaut General Aleksey Leonov, centre, who made the world's first spacewalk, with the author, right. Left is Dr Charles (Chuck) Berry, the man who said the President could not have dinner with the Apollo 11 astronauts the night before the launch. Margaret Turnill took the picture when they were all taking part in a Channel 4 TV programme re-enacting the moonlanding 30 years later.

the cosmonaut corps, but played a major part in designing the later Salyut space stations.

Next came two Soviet Air Force men, one with a taste for poetry and the other for painting. Their flight in 1965 proved that, given the courage, humans could leave their spacecraft and work free in space. The commander, Pavel Belyayev, had fought back after losing a whole year as a result of badly fracturing a leg during a parachute jump. He had been encouraged in his convalescence by his friend Gagarin, who visited him in hospital and finally accompanied him on his vital post-accident parachute jump, to check whether his nerve had gone. But he was to develop stomach cancer after the flight and die at the early age of 37.

Aleksey Leonov, who accompanied Belyayev, became the first man to wriggle outside a spacecraft and become a human satellite. It was only for ten minutes, but it proved that it could be done – and the Americans who were already capable of spacewalking but whose conservative policies had held back the first attempt, were quicker to benefit from the knowledge and soon overtook the Russians in 'EVA' – extra-vehicular activities.

Leonov was another cosmonaut who became a General. A convivial man undaunted by narrow Communist fears and restrictions, he made friends with US astronauts, notably Tom Stafford, and did much to make the joint Soyuz–Apollo flight possible in 1975. Leonov's vivid, impressionistic paintings following his brief spacewalk – revealing, incidentally, that it had been made possible by a telescopic airlock – provided a permament record of man's first reactions to this courageous venture.

Alan Bean, America's fourth man on the Moon, was another spaceman whose experiences inspired some dramatic art. The conclusion I had reached in early 1969, and forgotten until researching for this book, remains valid: the Russians devoted too much effort, in those pioneering space days, to propaganda successes. Had politicians like Khrushchev not interfered, they would at least have been the first to send men around the Moon – and probably the first man on its surface would have been Russian.

13 Final Rehearsals

Apollo 9

Despite the technically unqualified success of Apollo 8, NASA decided not to take for granted its ability to get men on the Moon in only three more flights, and made plans for five launches in 1969 – in February, May, July, September and December. Only five days after Apollo 8's splashdown, Apollo 9 was rolled out to the launchpad for the flight that was to try out the Lunar Module – the 'Lunar Bug' the astronauts called it – a weird spiderlike craft designed to deposit two men on the Moon and bring them back to the Apollo Command Module waiting in lunar orbit. For the public this mission seemed rather dull after the excitements of Apollo 8, and it was a near-impossibility to convey that it included far greater hazards than circling the Moon.

NASA was spurred on by the fact that the Washington celebrations of the circumlunar flight did not go unchallenged by the Soviets. On 14 January Soyuz 4 was launched with only one cosmonaut aboard, and a day later Soyuz 5 followed with a full complement of three. The world's first docking of two manned craft in space was then achieved; the Russians, perceiving the need to establish an alibi for their Moon failure, claimed to have created 'the world's first experimental space station'.

The expected crew transfers between the craft duly took place – but not in the way we expected. Apparently the Soviet scientists were not yet confident enough to dismantle and withdraw the docking probe between the two vehicles so that the cosmonauts could pass from one to the other internally. Instead an external hatch was opened on each of the docked craft, and two of the men in Soyuz 5 made a laborious 37-minute crawl with the aid of handrails over the outside from one hatch to the other. 'In effect', I reported, 'the mission had rehearsed the first emergency rescue in space.'

The Russians also made a first attempt to match the Americans' TV coverage of their activities. They still did not risk live transmissions, but showed the Soviet public launch pictures and a 'guided tour' of the spacecraft interior only an hour after these events had taken place. The whole mission was over in four days, but these rival activities succeeded in taking the edge off the American celebrations.

Just before my book *Moonslaught* went to the printers NASA provided more exciting material with details of the plans for Apollo 11 to bring back 50 lb of lunar rocks and dust. The astronomers and lunar scientists thought this would at last solve the mystery of the origin of the Moon, and to speed up that process arranged for samples to be made available to scientists around the world, including 14 British university groups. 'But not, I am glad to say, until both the astronauts and their samples have been put into quarantine for three weeks at Houston, Texas,' I reported. 'That will provide time to ensure that the samples don't include some unexpected germs that can breed on Earth. Some scientists consider that although unlikely, this must be regarded as a serious possibility.' The fear that extraterrestrial bacteria or other dormant life forms were lying quiescent just below the Moon's surface, awaiting the day when they could reach the Earth to spring to life and colonise it continued to provide us with many a good story. I was very knowledgable on the subject, for during quiet periods in the Apollo 8 mission we had had detailed briefings on 'The LRL' or Lunar Receiving Laboratory, by its excitable boffin manager, one Dr Persa Bell, and its Curator, Dr Elbert King. The LRL had just been completed at Houston at a cost of $12 million, and correspondents were allowed to visit it just before it was bacteriologically sealed from the outside world, complete isolation being ensured by 'negative pressure' – the maintenance of lower air pressure inside, so that any leaks would always be inwards.

The LRL had been created on the orders of a US Inter-Agency Committee on Back Contamination set up in 1958 to lay down the principles of a biological barrier system which would not only protect the Earth from lunar contamination, but also protect the lunar samples from contamination by earthly bacteria. The scientists took it all much more seriously than most NASA personnel; and the journalists, I fear, revelling in the pompous language accompanying it all, used the theoretical 'moon bugs' as a welcome source of light, mocking stories.

Meanwhile, an indication of NASA's growing confidence in the Apollo system was that one of its chief planners started worrying that the re-entry and splashdown of returning spacecraft was too accurate! If the spacecraft descending on its parachutes actually hit the aircraft carrier waiting with its helicopters and frogmen to recover it from the sea, the consequences, he minuted, would be 'truly catastrophic'. The sea provided a suitable cushion to receive Apollo's 9 tons, but the final descent was much too rapid for the astronauts to survive a direct impact with the unyielding armour-plated carrier. So he recommended that the recovery force should be located at least 8 to 16 kilometres from the targeted splashdown point.

This time, when the Apollo 9 countdown began a few days late at the beginning of March – there was a three-day postponement because crew members had sore throats and nasal congestion – I was not available to provide my usual first-hand launch coverage. It clashed with the imminent first flight of the Concorde prototype, and I gave that preference and went to Toulouse instead of Cape Kennedy. But as soon as Concorde had completed its brief maiden flight, there were urgent demands that I catch the first possible plane back to London to pick up space coverage for TV News, leaving the unfortunate Margaret to make the long drive with our car back to England on her own.

The Apollo 9 crew had spent two years working together in preparation for their mission, intended to rehearse the actual moonlanding in Earth orbit. They were to spend 10 days in orbit, matching the time needed for a flight to the Moon, lunar landing, stay and ascent, and return to Earth. James McDivitt, a member of the second astronaut group, was commander, having previously commanded the Gemini 4 flight in 1965, during which Edward White had made America's historic first spacewalk lasting 21 minutes. With McDivitt this time was red-haired Russell ('Rusty' of course) Schweickart, who was to share with him the risks involved in making the first flight in the Lunar

Module; while they did so the Apollo Command Module would be in the care of the handsome David Scott, prototype of *The Right Stuff*, as the astronaut heroes were later dubbed by Tom Wolfe.

The media, and especially the BBC, had not recovered from the excitement and cost of Concorde's maiden flight when, at 11 am the following day (3 March 1969) the second manned Saturn 5 took off, demonstrating its capability to place 125 tons in a low Earth orbit, and to send 50 tons from there to a lunar orbit. I missed surprisingly little as I travelled from Toulouse back to London: the crew was shaken about once more by the pogo effect when the Saturn 2's engines cut in, but they made no complaints about it; and David Scott had a little difficulty lining up again after separating from the S4B and turning around to extract the LM from the other end. One of the small RCS thrusters refused to work until switches had been recycled. And although they had all been warned by Frank Borman not to make sudden head movements when they started work, they all had space sickness symptons, despite taking medication. Schweickart was to suffer from it so badly that this was to be his only space mission.

Following a space mission from the confines of Television Centre was not my idea of reporting, although there were certain advantages. When the story was regarded as big enough, money was unlimited. Every minute of every available TV transmission was accepted, and I could sit in the recesses of the building peacefully watching what was going on while the cumbersome telecine machines revolved around me, recording the pictures for possible use later. I was much better off there than at Broadcasting House, which had to rely wholly upon audio transmissions – and sometimes better off than if I had been at Houston, where access to the incoming TV pictures, even if available, was accompanied by the chatter and activity of scores of rival correspondents. With my Apollo, Lunar Module and Saturn Press Kits on my knees, and my head full of briefings about personalities and problems, it was almost restful.

It was their third day in orbit when McDivitt and Schweickart carried out the first crew transfer via the docking tunnel of two spacecraft, rather than by the sort of external spacewalk used six weeks earlier by the Russians. When McDivitt first opened the hatch into the Lunar Module, he was assailed by a 'blizzard' of debris – small particles dropped by technicians which they had been unable to clear up when they had finished installing the systems. The two astronauts had a bad time making their first entry. While David Scott was busy dismantling the probe and drogue to clear the docking tunnel, and the other two were struggling into their pressure suits, Rusty vomited again, but managed to keep his mouth shut until he could reach a bag – thus just avoiding contaminating his all-important suit.

In the bulky suits they found it difficult to thread themselves through the 32-inch docking tunnel, but once inside they embarked on nine hours of checks and tests. Each had about as much standing room as one gets in a phone box, with the cover of the ascent stage engine sticking up waist-high between them. There were no seats or couches – there was not enough room – but they were dispensable anyway in the weightless conditions of space and the one-sixth gravity on the Moon. For manoeuvres, and later for the moonlandings, the astronauts were dependent upon body harness with automatic restraints.

Tasks for McDivitt and Schweickart included a 7-minute TV transmission, which was disappointing for viewers. The camera was in a fixed position, and little was to be seen except McDivitt operating switches and Schweickart keeping absolutely still because he was worried that any abrupt movements would cause him to vomit again. McDivitt experimentally throttled the LM's descent engines like a motorist trying out his new car; and satisfied that everything would work they returned to the Command Module for a night's rest.

Next day they were back in the LM for the next stage of the rehearsals. Some elaborate spacewalks to practise emergency procedures had been planned for Schweickart, but McDivitt was so concerned that he might vomit and choke inside his spacesuit that he decided that all such activities should be made less strenuous. However, the LM was depressurised and the hatch which would give access to the lunar surface was opened. This pulled inwards at the astronauts' feet, and Schweickart had to back out on hands and knees on to the LM's 'front porch' – a small platform at the top of a nine-rung aluminium ladder attached to the LM's front leg, to enable the moonlanders to descend 10 ft from the cabin on to the Moon. Secured by a nylon tether so that he could not float away, Schweickart stood up on the porch, anchoring his feet in 'golden slippers' which had been fastened there for him before the launch. Next David Scott depressurised the Command Module and opened his hatch.

The original plan had been for Schweickart, using handholds, to make his way there outside the craft, as a demonstration that if the Lunar Module failed to redock with the Apollo Command Module at any stage, this method could be used as a rescue procedure. (The Lunar Module, for the reasons described in the first chapter, had no heatshield, and astronauts using it could only return safely to Earth by transferring back into the Command Module.) We did not get the promised TV pictures of this particular drama, but Scott and Schweickart – who was feeling much better – filmed one another from their respective perches like tourists on a mountain-top before establishing that it was quite easy to clamber about outside. They then retreated inside their craft to stow the film for return to Earth. We did get some TV when they were back inside, and it came in while I was reporting during one of our main evening news programmes; for once I had a happy time 'ad-libbing' live commentary as we extended the news for 15 minutes. Normally such glamorous activities were claimed by the self-styled experts in Dick Francis's Apollo studio.

The EVA had been reduced to one instead of two hours, so that the crew members, and especially Schweickart, could conserve their strength for the key events of the whole mission: the separation of the Lunar Module for the period during which it would simulate the landing on the Moon, the ascent from it, and then – the manoeuvre upon which the astronauts' lives depended – redock with the Command Module. For these activities separate call-signs were needed, and it had been agreed that the LM with its thin legs and buglike body would be 'Spider', while the Command Module would be

'Gumdrop' because it had looked like a giant sweet – 'candy' in American parlance – when transported in its coloured foil wrappings. NASA bosses much preferred these rather childish names to the call signs like 'Faith' used by earlier astronauts and the dreaded possibility of headlines like 'America Lost Faith Today'.

The crew's fifth and most strenuous day in orbit was given an easier start when Mission Control agreed that McDivitt and Schweickart could enter the Lunar Module without helmets and oxygen hoses to start their checklist and preparations. The hatches between the Command and Lunar Modules were soon closed and checked, and David Scott punched the button to separate them. The LM hung on its capture latches; Scott punched the button again, and from their respective windows the astronauts watched the gap between them widening.

They took wonderful film of the dramatic manoeuvres in the following six hours, but there were no TV transmissions. By the time the film had been brought back to Earth, processed and released by NASA, TV news editors thought it was an old story, and gave it minimal exposure. I have always been baffled at the mental processes which led them – in my view – grossly to underestimate public interest in such pictures. There is probably truth in recent suggestions that this is because TV executives tend to have literary rather than scientific backgrounds! On the day, however, my radio colleagues rejoiced at the lack of TV while the Apollo 9 crew were doing their *pas de deux* in space, and made full use of the audio exchanges between Gumdrop and Spider to supplement their descriptive abilities. Stuck in the TV studio I triumphantly demonstrated it all with the help of the take-apart models presented to me several years earlier by North American Rockwell.

McDivitt used his small thrusters to make Spider 'pirouette' so that Scott could examine it from all angles through his windows, especially checking that its four legs were fully extended. For half an orbit Spider stayed within 2 miles of Gumdrop, so that if need be Scott could use his greater manoeuvrability to swoop back to the rescue. A series of test firings of the descent engine, which would lower Spider to the

Apollo 9. Left, the Command Module (CM) photographed from the Lunar Module (LM) 'Spider' right, which was in turn photographed from the CM. This first separation and re-docking was done in Earth orbit. David Scott, aboard the CM, said: 'You're the biggest, friendliest, funniest-looking spider I've ever seen!' Jim McDivitt and Russell Schweickart were in the LM. After they had re-docked, NASA was ready for Apollo 10, the final 'dress rehearsal' for the moonlanding. (NASA)

lunar surface and enable it to hover until a safe touchdown point was found, ended with the two spacecraft more than 100 miles apart. Then McDivitt jettisoned the Spider's descent stage – two-thirds of its 16 tons' weight – and started test-firings of the 3500 lb-thrust ascent engine, which would give the lunar astronauts a once-only chance of returning from the Moon. These test-firings were planned as 'a football manoeuvre', so that in an emergency Spider would drift back within 2 miles of Gumdrop. But once again there was no emergency, and McDivitt was able to redock Spider's remaining 5 tons with Gumdrop (now Apollo again) – with the comfortable knowledge that if necessary Apollo could now take over and fetch Spider.

Because the manoeuvres went so well, media interest in the remaining five days of the Apollo 9 mission dropped right away. During them a rather exhausted crew did little except rest, make Earth observations, and prepare for splashdown. That went well too, except that McDivitt, who as commander was the last to be lifted by helicopter out of the Command Module as it wallowed in a heavy swell, was dunked in the sea during the process.

Apollo 10

Apollo 10 had been rolled out to the launchpad two days before Apollo 9 returned, and there seemed little reason now why its mission should be another rehearsal. Since it was due to swoop within 25 miles of the Moon's surface, why not go all the way, and get the landing done? The obstacle was that its Lunar Module was not designed for a landing, being loaded with extra propellant for possible emergency manoeuvres, and it would have taken six weeks to switch the Apollo 11 Lunar Lander into Apollo 10.

Now that the Moon was within man's grasp, I began speculating about the possible date of a manned landing on Mars. With America's Mariner 6 and 7 spacecraft on Martian 'flybys', Texas University claimed to have found 'conclusive proof' of the existence of water on Mars, while Soviet astronomers refuted the claim, saying that the polar caps were mainly carbon dioxide, and did not consist of ice and snow.

With little oxygen in the very thin atmosphere, it had to be admitted that the prospect of finding any life-forms there was very poor; nevertheless NASA was talking of sending a robot to recover soil samples in the mid-1970s, and it seemed reasonable then to forecast a manned landing in the early 1980s.

My TV News masters had been so pleased at having a captive space correspondent available to them in the studio during Apollo 9, that once again evasive action was needed to enable me to resume first-hand coverage during Apollo 10. Happily the Daily Mail's Transatlantic Air Race solved my problem. I departed for New York as a participant in that on 4 May, and four days later started my Apollo 10 coverage from Houston. For me it was the happiest of all the space missions. I was accompanied by one of the nicest of the BBC's camera crews, Peter Matthews and Bill Norman, who would shoot to my scripts with quiet efficiency; and Peter Woon, former Daily Express air correspondent, whom I had recruited for the BBC, was now one of my bosses. He had been made editor of the News Division's first full-length TV news programme, and was eager to use film packages from me.

On my 54th birthday and on the back of a NASA handout exuding confidence with detailed preparations for the flights of Apollos 10–13, I wrote a script describing how Apollo 10 would be followed by two landings that year, 1969. For the first, the easiest and smoothest site had been selected in the Sea of Tranquillity, on the right-centre of the Moon as seen from Earth. Apollo 12 would land much further west, near Crater Flamsteed. For some reason, which proved to be prophetic, I did not mention Apollo 13; describing how the landings would become more ambitious I passed on to Apollo 14, which would attempt the first highlands touchdown near the apparently fresh Crater Censorinus.

Apollo 16, then due early in 1971, would tackle Crater Tycho, in the south-central area of the Moon – 56 miles in diameter, and probably over 2 miles deep. Apollo 17 would take a look at the volcanic domes of the Marius Hills area, in the north-west quadrant; Apollo 18 would also go north-west to Schroter's Valley, where reddish discoloration suggested that gas was leaking from below the surface. Apollo 19 would look at a mysterious 'gash', called Hyginus Rill, hundreds of feet deep in the centre of the lunar face – and an astronaut, it was planned, would go to the bottom of it, either on a tether or on a one-man 'flying platform'. Finally Apollo 20 would end the series with 'a frightening landing' inside Crater Copernicus, 60 miles wide and several miles deep, in the Moon's north-western quadrant.

Apollo 10, it was clear, would lay the foundations for a serious and satisfying exploration of the Moon. The crew – Col. Tom Stafford, 38, with two Gemini missions behind him; Commander John Young, 38, who had also done two Gemini missions; and Commander Eugene Cernan, 35, with just one Gemini mission – had the task of putting together everything done on the three earlier Apollo missions into a final rehearsal omitting only the actual touchdown and lunar lift-off.

While the Apollo 10 crew spent hour after hour in their simulators – Stafford and Cernan in the Lunar Module and John Young in the Command Module – linked with one another and Mission Control, going through over and over again their planned and unplanned activities around the Moon, there was plenty for me to do. In addition to a stream of TV film packages which the crew airfreighted back to London, I enjoyed providing a steady flow of radio pieces down the telephone. At von Braun's headquarters at the Marshall Space Flight Center at Huntsville I discovered they had invented a new word to explain why it was that Apollo 8's orbits around the Moon continually wobbled. The wobbles had occurred over the so-called 'seas', or flattened areas, and astronomers had decided that these areas, at least 12 of them, had been caused by massive lumps of material thudding into the lunar surface from outer space. These massive concentrations – shortened to 'mascons' – of material have the same effect as the Earth's magnetic North Pole, exercising a powerful attractive force on spacecraft passing overheard, and noticeably changing their orbit. Much work had gone into plotting these effects, and building in adjustments to Apollo 10's trajectory and orbits in order to neutralise them.

These adjustments I found had also been built into velocity and distance tables provided for mission controllers on the ground, covering every hour throughout the actual mission, and copies were readily provided. They proved invaluable for reference during the 10-day flight – though they did pose hazards for the unwary broadcaster. The first column showed GET, or Ground Elapsed Time, as shown on the countdown clock, which became a countup clock from the second of lift-off. That time was given against CDT, or Central Daylight Time, which at Houston was six hours behind London, or Greenwich Mean Time – unless of course it was winter, when it was CST or Central Standard Time. (In spring and autumn too there were periods when the clocks had been put forward or back in Europe or the US, but not in both so that the time difference was seven hours!)

When working at the Cape, it was necessary when using this table to remember that there was a one-hour difference between Texas and Florida, where we worked in EDT or EST (Eastern Daylight or Standard Time), with only a 5-hour time difference from London. The table gave the spacecraft's speed in FPS, or feet per second, which I usually tried to convert for my listeners into miles per hour. Two more columns gave the craft's distance at that time from Earth or Moon in nautical miles – and as most people thought in statute miles, another conversion was required. When the spacecraft went into lunar orbit, the tables dutifully provided its calculated velocity both in relation to the Earth and the Moon, again in FPS, and its distances from both, again in nautical miles. Flight Directors and astronauts occasionally added another complication by quoting distances or speeds in kilometres; because they were so used to working in the metric as well as the imperial system it was all the same to them.

Although they looked dull at first glance, these tables provided a dramatic guide to gravity effects. The spacecraft, as mentioned earlier, behaved much like a ball thrown into the sky. Having been hurled towards the Moon at a speed of 24 290 statute mph, the backward drag from Earth gradually slowed the spacecraft down until, after 69 hours and 30 minutes of flight, the speed had dropped to 2032 mph. It was then 235 196 statute miles from Earth, with another 18 352 miles to reach the Moon. That was the point at which lunar gravity became greater than Earth's, so that Apollo's speed increased again in the following 6 hours 15 minutes as it fell towards its target to 3483 mph, with the Moon only 116 miles away. But it was aimed not to crash on the Moon, but to swing past and around it, so that the braking rockets could be fired to slow it down to 2178 mph relative to the Earth, equalling lunar gravity at that distance and so placing it in a 'safe' lunar orbit of 90 miles. Apollo's velocity in relation to the Moon, however, was then 4344 mph and dropping as the orbit was lowered, and without explanation we tended to switch to that figure in our reports.

My *Daily Mail* and *Daily Mirror* colleagues frequently helped me with these conversions when I was drafting a hurried script. I found it all too easy to make mistakes, and never ceased to wonder that NASA's mathematicians never did the same – for any mistakes they made would be fatal for the crew. My problem was that I was often required to 'ad lib' during live broadcasts with this table in my hand. My mistakes would not cause loss of life, but could be fatal for me! To avoid errors in converting to British time I always wore two watches, a subject of mirthful comment for years among my American friends, who invariably identified me as 'the man with two watches'. When, all too often, I suspected that I was making mistakes over velocities and distances I found that the important thing was to make the mistakes with great authority and hurry on. The generous forgave the errors, and most people never noticed them. The mean-minded who pointed them out were usually those who had never made a live broadcast!

For light relief the camera crew and I drove a hundred miles south from Cape Kennedy (which had not yet reverted to its pre-assassination name) to Palm Beach. There we went aboard Ben Franklin, a 130-ton steel cylinder built by Grumman, makers of the Lunar Module. In this six 'aquanauts' were about to drift submerged for a month in the Gulf Stream, as it flowed 1500 miles from Florida to Novia Scotia. Their speed would be only 2 mph, but enlivened with observations through 20 portholes of the more regular users of the Gulf Stream swimming and 'talking'. The crew was led by the famous Dr Jacques Picard of

Switzerland, with three Americans and an Englishman from Portland's Underwater Weapons Establishment. Like astronauts they had to live in 'a closed ecology'. It made a colourful piece for TV News – the obverse of the moonflights, as it were; usually I tried to follow up with the results of such expeditions, but never heard any more about that one.

Back at the Cape, with four days still to go, the Russians, as usual, got into the story. Their Venera 6 and 7 spacecraft, which had been launched 127 and 120 days earlier, reached Venus and descended on parachutes into its thick atmosphere. Venera 7 reached the surface and transmitted for 23 minutes before succumbing to an atmospheric pressure 90 times that of Earth's. Surface temperatures of 475 °C were recorded. This success led me to speculate that Soviet scientists might be planning to bypass the Moon and send cosmonauts to the planets.

Boeing scientists told me that a feasibility study done by them showed that six Saturn 5 launches could assemble a 1000-ton spacecraft in Earth orbit; it would carry six astronauts to Mars, and by using a nuclear-powered upper stage called Nerva, development of which was then well-advanced, could substantially reduce the 2–3 years' round trip time generally thought to be necessary. But NASA as usual took little interest in Soviet activities; all their energies were concentrated on Apollo 10, with the crew taking and passing their final medicals.

While this was going on I did an apologetic piece reporting that when they entered and separated in lunar orbit we would have to call the Command Module 'Charlie Brown' and the Lunar Module 'Snoopy'. Once again the astronauts had fallen back on Schulz's Peanuts cartoons for their call-signs. Most people seemed to approve, but Julian Scheer, NASA's public affairs chief, shared my own views and told headquarters that 'something a little more dignified' should be picked for the Apollo 11 landing. However, I was able to point out that 'Snoopy' was quite appropriate when you considered that the final flight plan for Stafford and Cernan was to swoop down to within 9 miles of the lunar surface and spy out the suitability of the surface of the 'Sea' of Tranquillity for their successors' touchdown:

> And if you think Tom Stafford and Gene Cernan may be tempted to make an unofficial landing, you'll be wrong. If they do touch down, either accidentally or on purpose, they won't come back. The ascent half of Snoopy contains only half the amount of fuel needed to get back into lunar orbit from the Moon's surface – a decision taken to make it more manoeuvrable for this mission, when it'll be wandering about, at times 350 miles from Charlie Brown, its parent spacecraft, orbiting the Moon much higher up.

The much-maligned Chuck Berry and his medical team came up at this time with a very effective, and for once popular, way of minimising space sickness just after launch, which has been used by astronauts ever since. Instead of insisting on isolation, the doctors sent the crew off in their T38 jet fighters to spend a hour doing aerobatics, climbing vertically and then going into parabolic dives to give brief periods of weightlessness. This shock treatment, it was theorised, would prepare the inner ear and act as a sort of inoculation when the real flight began. For the astronauts, it was a period of joyful relaxation after long hours in the simulators. A less popular, additional remedy was to provide them each with a one-pint plastic jug. Stomach troubles and flatulence had been caused on previous flights, it was decided, by hydrogen gas in the drinking water produced as a bonus by-product of the British-designed fuel cells. Before drinking the water the astronauts were instructed to fill the jugs and set them spinning in their weightless cabin, to separate the suspended gas from the water. This, however, proved to be a theory that did not work in practice. In orbit the gas did indeed settle at the 'bottom' of the spinning container; but as the astronauts drank, the gas remixed with the water, giving them stomach pains and cramps, so that they dreaded using it to reconstitute their dehydrated foods, and consequently ate very little.

With two days to go, Stafford, Young and Cernan spent much time studying earlier pictures of the Moon and its topography, and playing handball in the astronauts' gymnasium. Denied any views of them, we were briefed on the new 13 lb colour TV camera they would be taking with them. It had been developed at a cost of millions of dollars for classified military and underwater work. We tried, and in my case failed, to understand how the TV transmissions, if received first in Spain or Australia, would be in monochrome, and would have to be bounced by satellite to Mission Control to have the colour reconstituted. It was, and sometimes still is, most frustrating for TV people having to report live on rather dull monochrome pictures as they come in, knowing that when they are reissued in glorious technicolour half an hour later the newsdesks will no longer be interested in letting the public see them.

Our advance stories had covered the mission plan in such detail that when it was actually in progress there was every excuse for dwelling on the things that did *not* go according to plan. It was a lunchtime launch at the Cape – a luxury for the newsmen, and ideal for hitting peak evening times on British radio and TV, but hard work for the newspapermen, who had to move fast to meet Fleet Street deadlines. Stafford, Young and Cernan had an eve-of-launch dinner with Vice President Spiro Agnew, a impressive-looking man, but soon to be publicly disgraced. They had a big meal – shrimp cocktail, roast beef, strawberries and cream.

We were also informed that, as another step towards making space travel 'normal', they were taking 'real' bread to eat during the flight. After ten hours in bed Gene Cernan had half an hour with his Catholic priest before joining the other two for a steak and egg breakfast, with ten more guests, including astronauts and the head of NASA. The Apollo 11 team – Armstrong, Aldrin and Collins – was among the VIPs, well away from the media, watching the launch. With them too was His Excellency John Freeman, Britain's Ambassador to Washington, for the fashion had begun of not giving that appointment to career civil servants. Freeman's wife, Catherine Dove, who had transferred her affections from my BBC colleague Charles Wheeler, was not with him.

This time I was allowed to describe the launch into Radios 1 and

2, while in London Larry Hodgson's Apollo Studio did it for Radio 4, and Dick Francis's TV Apollo team took over BBC 1 from the far end of the Press Stand. But at least I had something to do as I sat in Seat No. A13 between Ronnie Bedford and Angus Macpherson, and I had enjoyed my share of the glory in all those 'run-up stories'. The final stages of the countdown were marred by nothing more serious than screams from female space groupies as a 3-ft snake invaded the grass area in front of the Press Stand, and I wish we had had some idea of the drama inside Apollo 10 when both the first and second stages of Saturn 5 shook the crew backwards and forwards so violently as a result of fuel sloshing that Cernan admitted later that he wondered about abort procedures. But they said nothing over the radio – mostly because speech was so difficult anyway – and the Flight Directors pressed on with the mission and kept quiet about it in their briefings.

Just after TLI – translunar injection – we had marvellous colour TV pictures, the first from space (before that we had had to make do with monochrome until colour film and photographs were brought back to Houston) sent from the Command Module as it separated from the Saturn 4B, turned around and then withdrew Snoopy, the Lunar Module, from the 4B's far end. After that we of the media chased off to Melbourne airport to catch the chartered Press plane to Houston, while Stafford, Young and Cernan – 'The talkiest crew anyone can remember' according to Mission Control – relaxed. Gene Cernan had told the media three weeks earlier: 'The translunar coast is sort of like two and a half days of sheer ecstasy waiting for that 24 hours of stark terror that we're going to be confronted with.'

Cernan certainly enjoyed himself *en route* to the Moon, sending back unscheduled TV transmissions, including what I described as 'the first fabulous colour pictures of what our own planet looks like to Martians and Moonmen'. It was the first astronaut crew, I pointed out, that really wanted to share its glamorous journey with ordinary people around the world; but the TV networks were very irritated by their unscheduled transmissions when they had told NASA they wanted them to be carefully scheduled at prime viewing times:

> Each network wonders and worries whether its competitors will break into advertising and ancient Westerns, or just stockpile the pictures for the future. Certainly the last views that Cernan sent of the Earth, showing it with a nasty pimple on the right-hand side – 'Could that have been Everest?' I wondered – have not been seen in America. And in England, you would have been in bed anyway!

Very inconveniently Apollo 10 was 11 minutes late arriving at the Moon because its initial trajectory had been so accurate that only one instead of the expected four mid-course corrections had been necessary. This was very satisfactory from a mission point of view, because hoarding propellant was a major concern; but for newsmen it meant that all our flight plans needed correcting if we were not to get in a muddle!

In due course the 42-ton Apollo, with Service Module and Lunar Module attached, disappeared around the 'leading edge' of the Moon – in other words, passed in front of it, in the direction it was travelling. The crew embarked upon the period of 'stark terror' light-heartedly referred to by Cernan. While we waited for them to reappear they turned the spacecraft so that the main propulsion engine faced forward, and lightened it by 13 tons – the amount of propellant burned off – by firing the engine for 6 minutes. That put them in an initial egg-shaped 185 × 68 miles lunar orbit, close enough to the planned 195 × 69 miles. They emerged joyfully saying they could see the lunar surface at last, with Stafford promising to send us pictures later. This was received all the more eagerly by astronomers, for observers in California, New Mexico and Spain had all reported what they thought was some volcanic activity in Crater Aristarchus, 400 miles to the right of Apollo's orbital path.

In England my reports led the BBC news bulletins despite the fact that the contest between Prime Minister Wilson and the dying Smith regime in Rhodesia was reaching its climax, and despite Technology Minister Wedgwood Benn reporting that Concorde costs had soared to £730 million, with accompanying hints of possible cancellation. On the early morning *Today* programme, however, which still took a light-hearted look at life for breakfast-time listeners under its first presenter Jack de Manio, I was not called in until after a piece from Plymouth about a musical watering can.

While the crew were opening up the docking tunnel into Snoopy, NASA gave us a list of 50 unofficial names which the astronauts had drawn up in advance for identifying landmarks around the Tranquillity Base landing site: there was Fay Ridge for Stafford's wife, and Weatherford for his home town; Barbara Mesa for the wives of both Young and Cernan; St Teresa for Cernan's 6-year-old daughter, and Mt Marilyn, reserved for the now famous twin lunar peaks. We pointed out to NASA officials, who did not want to hear, that no lunar landmark had yet been named after the late President Kennedy, who had inaugurated the whole expedition eight years earlier. I suggested in *Today* that the most fitting memorial to him would be to name the landing site after him – but neither then nor later did anyone take notice! Even more after his death than in his lifetime, John Kennedy's name evoked very mixed reactions; he was much less of a hero in the US than, for instance, in the UK.

With Apollo's orbit circularised at 69 miles the crew gave us the promised colour TV tour of the surface. With the adrenalin flowing, all three were free of space sickness, and only Gene Cernan had taken pills – two aspirins because his 'athletes' feet' were bothering him.

The astronomers were reassured by the crew's conviction that they could see evidence of volcanic, as well as impact activity. The crater of Langrenus, 90 miles in diameter, with a 7-mile-high central peak, was breathtaking. The Sea of Tranquillity could be seen to have plenty of clear landing areas, although it was dotted 'with basins that looked like the sinking sand in an egg-timer'. Despite the unmanned Surveyor landings, there were still some fears that the surface dust might be so soft in some areas that the astronauts, when they touched down, would be swallowed up in it.

An unexpected result of the main engine firings to circularise the orbit was that when Cernan checked out Snoopy he found himself in what he called 'a snowstorm'. It consisted of plastic crumbs shaken from the insulation system which they were in danger of breathing into their lungs. On advice from Mission Control most of the crumbs were trapped by placing wet tissues over the air filters. The crew cut short a sleep period so that they would be ready in good time for their fifth day in space and the all-important eight hours of separation.

As I sat at my microphone, watching Houston's TV screens filled with a slow-motion background of the rotating Moon, listening intently with one ear to the three-way exchanges between the astronauts and Mission Control, and only half-listening with the other ear to the BBC with its stream of excited programme requests, this was one of the most memorable events of the whole Apollo programme. Stafford and Cernan became concerned to discover, when they were ready for undocking, that the Lunar Module had slipped 3.5° out of line with the docking module at the latching point. They feared that by separating in this position some of the latching pins might be sheared off, making it impossible for them to redock when they returned. After consultations at Mission Control, they were assured that there was no danger so long as the misalignment was less than 6°.

In the unusually emotional exchanges before the separation there was recognition of the risks involved: 'We'll see you back in about six hours,' Tom Stafford told John Young, now alone in Charlie Brown. Gene Cernan urged Young to watch over them, and come to their rescue if the need arose.

> . . . We've been watching the most extraordinary colour pictures of Snoopy, the lunar module, gradually drifting away from the parent craft Charlie Brown. John Young in Charlie Brown started this most dramatic eight hours with a gentle push from his small thruster engines. The colour camera is mounted on the command craft in a fixed position while the astronauts get on with their work, and we've been watching them drift slowly apart: only 50 feet at first, and we could see every detail of Snoopy – its landing platform, the ladder on the leg which the astronauts will descend to the Moon

next July, and a flashing red beacon which Houston has just asked them to turn off because it's dazzling the camera.

Now we can see Snoopy disappearing in the centre of our screen – perhaps half, three-quarters of a mile away. And very shortly Stafford and Cernan will press the button in Snoopy that gives the computer the go-ahead to descend. Two bursts at different throttle settings will lower Snoopy's speed, which was 3600 mph, just sufficiently to start an hour-long drift down to within nine miles of the lunar surface, causing it to sweep just above the Sea of Tranquillity. And at that point they'll test out the radar landing system, the only part of the whole moonlanding project not yet tested; and within the hour, America will know for certain whether this project should be successful in achieving a moonlanding in mid-July . . . Stafford and Cernan must be longing to land now. But it would be fatal if they did.

Few of us moved from our seats even to visit the lavatory. Five hours later I was reporting that Snoopy had successfully completed its second pass, and had started the series of manoeuvres intended to culminate in redocking with Charlie Brown after three more hours.

Snoopy was at one time so far away from Charlie Brown – about 350 miles – that the two, though still able to talk to Earth, lost radio contact with one another. But Mission Control was able to pass messages between them until the two orbits drew closer together again. It was John Young, in Charlie Brown above, who was first to announce the success of the descent manoeuvre. 'They're down there among the rocks, rambling about among the boulders right now!' he said.

In fact Snoopy's two passes over the Moon were not quite as low as planned: ten and a half miles the first time, thirteen and a half the second. Approaching the second low pass over the Sea of Tranquillity Tom Stafford could be heard complaining bitterly that he couldn't get his film camera to work. Cernan too has had camera problems. During the two 20-minutes passes they were hoping to get close-up pictures which would be vitally important in pinpointing an actual landing spot . . . A problem with the rendezvous radar was resolved only minutes before Snoopy went behind the Moon to make the critical descent engine burn. When it re-emerged the communications system was swamped with jubilant cries from both Stafford and Cernan. 'We just saw Earth rise, and it had to be magnificient', they said.

Half an hour after midnight, London time, came another dramatic moment in these eight tense hours. Snoopy was split into two halves. The descent stage was jettisoned, for that's the part that contains not only the descent engine but acts as a launch pad for the return flight from the Moon. This time it was jettisoned during flight and abandoned in lunar orbit. Stafford and Cernan, firing their ascent stage engine for the first time – a tiny rocket with a mere 3500 lb thrust – had to make an evasive maneuver to avoid collision, and then climb away to rejoin Charlie Brown at the 69-miles level. The manoeuvres seemed both endless and complicated but at Mission Control tensions gradually relaxed as one followed another with complete precision.

Pressing our headphones to our ears, this is how we heard it all, starting with Snoopy's ascent burn:

SNOOPY: 2 minutes to the burn.
CHARLIE BROWN: Roger.
SNOOPY: And Charlie Brown, our charts agree very closely so we're go.
CHARLIE BROWN: Roger, my numbers agree with your numbers. Then I guess we're all in agreement, then let's go. 119 [seconds] to the burn.
SNOOPY: Roger.
CHARLIE BROWN: Mark, one minute to the burn.
SNOOPY: Roger.
CHARLIE BROWN: 35 seconds, DSKY blank. ['Disky', as it was called,

was the display and keyboard for the Apollo guidance computer, with a 19-button keyboard to enable the crew to instruct or interrogate it. The figures continually read up by Mission Control were computer instructions to be fed in.]

SNOOPY: How many seconds, John?
CHARLIE BROWN: 5, 4, 3, 2, 1, burning.
SNOOPY: We're burning.
CHARLIE BROWN: Go to it.
MISSION CONTROL: We copy.
SNOOPY: We got 15 to go.
CHARLIE BROWN: Right.
SNOOPY: Burns complete.
CHARLIE BROWN: Roger, good show.
MISSION CONTROL: Snoop, Houston. We see you trimming. Good show.
SNOOPY: Okay, zero 1 and 1 tenth.
MISSION CONTROL: We copy, Snoop.
SNOOPY: And Snoopy's pitching back up to acquire. Houston, this is Snoop. You can't believe how noisy those thrusters are.
MISSION CONTROL: Roger, 10. We can't even imagine.
SNOOPY: It sounds like being inside a big rain tub with about two inch hail beating all over you.
CHARLIE BROWN: Hey, babe. Here's where . . .
SNOOPY: Okay, I'm pitching up to give you radar target here [Garbled]
SNOOPY: Okay Houston this is Snoopy. We have solid [radar] lock and first update appear real good.
MISSION CONTROL: Roger, Snoop, we copy. We got 4 minutes 50 mark to LOS [loss of signal] for you.

Both spacecraft, still about 27 miles apart, disappeared behind the Moon for the 16th time. There was time for us to snatch a cup of tea knowing that when they reappeared 47 minutes later they should be station-keeping within a few feet of one another:

SNOOPY: Okay, you ready to dock?
MISSION CONTROL: Snoopy, this is Houston. How do you read me?
SNOOPY: Looks like it's good.
SNOOPY: Hey Joe [Mission Control] we're about ready to dock. Stand by.
MISSION CONTROL: Very good.
SNOOPY: Don't call us. We'll call you.
MISSION CONTROL: Roger, out.
SNOOPY: Okay John, you're in to about 5 feet. Looking beautiful.
CHARLIE BROWN: How far?
CHARLIE BROWN: Twenty.
SNOOPY: I captured? [Young had slid the Command Module's probe into the centre of the Lunar Module's drogue – first stage of the docking procedure.]
CHARLIE BROWN: Yes, you are on.
SNOOPY: I got capture, John. Fire when you're ready. [Stafford had to fire his thrusters to ram the Lunar Module forward to cause the capture latches to close.]
CHARLIE BROWN: [Cernan talking to Stafford] Everything looks good here, Tom.
SNOOPY: Okay here. Oh, we look good.
CHARLIE BROWN: Yell when there's a rock in the cabin, babe.
SNOOPY: Oh, we got that one, all.
SNOOPY: We got them [the latches] John. We heard them in there.
SNOOPY: Yep. Hello, Houston, Snoopy and Charlie Brown are hugging each other.
MISSION CONTROL: Roger. We heard them down here.
SNOOPY: Okay, let's stay – let's stay in our helmets, babe, until we get this thing squared away.
SNOOPY: Okay John. That was beautiful. Just beautiful, babe.

When the moment of docking came, the only sign of emotion in Mission Control was that one of their huge TV screens was suddenly filled with a colour cartoon of the original Snoopy saying to his master: 'Smack on target, Charlie Brown!' For several hours after that the exchanges between the astronauts and Mission Control continued unbroken. Because there had been problems with the cameras and with oxygen in Snoopy, Stafford and Cernan were asked to bring the cameras through the docking tunnel for return to Earth, as well as dismantling and bringing back the lithium hydroxide canister for inspection. That caused some dismay; where was this large object to be stowed in the already crowded Apollo Command Module? Mission Control asked for time to consider that, and finally said it would have to go under the left-hand couch in the sleeping bag. The crew, exhausted but happy, settled down to nine hours' sleep, but in Britain my programmes, as usually happened, were just going on the air. I prepared the ground for a bit of rest myself in a piece that led the *Today* programme:

> All that Stafford, Cernan and Young have to do now is to get back from the Moon. It's so simple that most of the correspondents here at Houston feel the story's really over. That fact – that a quarter-million miles' journey from the Moon now has no terrors, is the best proof that experimental spaceflight has now settled down into space exploration.
>
> It's a funny thing about these Apollo crews. Every one seems to be better than the last. Certainly this team has been no exception. Only over the cameras have there been a few tantrums. Stafford and Cernan both got angry because of the trouble they had with them during their low sweeps over the Moon. 'You have to kick it and stamp on it to make it work,' Cernan said of his. Stafford's comments about his jammed magazine filled practically the whole of page 365 of Mission Commentary, which is transcribed for us, just like Hansard in the House of Commons, as it goes along.
>
> I was a bit relieved to discover that these men were human when John Young in Charlie Brown radio-ed a warning to Snoopy that they were transmitting. The transcript alleges that the tape was 'garbled' after someone in Snoopy began speaking freely about some

> valves. That was when Snoopy was being asked to give Houston a POO – pee double oh, which stands for Program Zero Zero. It means cleaning out the computer, so that fresh data and instructions can be fed into it.
>
> And that's really what all this is about. But it's certainly a consolation to me to know that they'd never get back from the Moon, let alone around it, without the computers. If they had to do the sums themselves it would take them as long as it does me to sort out the difference between Ground Elapsed Time, Central Daylight Time, Eastern Daylight Time and Greenwich Mean Time. Oh, and British Summer Time, of course.

Next day, with Apollo 10 doing the last eight of its 30 orbits around the Moon, John Young was busy with his radar measuring 'targets of opportunity' selected by astronomers from the Apollo 8 pictures, to obtain the exact heights of mountains and crater rims, so that later crews would not crash into high ground. Our peace was shattered by indignant calls from BBC and London paper newsdesks saying that the agencies were carrying stories that Snoopy had gone out of control and nearly crashed while near the lunar surface. We had heard Gene Cernan exclaim 'Son of a bitch' at about the time Snoopy was due to jettison the descent stage, but thought it was a continuation of the complaints about the cameras. But when we summoned Glynn Lunney, our favourite Flight Director, he admitted that a switch position had been omitted from the checklist, with the result that just after the separation the ascent stage, then 35 miles above the Moon, had started gyrating wildly because its radar was 'looking' for Charlie Brown. Tom Stafford had quickly taken over manual control from the computer and it was three minutes before Snoopy was stabilised – a long time when you think you might be crashing into the Moon. But having missed the story myself, I followed journalistic practice by knocking it down; it had NOT been a near disaster, and Snoopy was never in real danger of crashing, I said quite truthfully! This, however, was not good enough for TV News who demanded a dramatic voicepiece.

As soon as John Young had fired the main engine to increase speed and break out of lunar orbit for the 54.5 hour journey back to Earth, Mission Control issued the preliminary flight plan for Apollo 11's moonlanding. Lift-off would be on 16 July, with Neil Armstrong and Buzz Aldrin due to touch down on the lunar surface on Monday 21 July, they said. At 5.12 am British Summer Time Armstrong would clamber out on to the 'porch', and the first thing he would do would be to pull a cord which would deploy a TV camera so that the whole world could watch as he clambered down the ladder to become the first man to step on to the Moon.

14 The Eagle Soars

Preparations for Apollo 11

The Apollo 10 crew were still only half-way back from the Moon – but so accurately on course that they finally splashed down within 3 miles of the recovery ship standing by off the Samoan Islands – when they were told that Armstrong, Aldrin and Collins had already done a rehearsal of the Apollo 11 splashdown. Stafford's crew must have been thankful that they were to be spared the ordeal facing their successors.

Armstrong's crew had to practise spraying one another with mustard-coloured disinfectant to kill off any optimistic moonbugs hitching a ride with them, and then don their 'BIGs' – biological isolation garments – to ensure that they had no contact with the Earth and its atmosphere before they entered the quarantine facility. And the space doctors, still dreaming up nightmare scenarios, had decided that the 18 volunteer doctors and debriefers who would share the three weeks' quarantine of the returning moonlanders would themselves have to enter the quarantine facility three weeks in advance. That was to ensure that any infections they might take into the quarantine facility would show up before the astronauts arrived and eliminate any chance that earthbugs could cross-infect the astronauts and link-up with their hypothetical moonbugs.

The astronauts appeared to treat all this with amused acceptance, and as part of the price they had to pay for taking part in the great moonrace. And just as professional soldiers traditionally respected their soldier enemies, so the astronauts and cosmonauts began to express public recognition of each other's achievements. During their return flight, Apollo 10's commander, Tom Stafford, having listened respectfully to President Nixon saying that he wished he was young enough to make a spaceflight, asked that the crew's congratulations should be sent to the Soviet Union upon the successful landings of their unmanned craft on Venus. A few hours later Moscow Radio reciprocated by reporting that Apollo 10 had safely splashed down, describing the flight as 'a great space success'.

Dr Thomas Paine, the only NASA Administrator who enjoyed talking to the space correspondents until Admiral Truly's brief tenure 20 years later, was telling us within an hour of Apollo 10's return that Apollo 11 would lift off on 16 July. Paine, a materials engineer from General Electric, had been promoted by President Johnson to Administrator, the top NASA job, while Apollo 9 was carrying out the final moonlanding rehearsal. Changes in NASA's leadership, only nine months before Apollo 11, went unnoticed in Britain, and aroused little interest even in the US. It was 1997 before I learned the full story when reviewing *Aiming at Targets* by Robert Seamans, who for three years from 1965 to 1968 was No.2 in the decision-making triumvirate which made the moonlanding possible. Seamans resigned from NASA in 1968 when he got tired of trying to keep the Apollo programme on track in the face of President Johnson's 'barnyard vernacular' and disputes with his boss, Jim Webb.

Soon after Seamans's departure, on 16 September 1968, Webb himself told Johnson he thought of resigning in the near future – and found himself hustled out of office the same day! 'I've been thinking along the same lines,' replied the President. 'Let's step outside and tell the Press that you are leaving, effective immediately!'

Paine brought fresh air and enthusiasm to the top job. He told us that the Apollo 11 lift-off date depended only upon supporting evidence from the photographs brought back by the Stafford crew. And despite complaining about their cameras, Stafford's team had brought back 2000 still photographs and 1800 feet of film, described by NASA technicians as of excellent quality. Paine added that America was not just aiming at the Moon:

> While the Moon has been the focus of our efforts, the true goal is far more than being first to land men on the Moon, as though it were a

> celestial Mount Everest to be climbed. The real goal is to develop and demonstrate the capability for interplanetary travel. With some awe, we contemplate the fact that men can now walk upon extra-terrestrial shores. We are providing the most exciting possible answer to the age old question of whether life as we know it on Earth can exist on the Moon and the planets. The answer is Yes!

Within a few days of my return to England after a month away Cosmonaut Aleksey Leonov revived speculation that the moonrace might still be on. With hindsight, he ought to have known better, and probably did. But he possessed a mischievous sense of humour, and was probably exercising it when he told some Japanese journalists visiting Moscow that Russia planned to land a man on the Moon either that year, 1969, or the year after. And he assured his visitors that he was certain that pieces of moonrock recovered by Soviet cosmonauts would be on show in the Soviet Pavilion during the 1970 Japanese World Fair in Osaka.

Tom Paine, passing through London on his way to the Paris Air Show the day after this report, told me he found it hard enough to keep up with America's space programme, and had no idea what the Russians were up to. But it was in that interview that Paine first disclosed that American scientists thought that lunar rocks, when brought back to Earth and exposed to our oxygen atmosphere, might burst into flame. That was why they would be kept in vacuum boxes and opened in a vacuum chamber, 'to make sure we don't get into any situations we don't understand'.

Always full of ideas and enthusiasm, Paine talked of assembling over a 10-year period a space station capable of containing 80 astronauts by the mid-1980s. He also wanted to send an Apollo spacecraft to spend 28 days in polar orbit around the Moon so that the lunar globe could be mapped as thoroughly as Earth. Rather wistfully he added that he did not think that America would be ready to send men to Mars before the 1980s – an unfulfilled target long passed when he died in the early 1990s.

Four weeks after leaving America – half of which was spent in Paris – I was back at Houston with a camera crew covering the final preparations for the moonlanding launch. There was no shortage of stories for either radio or TV, and the first came with a NASA announcement issued within minutes of my arrival. As mentioned earlier, because neither the US nor the Soviet Union had felt confident of reaching the Moon first, they had long since agreed that whichever it was would not attempt to claim the whole Moon as national territory. Despite its name, the National Aeronautics and Space Administration was not nationalistic in its outlook – for the very good reason that many of its most brilliant people had been recruited from other countries – notably Germany and Britain.

But NASA's proposal that the US should therefore adopt a magnanimous international approach by having Armstrong and Aldrin erect the blue United Nations flag on the Moon as soon as they stepped upon it was firmly over-ruled. It must be the Stars and Stripes, insisted Congress, since the whole idea of spending all that money to get to the Moon first was to demonstrate to the world that the United States was superior to every other nation. There was a last minute hustle, I reported, to prepare such a flag on a telescopic flagpole. (Years later I learned that a NASA employee had been despatched to the local store, where he paid $5.50 for the flag which became the first to be planted on the lunar surface.)

It was hoped, but with no certainty, that Armstrong would be able to thrust the aluminium pole into the lunar surface; and, as there was no wind on the Moon to stop it drooping in the most dismal way, the flag would have to be stretched out on a telescopic wire frame. A committee set up by Paine instructed that no other flag should be unfurled or left on the Moon; but miniature flags of all the United Nations and of the 50 states of the US, its territories and the District of Columbia, would be stowed in the Lunar Module and brought back to Earth. In addition two full-sized US flags which had flown over the two houses of Congress would be carried in the Command Module.

We were also shown a copy of a cylindrical steel plaque which had been affixed to one of the Lunar Module's spindly landing legs. 'President Nixon's plaque', we were told, would be 'a permanent artefact on the lunar surface'. Surprisingly, it was much less controversial than the flag had been. Engraved at the top were the Earth's two hemispheres, showing continents and oceans, but no national boundaries. Then came the words:

> HERE MEN FROM THE PLANET EARTH FIRST
> SET FOOT UPON THE MOON JULY 1969 A.D.
> WE CAME IN PEACE FOR ALL MANKIND.

Beneath were facsimile signatures of the three Apollo 11 astronauts, with that of President Richard Nixon, 'President, United States of America' in modestly small letters at the bottom. Having done my voicepiece about this, I appended a note for the anchorman pointing out that 'an artefact' could be defined as 'a product of human workmanship, especially of simple primitive workmanship'. I never learned whether this was regarded by the presenter as helpful or offensive.

The space correspondents gathering at Houston were all intrigued as we followed a great power struggle in progress between Deke Slayton, the Chief Astronaut, trying to protect the hardworked Apollo 11 crew from being distracted from their rehearsals by media demands, and Julian Scheer, the dynamic and powerful Assistant Administrator for Public Affairs at Washington. Scheer, says the official NASA history, 'wanted the public to see the pilots as human beings, to foster a better understanding of their training and goals'. Deke was appalled at Scheer's demand that each crew member should spend at least a full day with each of the networks – and if the media wanted to see Armstrong-the-man in his home state of Ohio, rather than Armstrong the spaceflight technician, he should be willing to go to Ohio for them. Scheer also wanted the wives of the crew to attend 'a tea' given for the women of the Press corps. 'Homes and wives are personal', snapped Slayton, 'and landing on the Moon does not change that.' The mission, he pointed out, would not go anywhere on schedule if he could not keep the crew working instead of entertaining the press.

Scheer had also been asked by Nixon to add a mention of God in the wording of the landing plaque. Scheer disapproved of this, and the request was quietly forgotten.

Despite having a TV crew on my own back, the US TV networks were so crudely predatory that my sympathies were wholly with Slayton. Nevertheless, Deke had to make concessions, and I was among the beneficiaries. Media access to training sessions such as rehearsals of lunar activities was provided on a pool basis; but this meant that reporters, cameramen and photographers had to form a committee, and nominate say five of their members to cover these occasions, and then make their personal reports and pictures available to everyone else. Producing radio and filmed TV reports was already keeping me busy for 18 hours a day, so I stayed out of those activities; the bonus came when Deke was forced to make the Apollo 11 crew available for a news conference only 11 days before lift-off. The doctors, who had been trying to keep them isolated from possible coughs, colds and other germs, were horrified.

'Neil Armstrong, Edwin Aldrin and Michael Collins will be sitting in an improvised wind tunnel when they face us at 7 pm British time tonight to answer questions about their historic flight', I reported:

> For two days workmen have been busy erecting a plastic screen and special ventilators to ensure that air currents flow *from* the astronauts towards the newsmen. Smoke tests have been made to check that the system works well, and that while we will get the full benefit of any germs the astronauts may be breathing out, our germs can't possibly reach them. And when they do their TV interviews afterwards, there'll be a plate-glass screen betwen them and the interviewers . . . Having successfully completed the major countdown rehearsals at Cape Kennedy, they have flown here to Houston to spend their last week-end at home with their families for the next six weeks.

Despite the draught, we sat in comfort in the glossy new auditorium at Houston's Manned Spaceflight Center – a quarter of a century later it is still a theatre of which any medium-sized town could be proud – for the main news conference. But that was not good enough for the American TV networks, so a subsequent 30-minute interview was arranged in the quarantine laboratory to be conducted by four interviewers. They were to be separated by a glass screen from the interviewees, while the cameramen and technicians required to be on the astronauts' side of the screen were ordered to undergo physical examinations beforehand, wear face masks and undertake not to approach nearer than 20 ft from the astronauts! This caused so much ill-feeling that the interview was postponed until 36 hours before lift-off and then took place with interviewers and interviewees 15 miles apart over a closed-circuit TV link!

London took a 'live feed' of our evening news conference, so I was able to sit quietly through it, contributing my questions, and then describe it for the late night news. Julian Scheer, I had been relieved to discover, had shared my dislike of the cartoon-character call-signs selected by earlier crews, and behind the scenes had urged that more dignified names were chosen for the lunar landing 'to be witnessed by all mankind'. It was only afterwards, however, that I learned that Armstrong had delegated this task to Michael Collins, accepted as the intellectual member of the crew.

'Inevitably, America's moonlanding mission is taking on a more patriotic flavour every day,' I reported that night:

> Neil Armstrong, who'd clearly been waiting for a question on the subject, told us that the Snoopy and Charlie Brown call signs used on the Apollo 10 mission would be dropped this time. Instead, the lunar spacecraft will be 'Eagle', America's national symbol, while the command spacecraft waiting in orbit to bring Armstrong and Aldrin home again will be 'Columbia' – to Americans a symbol of liberty and justice. And Colombiad was also the name of Jules Verne's imaginary spacecraft which also took off for the Moon from Florida. Armstrong hinted, too, that they had an unofficial name that they'd apply to the lunar landing site, but didn't tell us what it was.

These three Apollo men are much quieter and less voluble in temperament than all of the earlier Apollo crews. Armstrong, fair-haired, looking much younger than the other two – although in fact all three were born in 1930 – was tense at first but soon thawed out. Asked whether he'd thought of something historic to say when he stepped on the Moon, he said he hadn't. Aldrin , who's always known as 'Buzz', broke the tension by breaking in: 'I hope he doesn't slip, that's all'. Aldrin, who has thin, gingery hair, declared that they had the utmost confidence in total success, and both he and Armstrong said they'd spent little time thinking of what they'd do if they failed to get off the Moon again. Michael Collins, who is dark and has the readiest smile of the three, was very serious when he explained he'd be able to wait for up to two extra days if they did have difficulty getting back from the lunar surface. But Collins got a laugh when, asked what precautions were being taken to prevent the astronauts picking up germs from their families during this last week-end at home, replied: 'My wife and children have signed a statement that they have no germs.'

We got deeper into their problems when Buzz Aldrin explained that the most difficult part of their whole mission was dealing with the unexpected. Their fate would depend on correctly interpreting the things that haven't been thought of in advance. The biggest moonrocks they expect to be able to bring back on this first flight will be no bigger than golf balls. And, said Armstrong, unfurling the Stars and Stripes won't be the first job he'll do on the Moon. That will have to wait nearly an hour until Aldrin has joined him, got used to working in one-sixth gravity and is able to help to push the aluminium flagpole into the Moon's surface.

Mike Collins had almost the last word: asked if it wasn't pretty frustrating for him to go so near the Moon but have to wait in orbit while the other two made the landing, he replied: 'I'm not a bit frustrated. I shall go 99.9% of the way and that suits me.' And Armstrong chipped in to point out that Collins had a giant job to do in looking after them.

These very visible and rather risible efforts to protect the astronauts from infection brought to a head a story which I had started a week before. Chatting with Chuck Berry about the unspoken criticism of the fact that Vice President Agnew had had a pre-flight dinner with the Apollo 10 crew, Chuck had let drop that President Nixon had been invited to have dinner with the Apollo 11 team the night before lift-off – and that in his view this would seriously prejudice his 'preventive medicine policy'. It was not just a question of ensuring they did not develop colds during the mission, but of being sure that any infections they brought back came from the Moon, and not from pre-flight infection.

When I raised this with NASA executives, the query was passed to and fro, and finally the answer came back that Dr Berry appreciated that his view was only a recommendation. 'Top NASA management' had decided that Messrs Armstrong, Aldrin and Collins would be sufficiently isolated if they had their dinner party in their crew quarters at the Cape in specially controlled airflow conditions. So the ambitious Frank Borman, who had resigned his astronaut appointment to become Director of Manned Space Stations as well as the President's 'adviser' on space, and was just off to Russia on a goodwill mission, had been told to go ahead and invite Nixon, who had promptly accepted. 'With the Apollo programme almost over, NASA management is hoping to persuade the President that a long-term follow-on space programme, costing perhaps £10 000 million pounds over ten years, should now be started . . . Not for the first time in the space business, medical expediency has taken second place to political expediency.'

My story of 'The Man Who Told the President he Couldn't Come to Dinner' hit the headlines in newspapers, on radio and TV, until finally the White House announced that Nixon had decided not to attend after all. And when Borman returned from Moscow the simmering resentment between the astronauts and doctors (which had its origins in the refusal to allow Deke Slayton to fly in 1962 because of a heart irregularity) blew up into a public row which was to sweep our friend Chuck aside. An angry Frank Borman told us at a news conference that medical advice should be kept private, and not made public, and recalled that two weeks before his own moonflight he went to a White House dinner with pride:

> The person who would isolate the astronauts in the last few days, particularly from the President of the United States, does not understand the entire psychology of the pilot. A visit from the President the night before launch does a lot more than just offer encouragement. It would be a very fine gesture.

Would he try to get the decision reversed? 'How can you get it reversed now? If one of them even sneezed on the lunar surface, it would be because President Nixon was there!' Borman added, for good measure, that one of the reasons that Russia had fallen behind in the space race was that the Soviet cosmonauts had suffered from their medical community even more than their American counterparts.

The astronauts never forgave Chuck Berry. Five years later Michael Collins wrote sarcastically in *Carrying the Fire* that though Berry 'billed himself as the astronauts' "personal physician"', they seldom saw him. Once we had broken the story, he pointed out, the President had no option but to cancel, for if an astronaut did fall ill, Nixon was bound to be blamed; even if nothing happened, 'he could be charged with callous disregard of the professional advice of our personal physician' and went on:

> Had we been operating behind a germ-free barrier, this [ban on the dinner] might have made a modicum of sense. But we were in daily contact with dozens of people, and one more – especially one whose physical condition is subject to constant checking – wouldn't have made a particle of difference.

As an admirer of both Collins and Berry, I was sorry to read that; for it was we of the media who had billed Berry as 'the astronauts' personal physician', and that is exactly what he had been right back from Mercury days. Our exchanges of views and jokes with him over the years, including his amusing medical analysis of the possibilities of love-making in weightless conditions, had helped to provide stories which kept up public interest in the space programme.

After this Berry soon became much less visible to the media, and although he was still around during the Apollo 13 drama, he left NASA for other appointments soon after – although his son, Dr Michael Berry, subsequently joined the NASA medical team. Collins himself provided justification for Berry's attempts to prevent the astronauts getting over-tired by admitting elsewhere in his book that by launchday he had 'ticks' in both eyes as a result of the stress of continuous training.

So, during those last few pre-flight days there was in fact more concern about protecting the health of the astronauts than about the five million parts in the launch vehicle – all made, as the astronauts were fond of saying, by the contractor who had put in the lowest bid.

In July 1999, taking part in a Channel 4 TV programme re-enacting the 30th anniversary of the moonlanding, I had a happy re-union with Chuck Berry, also taking part. I had with me a copy of Michael Collins's *Carrying the Fire,* and asked the other participants to add their comments to Collins's own good wishes, dated May 1976. Chuck wrote: 'It is a real pleasure to be with you again on such a memorable anniversary program – the 30th. You know my view about Mike's comments re the medical program – it's astronaut Bullshit and surprising he was not aware of efforts being made to protect him.'

Chuck also surprised me by saying that in his view no one had yet achieved sex in space – despite NASA having had one married couple, Mark Lee and Jan Davis on the eight-day STS-47 mission, and, on the Russian side, the delectable Svetlana Savitskaya having clocked up 19 days on two spaceflights!

Other participants in the TV programme included Gerry Griffin, Apollo 11 Flight Director, and Cosmonaut General Aleksey Leonov, first man to walk in space, who repeatedly called for re-supplies of whisky and vodka, and signed his name under that of Collins with a huge flourish. He confided that, when in space, he resented instructions by Moscow Mission Control, referring to the controllers contemptuously as 'Big Bananas'.

The superstitious – and there were many in the space business – thought that the fate of Bonney the adolescent monkey was a bad omen

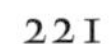

'Astromonk' Bonney, who died 12 hours after being brought back from space a few days before Apollo 11 was launched. His flight was cut short after 9, instead of the planned 30 days, in orbit. With 8 sensors in his head, including one in each eye – monitored via the box on his head – Bonney refused to co-operate. Project Biosatellite was unpopular with the astronauts as well as the public. (Associated Press)

for the moonlanding mission. Bonney had been sent into orbit at the end of June 1969 on what was intended to be a 30-days 'long-duration' flight. Amid protests in which the astronauts as well as the media had joined, the space doctors had planted 24 censors in Bonney's body – eight of them in his head, including one in each eye – in the expectation that they would get information about the effects of long-term weightlessness that they could never get from a man because, of course, they could not treat a human body in the same way. Bonney's flight was the third of four planned missions in Project Biosatellite, and the idea was that he would be sitting quietly in orbit, earning his food twice a day by punching buttons on a ' fruit machine' – which he had enjoyed doing on the ground – forgotten by the media while the Armstrong crew visited the Moon.

Now that NASA was making such progress with manned

spaceflight the continued used of animals in this way seemed pointless, and many of us had said so. Bonney apparently agreed with this view, for once in orbit he got lonely and depressed sitting there strapped into his 7 ft long spacecraft. Since he refused to co-operate, the spacecraft was brought back after nine days and all the newsmen at the Cape, looking for a story to keep them going, covered in detail his sad end. The aircraft sent to catch him as he parachuted down in a descent capsule missed him, but a helicopter fished him out of the Pacific 150 miles off Hawaii. He was placed in intensive care and fed intravenously because he had eaten little and drunk nothing at all for days. 'He did well for a while and then suddenly passed away,' a public affairs spokesmen announced. NASA's press office got busy finding other stories to divert attention from this little drama, and future Project Biosatellite missions were quietly cancelled as an 'economy cut'. For many years afterwards – in fact, until well into the Shuttle programme – NASA's medical people put their animal experiments, with which they still persist, on Soviet spacecraft, because that attracted much less attention.

There were in fact plenty of other stories. Every BBC radio and TV programme wanted something in every edition; and a small bonus came my way as a result of the conflict between Julian Scheer in Washington wanting to wring every ounce of political advantage out of the mission, and Houston wanting to shield the crew from the media. Scheer, as recounted earlier, had got rid of Colonel John 'Shorty' Powers, the first official countdown commentator, and replaced him with a journalist, Paul Haney, at Houston. But now Haney too had been diverted by the bright lights and become more interested in being the famous 'Voice of Apollo' than in the duller side of his job as head of Public Affairs. When Scheer tried to move him out, Haney suddenly resigned, and was promptly signed up by Britain's growing and increasingly competitive ITN. Haney was an unpopular man, especially at the Cape, and when he re-appeared there in his new role, letting it be known that he intended to 'give the BBC a bloody nose' all doors were suddenly opened to me – accepted for the past ten years as a 'Cape Vederan'.

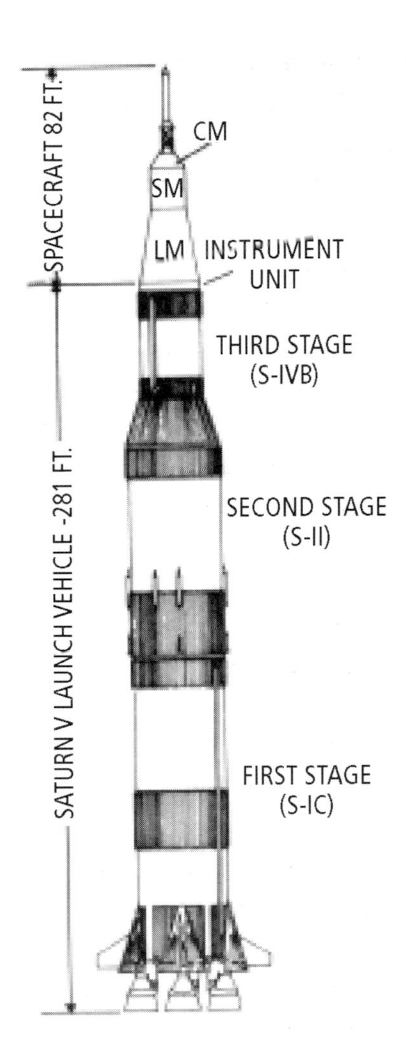

The Saturn 5 launch rocket plus spacecraft. (NASA)

What seems a small privilege now was a 'big deal' then: I was allowed to take our camera crew to the top of the 363 ft Saturn 5 rocket, filming a descriptive piece as we rode in the astronauts' elevator; then we tried out the access arm to the Apollo 11 hatch, shared with our viewers the spectacle of the Banana and Indian Rivers flowing around Merritt Island to the Atlantic, and followed the emergency route to the nine-man trolley providing escape down the frightening slidewire if the rocket caught fire. That I did not try.

> 20 seconds to decide: touch down, or get out by firing the ascent engine?
>
> At 5 ft the wire probes dangling from three of the landing pads touch the surface; a light flashes in front of Armstrong. He's still got time to abort – but not much. If he's satisfied, he tries to wait one second – and then cuts off the rocket engine.
>
> The 12-minute burn has halved Eagle's weight of 16 tons – so now there's only 8 tons to drop the last 5 ft on to the Moon – and of course that's only one and one third moon-tons. The four landing pads, each 3 ft in diameter, crunch down and should easily absorb the shock.
>
> If it doesn't seem safe, there's a three-minute window for immediate take-off. But having got there, Neil Armstrong and Buzz Aldrin will certainly want to stay.

The atmosphere in the astronauts' living quarters at the Cape when seven doctors, headed by Chuck Berry, gave the crew their final medical examinations to pass them fit for flight must have been extremely cool. With four days to go, Florida's police were planning to handle 350 000 cars, 2000 private aircraft and an armada of boats converging on the Cape to watch the lift-off. Dr Ralph Abernathy, who had taken over from the assassinated Martin Luther King, was also arriving with a mule and wagon train, to lead a silent protest demonstration against the cost of the mission on behalf of what were called the 'Moonport's Hungry People' – 7000, both black and white, allegedly underfed and living in the area. Tom Paine skilfully defused the protest by meeting Abernathy and inviting 10 of the protestors to join spectators on the VIP site for the launch.

The 93-hours' countdown, spread over five whole days because it had so many precautionary 'hold' periods built into it to provide opportunities to clear technical snags without delaying the launch at the end of it, had been under way for two days when the Russians once again achieved a dramatic intervention. They launched Luna 15 towards the Moon, starting us on instant speculations that, having been defeated in the manned contest, Soviet scientists were making a final bid to land a robot on the Moon to gather a few ounces of lunar soil and return them to Earth just before the Apollo astronauts got back with their much more significant haul. But even a few grains would snatch for the Soviets the prestige value of being first. Luna 15, we learned, was huge – 5700 kg, four times the size of earlier Lunas – confirming that it was carrying a lander with automatic scoops, and an ascent stage mounted on top containing enough propellant for the return journey.

I sought out Wernher von Braun with my BBC microphone, and asked him how he would feel if at this late stage the Russians succeeded in snatching lunar samples ahead of Apollo. 'Frankly', he said, 'I had always expected they would do something to beat us to the Moon. I am personally convinced that they tried to land a man there ahead of us, and this of course is the second-best thing they can do . . . This is undoubtedly quite a challenging mission, to soft-land a spacecraft on the Moon and scoop up a sample of lunar soil and fly it back to Earth. It would show again that we have very competent competitors in this race . . . I am even quite glad that we are not alone in doing this thing. Every now and then people ask us: Is this thing really worth doing? Now they have a lot of smart people in the Soviet Union also, and if this thing was merely Army-Navy football they wouldn't do it, you know. They're committing a great deal of their national resources to it, so it confirms to us that we seem to be doing what is really a sound, sound thing.'

In a way, added von Braun, he wished them 'full success', even if it would 'take a little bit off our programme – but not too much really'. When I pointed out to him that success for the Russians would be claimed as justifying their position by the many vocal American scientists who believed that robots could explore space more cost-effectively than men, von Braun brought a gale of laughter from those listening to my interview when he replied: 'What would have happened to America had the Spaniards sent an unmanned sailing ship across the ocean?' The Russian mission, he added, was not an impulsive one; to have the hardware ready it would have had to be planned eight years earlier.

The docking system. Once the probe is inserted, 12 latches lock the probe and drogue together. The astronauts must then dismantle the system to pass through the tunnel and became adept at doing so despite the confined working space. (NASA)

My NASA friends observed my battles with my own bureaucracies as amusedly as I followed – and reported – theirs; so there was a good deal of comment when, with two days to go, a team of engineers arrived from BBC New York to install all the equipment, both at the Cape and Houston, deemed essential for the high-powered TV producers, assistants, secretaries and commentators. I was not invited, nor did I volunteer, to join their new set-up and benefit from their facilities; they set up camp among the other TV groups at the far end of the Cape's Press Stand. As before I was left undisturbed among the general mix of radio and newspaper people in my central, front row seat, A13, between Angus Macpherson and Ronald Bedford. It had acquired a comfortable familiarity.

By this time I had sent off the last of my film packages for TV News, including 'astronaut profiles' for use on the evening before launch – a 'wildtrack', or filmed voicepiece over which NASA film of astronaut interviews could be laid – and an interview with our favourite blonde, 25-year-old Poppy Northcutt, the first woman to qualify for an operational job in Mission Control at Houston. Having done much

Apollo 11 crewpatch. It was Armstrong's idea that the Apollo 11 patch should bear no names of the astronauts. He wanted it to represent all the scientists, astronauts, technicians, recovery crews and others who had participated in the moonlanding programme. The only other Apollo crew patch that bore no names was Apollo 13, and that was due to the last minute change when it was decided that Ken Mattingly might develop measles. (NASA)

of the arduous work computing the necessary trajectories to and from the Moon, her task now was to sit at her console monitoring Apollo 11's progress, always anticipating possible problems and computing ways of telling the astronauts how to solve them. I said a reluctant goodbye to my camera crew, Keith Skinner and Sim Harris, no longer needed now that TV News could steal pictures from the costly satellite links installed by the invading army, and settled down to do my final progress reports.

The pre-launch news conference, without which no mission could possibly take place, was a packed affair, full of tension and attended by the 3000 real and avowed newspeople standing around the Press Stand. The only news it yielded was an admission by Christopher Columbus Kraft, the Director of Flight Operations, that if either Armstrong or Aldrin broke an ankle on the lunar surface, it would probably be impossible for the other one to haul him up the nine-step ladder back into the Lunar Module for return to Earth.

We were given only sketchy details of the astronauts' pre-flight dinner. After all the fuss about keeping them germ-free they sat down at a table for 10 to enjoy boiled sirloin and buttered asparagus followed by fruitbowl. The backup and support crews, who included the

Apollo 11: *Flight's* aviation cartoonist, the late Chris Wren, marked the occasion with a cartoon about the BBC Aerospace Correspondent, normally invited aboard inaugural flights of new aircraft: 'He's pretty sick about the Apollo 11 moonlanding. First inaugural he's not been invited on!' (Chris Wren)

Apollo 8 astronauts Lovell and Anders, were there with Deke Slayton, the Chief Astronaut. What we were not told was that Tom Paine had had a quiet dinner with them the evening before that, and according to both Aldrin and Collins in their subsequent books, told them not to take chances, with an assurance that if they decided to abort and come home he would make sure they got the next flight for another try at the landing. This, as Collins explained, 'removed the obvious risk of letting our desire to be first on the Moon cloud our judgment in analyzing hazards'; but I have always doubted very much whether either Deke Slayton or later Apollo crews would have honoured that assurance, which went against the whole impersonal system they had developed of leap-frogging crew selection.

Despite the fact that the British public was deeply absorbed in the moonlanding attempt, the Government itself, then as always, remained aloof, justifying its implied policy of regarding the whole expedition as a waste of money. Britain's official representative was the Permanent Secretary to the Ministry of Technology; the Ambassador to Washington, the former TV interviewer John Freeman, who had attended the Apollo 10 launch, this time turned down a formal invitation. A British Embassy spokesman was much quoted for his explanation that 'when you've seen one Apollo launch, you've seen them all'.

Dawn on 16 July 1969, like most midsummer days in Florida, was hot, stuffy and windless. Sliding up from the Atlantic, the Sun slanted maddeningly into our eyes as we looked east at the launchpad; it was several hours before its effect on the sea was to provide a cooling breeze. On the VIP stand, well away from the Press, the leading Americans were ex-President Lyndon Johnson and the soon-to-be-disgraced Vice President Spiro Agnew. I recorded that Colonel Charles Lindbergh, first man to fly solo across the Atlantic, was there to see the start of an even greater adventure than his own; but Lindbergh watched from a private and secluded enclave provided for astronauts. Less concerned about being noticed among the VIPs was Germany's Hermann Oberth, at 75 the last survivor of the three acknowledged originators of rocketry. The other two, Russia's Konstantin Tsiolkovsky and America's Robert Goddard, had been dead for many years.

Arthur C. Clarke, the space writer whose vision of communications satellites in space fiction played a part in helping them to become a reality, wandered restlessly about the Press site, during breaks from appearing with Walter Cronkite on CBS. Maybe he would have preferred to be on the BBC team. Periodically he stopped to rest his chin on our front-row desks and to utter a marvellously quotable sentence – rehearsing it no doubt for his American audience! As the countdown clock neared zero, it was: 'This is the last day of the old world.' Shortly after Apollo 11 rose shatteringly from the launchpad exactly on time – 9.32 am Cape time, 2.32 pm London time – Clarke returned to tell us: 'At lift-off I cried for the first time in 20 years, and prayed for the first time in 40' – and hurried off to the CBS studio next door to the Press Stand.

15 The Eagle Swoops

Apollo 11 lands on the moon

After 12 years covering every step in NASA's steady progress towards launching men to the Moon, the last thing I expected was that when the moment came I would sit watching with absolutely nothing to do. But there I was at the critical moment, slowly turning lobster red in a temperature of 103 °F in the direct sunshine of high summer, and just staring.

We had had a few seconds of anxiety during the Apollo 10 lift-off, for the whole Saturn 5 column had become enveloped in brownish smoke, and only when the needle-point of the escape tower broke through the top could we be sure that the vehicle was safely on its way.

The Apollo 11 lift-off was quite different. Twin fan-shaped columns, consisting mostly of brilliant white steam, formed on each side of Pad 39A, and there was a breathaking view of the stack rising with majestic symmetry on its column of flame between the wings of the fan. After the 'Good Luck and God Speed' exchanges at minus 3 minutes 45 seconds, Jack King, the Voice of Apollo-Saturn launch control, talked us non-stop through the last three minutes with the computers running smoothly on automatic sequence:

> Two minutes 10 seconds and counting. The target for the Apollo 11 astronauts – the Moon. At lift-off we'll be at a distance of 218 096 miles away. Just passed the 2-minute mark in the countdown, T minus 1 minute 54 seconds and counting. Our status board indicates that the oxidizer tanks in the second and third stages now have pressurised. We continue to build up pressure in all three stages here at the last minute to prepare it for lift-off. T − 1 minute 35 seconds on the Apollo mission, the flight that will land the first man on the

> Moon. All the indications coming into the Control Center at this time indicate we are GO.
>
> One minute, 25 seconds and counting. Our status board indicates the third stage completely pressurised. Eighty second mark has now been passed. We'll go on full internal power at the 50 second mark in the countdown. Guidance system goes on internal at 17 seconds leading up to the ignition sequence at 8.9 seconds. We're approaching the 60 second mark on the Apollo 11 mission.
>
> T−60 seconds and counting. We have just passed T−60. 55 seconds and counting. Neil Armstrong just reported back: It's been a real smooth countdown. We have passed the 50-second mark. Our transfer is complete on internal power with the launch vehicle at this time.
>
> 40 seconds away from the Apollo 11 lift-off. All the second stage tanks are now pressurised. 35 seconds and counting. We are still GO with Apollo 11. 30 seconds and counting. Astronauts reported: Feels good. T−25 seconds. 20 seconds and counting. T-15 seconds, guidance is internal – 12, 11, 10, 9, ignition sequence starts, 6, 5, 4, 3, 2, 1, zero, all engines running...
>
> LIFT-OFF. We have a lift-off, 32 minutes past the hour. Lift-off on Apollo ll. Tower cleared.

Mission Control, Houston (now John McLeish instead of Paul Haney, who had resigned in a huff and gone to work for ITN) instantly took over: 'Neil Armstrong reporting their roll and pitch program, which puts Apollo 11 on a proper heading. Plus 30 seconds...'

Every telephone and microphone except one on the Press Stand was busy: American, Australian, British, French, German, Japanese, Spanish voices in a frenzied babel of description. Once again I was suprised to note that most people were crying, and those with nothing else to do were punching the air, and yelling GO, GO.

It was a hearteningly unifying moment: everyone wanted Apollo 11 to achieve success. Even the worriers like myself stopped fretting about the many possibilities of an abort, and whether in my case I would have the necessary instant recall of all the information I had tried to absorb from scores of Press handouts issued by every company which had contributed some subtlety to this 3000-ton voyager to another world.

The reason I was unemployed at this supreme journalistic highlight in my career was that the BBC's 'Apollo Studios', at Shepherd's Bush for TV, and at Broadcasting House for radio, had insisted on taking over the launch coverage. The studios were packed with 'experts' who had to be given something to do, and once again it had been a case of 'Leave the facts to us, Reg. We'll just come to you with the question: "What's the atmosphere at the Cape?"' Since most of the experts had never been near America's space programme, I was well aware that they would call upon me for tougher explanations than that if anything went wrong.

But nothing did go wrong. And although it was only for a short time that I could not report on the culmination of this story that had occupied my mind and time for 12 years, I sat there with a bleak sense of loss as the Saturn 5 first stage dropped off, and the flame of the second stage engines became a starlike point a hundred miles away. I was wearing headphones linked to the telephone beneath the narrow desk on which was piled my portable typewriter, a selection of Press Kits, and the Flight Plan. Suddenly I became aware that the 'programme feed' of what BBC radio was transmitting had fallen silent. I realised that Larry Hodgson's team had momentarily exhausted their repertoire of eulogies, and had switched to me: 'What's the atmosphere at the Cape, Reg?'

It was quite an effort:

> Hello London, and as you probably gathered, just about the most emotional launch we've ever seen. Apollo 11 has just disappeared from my sight, looking rather like a tiny tadpole in the sky . . . a very emotional crowd of pressmen here, I've never seen them clapping and applauding and shouting Go like they did on this occasion . . . But after all, this is America's twenty-first manned launch, and the

> world's first real ['Fluff' noted the BBC's recording angel, when typing out the 'as broadcast' transcript] exploration flight . . . I'm trying to look at my clock, it's now about five minutes into the flight, so we're seven minutes away from Earth orbit. A hundred miles circular orbit is the aim. Two orbits after that, when they've checked out the systems and got ready for translunar orbit injection, a journey of 65 hours to the Moon, of course . . . they're so much ahead of schedule they're hoping to try out the colour TV camera at the end of the first orbit, which should be about 4 pm British time . . . It was Arthur C. Clarke [it must have been a bad line, for that was transcribed 'after Steve Clark'] who put the launch into perspective for me. He wandered past my seat . . . just now and said 'This is the last day of the old world.' And I think that's how we all feel . . . Back now to London.

It was the biggest moment, Wernher von Braun told us in his guttural German accent at the post-launch news conference an hour later, 'since life first crawled out of the slime'. From Earth orbit Armstrong reported back more prosaically: 'We have no complaint with any of the three stages. That ride was beautiful.' His appreciation was enhanced by the fact that for the first time there was no 'pogo' bounce.

Because this was the third time it had been done, no one was much surprised when the Saturn 4B third stage was relit, and fired Apollo 11 out of Earth orbit and into a translunar trajectory at 24 250 mph – within 3 miles an hour of the planned speed. But with the evening news programmes still running, I waited at the Cape until Collins had successfully completed TD&E – transposition, docking and extraction – which involved undocking Apollo from Saturn 4B, then turning around and nosing back to dock with the Lunar Module, then pressurising it from the Service Module tanks, and extracting it from the S4B. After that came an evasive manoeuvre to eliminate any possibility of collisions as the two craft continued – Apollo to pass in front of the Moon, and the S4B behind it.

That done, the crew could settle down two hours early for their first rest in space – and newsmen like me could carry out our own well-rehearsed extraction, transposition and docking manoeuvre: the scrambled transfer from the Cape to Mission Control at Houston. We envied the crew, for this had to be done in our sleep period, and it was a far more hysterical operation: checking out laden with Press Kits, etc., making it in time to the airport to catch our special plane, then literally racing 50 miles in our hired cars to the three hotels at Nassau Bay. They were much less friendly and efficient than those at Cocoa Beach, and despite our reservations, confirmed and reconfirmed for months past, they were as always overbooked and covered it up by denying that last arrivals had ever made those bookings. It worked well for the hotels, forcing some newsmen to share rooms, which meant that neither got any sleep because of the other's long-distance phone calls.

The three hotels were a one-minute drive or a five-minute walk across the new 'NASA Road 1' to our tables in the enlarged News Centre at the Mission Control. TV screens gave us a continuous view of what was happening in Mission Control, unless the crew was transmitting TV, when that was fed to us instead. Flight directors and technicians briefed us every time they changed shifts, NASA managers, scientists and astronomers held news conferences whenever they thought they had something to say, and made themselves available to us on an individual basis more or less on demand. Seldom have newspeople been more spoiled.

My table was between Canadian and French radio stations. A BBC New York engineer had spared half an hour away from Richard Francis's team to help me hook up my Uher tape recorder to Mission Control transmissions as well as my telephone, so that I could record exchanges between the astronauts and Mission Control, and also take a feed from the news conferences, and then – if I had time to find and mark the excerpts – feed bits of 'actuality' into my regular news reports. The disadvantage was that when I wanted to take my tape recorder to an interviewee elsewhere I had to unhook it all and then hook it up again with frenzied fingers and then dial myself through when I was overdue for broadcasts. 'This is the day I won't make it!' I would tell

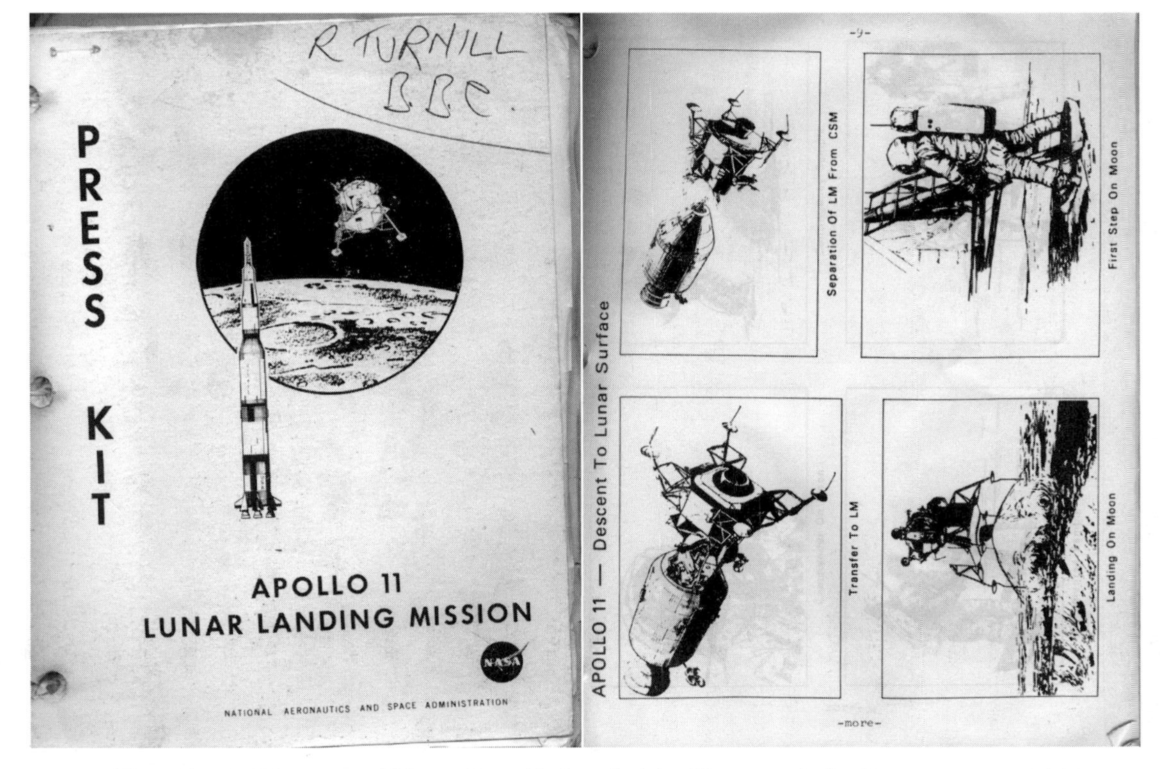

The author's all-important 250-page Press Kit. Page 9, right, illustrates the landing sequence – which Armstrong and Aldrin carried out perfectly on the day. (NASA)

myself time after time as I inserted the back of a handout into my typewriter with the clock showing only a minute or two before the next live or recorded broadcast. I always felt on show and rather foolish, and it was a long time before I realised that those surrounding me envied the demands made upon me and sympathised with my efforts to meet them.

At Houston my desk was some distance from my *Mail* and *Mirror* friends, whose neighbours were mostly drawn from the 119 Japanese journalists who spent much of their time translating sections of the Press Kit into Japanese at the tops of their voices to Tokyo, interpersed with much 'Hoy hoy-ing' and 'Ah-so-ing', breaking off periodically to demand in poor English from their neighbour explanations of what certain sections meant. Once again, NASA public affairs officers were not above pointing out to them the regular space correspondents and telling them to come to us for explanations, indifferent to the fact that in my case I was frequently broadcasting live to several million people in Britain.

We were not unsympathetic, for while we had built up to this mission step by step with the astronauts, the Japanese reporters had mostly been despatched by their offices without much knowledge of English and with no knowledge of the space programme. But they were soon to catch up and surpass most Western journalists in their knowledge of and enthusiasm for space exploration.

The Apollo 11 crew gave us several TV transmissions during their outward journey, demonstrating the quality of their new colour cameras by taking Mission Control and the rest of the world with them as they opened up the docking tunnel and checked out the Lunar Module. Like any amateur photographer, Buzz Aldrin inquired 'Do you think F22 will be all right?' and Mission Control pronounced themselves well-satisfied with the clarity of the picture. It was so sharp that close-ups enabled viewers to read some of the controls and instruments that would guide the astronauts to the touchdown. A close-up of the 'Abort' button was accompanied by the comment: 'We're going to tape that over.'

This improvement even on Apollo 10's colour TV transmissions was to mislead an expectant world into believing that it would share the actual moonlanding just as vividly. Only the Command Module was equipped with a colour TV camera. The Lunar Module carried only black and white TV and film cameras – a decision which had evoked furious protests from Maxime Faget, responsible for much of the spacecraft design. Early in 1969 he had submitted a written protest that it was 'almost unbelievable' that the culmination of a $20 billion programme was to be recorded 'in such a stingy manner'. George Low, the NASA chief concerned, had replied that some of the scientists had insisted on black and white film because it had a higher resolution than colour film and that it was felt that, with no atmosphere to absorb the solar energy in the ultraviolet, colour film might not be effective on the lunar surface.

Only one mid-course correction was necessary as Apollo 11 cruised smoothly towards its target – a 3-second firing of the small thrusters to ensure that they passed only 69 miles ahead of the Moon. Their previous course placed them 200 miles ahead. Now that there was little for the crew to do, Mission Control began updating them with world news – beginning with the progress of Luna 15, *en route* to the same destination ahead of them.

They listened without comment to that and a report on protests in the House of Lords about a US submarine exploring Loch Ness with the intention of using a dart gun to snatch a skin sample of its alleged monster. Somewhat embarrassed by the silence, Mission Control added somewhat diffidently 'We think you're doing a great job out there.' Armstrong replied drily 'Thanks'. He always took the view that others were working just as hard without having so much fun; he may have had in mind the 12 men already sealed inside the Lunar Receiving Laboratory, ready to share with the crew their strict 21 days' isolation starting from the moment when it was hoped they would lift safely off from the Moon.

There was in fact increasing concern among NASA scientists at the possibility that the Soviet commands and communications, flowing to and from Luna 15, would clash with the even more complicated transmissions with Apollo – especially when the Command Module and Lunar Module were separated, with their safety depending upon perfect communications between one another and Mission Control. This was reflected in the next news transmission to the crew late on 17 July:

> Britain's big Jodrell Bank radio telescope stopped receiving signals from the Soviet Union's unmanned Moon shot at 5.49 BDT [British Daylight Time] today. A spokesman said it appeared that the spaceship had gone beyond the Moon. 'We don't think it has landed' said the spokesman for Bernard Lovell, director of the Observatory.

News items following this included Vice President Agnew demanding that a man be sent to Mars by the year 2000, and an announcement by immigration officials that 'hippies' would be refused tourist cards to enter Mexico unless they first took a bath and had haircuts.

The crew were not told at that stage that Frank Borman had been given the task of phoning his recently-made friends in Moscow for detailed information about Luna 15 and its operations. He got through to Academician Keldysh, President of the Academy of Soviet Sciences, but the line was so bad Borman had to ask for information to be cabled. Keldysh, always co-operative with Western scientists, sent it at once: Luna 15 had been placed in a 179 × 73 miles lunar orbit, and would stay there for two days. It would be gone by the time Apollo 11 arrived, leaving us to assume that the Russians hoped that by then it would have visited the lunar surface, scooped up a soil sample, and be on its way back to Earth. Poppy Northcott and her computer department worked out how close the spacecraft might be when and if they passed, and reported that there was only a billion-to-one chance that there would be any problem.

At Houston it was lunchtime Saturday when Apollo 11 finally passed in front of the Moon at a distance that resulted in lunar gravity, with the help of a 6 min 2 s firing of the main propulsion engine, pulling it into a safe orbit around it. The weight of the vehicle as it disappeared behind the Moon was 96012 lb; when it reappeared the long burn had reduced the weight to 72004 lb, and its lunar orbit was 70 × 196 statute miles – incredibly within a tenth of a mile of the flight plan at the high point, or apocynthion [the word deriving from Cynthia, the Roman goddess of the Moon], and 1.5 miles at its lowest, or pericynthion. It was another triumph for those who, months ago, had calculated all the required trajectories, and the angles and attitudes and seconds of propulsion needed to change them.

A later, much shorter burn, circularising the orbit to 62 × 75 miles, meant that the landing on the Moon's surface should take place during Sunday night's 'prime time' TV in America; but the actual moonwalk would not start until 2 am on Monday in the US and 5–6 hours later in Europe. Another news summary to the crew informed them – and me, since I had not been told – that the BBC was considering transmitting a special radio alarm to wake people up in case the moonwalk took place early. I rashly assured my listeners that if the touchdown was completed according to the 184-page 'Lunar Surface Operations Plan', which had now worked its way to the top of my cluttered desk, Armstrong and Aldrin would adhere to the plan to have four hours' rest before venturing on to the surface just after 7 am in Britain.

The taciturn Neil Armstrong almost sounded excited as he got his first good view of the landing approach. 'The maps and pictures brought back by Apollos 8 and 10 give us a very good preview of what to look at here. It looks very much like the pictures, but like the difference between watching a real football game and watching it on TV – no substitute for actually being here!'

The crew's first attempt at having a meal in lunar orbit was interrupted when Mission Control passed on the astonomers' request that they should look for what appeared through their telescopes to be a 'transient event' in the vicinity of Crater Aristarchus – just within view of their horizon, 394 miles north of their track. Somewhat inconclusively the crew said the area seemed to 'have a slight amount of fluorescence to it' but there was no colour. Rather than the volcanic activity of the astronomers' dreams, it seemed to be earthshine on the wall of a crater. Sharing their progress with Mission Control by means of the TV cameras, the crew was naturally more interested in talking their way through the powered descent the next day, picking out the landmarks that would help to guide them – with Mount Marilyn, the ignition point for the start of Eagle's descent, prominent among them.

All three astronauts, and especially Buzz Aldrin, noticeably enjoyed that day in lunar orbit, getting ready for the descent. He and Armstrong spent much of it powering up the Lunar Module, and for the first time, as the communication systems were checked out, the call-signs Eagle and Columbia were used so that Houston could call them separately. Yet again, in case of an emergency in which the crew lost contact and had to make their own decisions to return to Earth, an updated programme for TEI, Trans Earth Injection, was read up:

> Rog Columbia. Here we come with the TEI 11.SPS G&N, 37200 minus 060 plus 047.Noun 33, 098, 05, 2422 plus 41448 plus 03719 minus 02422. Roll is NA, pitch 020, the rest of the pad is NA.Set stars are NA. The ullage is 2 quads, correction, 2 jets for 16 seconds. Use Bravo and Delta. In a comment, the undocked present CSM, correction, this is upfront. TEI 11 is undocked. Present onboard weight of CSM is 37200 pounds. About 50 Alpha on your dap. Over.

I found the numbers difficult to hear. A quarter of a million miles away Mike Collins calmly read them back without an error.

On the Saturday night, the crew were supposed to have nine hours' sleep. Mission Control assured us periodically during what was a very short night for us that they were in 'a deep sleep', but they did not in fact sleep well, and were up and active making their landing preparations long before Mission Control sent them the routine 'wake-up' call – omitting, on this occasion, I was glad to note, the usual jokey music. Unlike the earliest astronauts, they were not required to report bowel movements, but they did have to report that the commander had had 5.5 hours sleep, the Command Module pilot 6 hours and the Lunar Module pilot 5 hours.

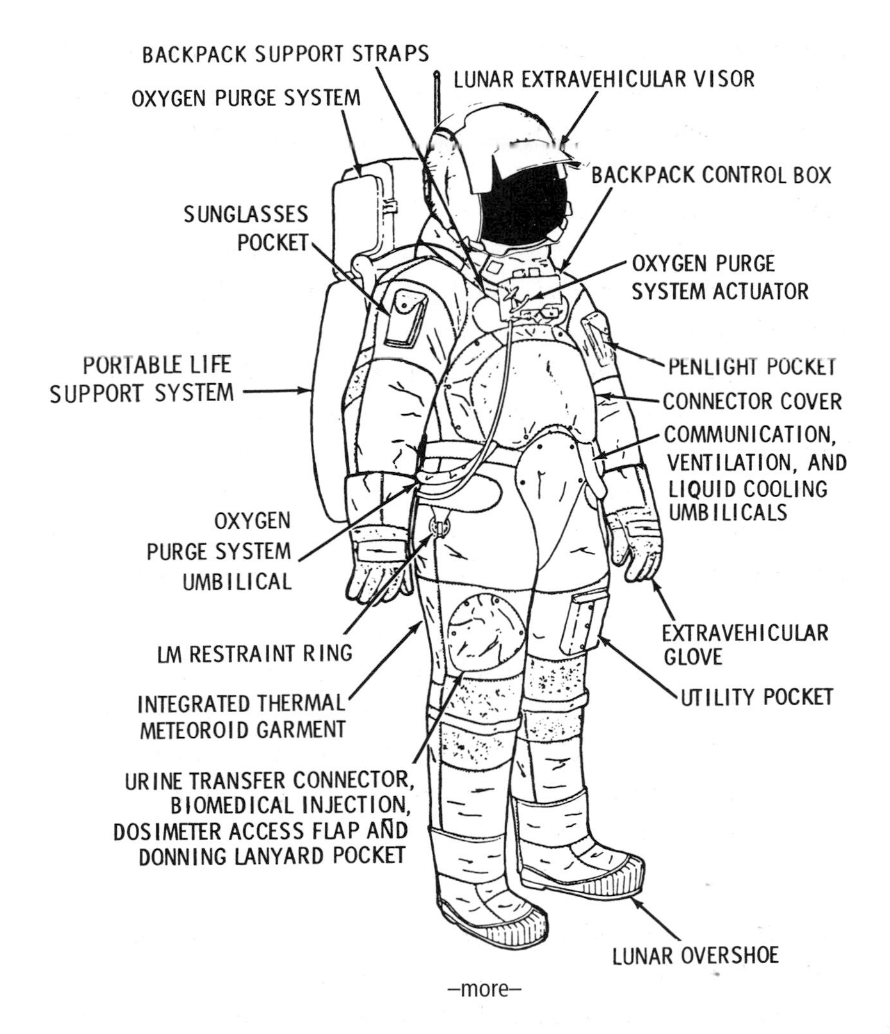

Fashions for 1969: the well-dressed man on the Moon. (NASA)

There were still two lunar revolutions to go before separation started on the 13th, so while the astronauts had breakfast Mission Control did their best to relax them with more 'morning news briefs'. President Nixon's Sunday service at the White House would be dedicated to the mission, they were told. Maybe I was wrong when I thought

I detected a note of irony as Capcom (the Capsule Communicator) added: 'Frank Borman is still in there pitching and will read the passage from Genesis which was read on Apollo 8 last Christmas.' If that was the sublime part, the ridiculous quickly followed:

> Among the large headlines concerning Apollo this morning there's one asking that you watch for a lovely girl with a big rabbit. An ancient legend says a beautiful Chinese girl called Chango has been living there for 4000 years. It seems she was banished to the Moon because she stole the pill for immortality from her husband. You might also look for her companion, a large Chinese rabbit, who is easy to spot since he is only standing on his hind feet in the shade of a cinnamon tree. The name of the rabbit is not recorded.

'OK, we'll keep a close eye for the bunny girl,' commented Buzz Aldrin.

I always suspected that the astronauts were somewhat bored and irritated with the rather trivial news items selected for them, and that these were really meant to attract media attention. With the crew's thoughts focused on the trials ahead of them, Mission Control – Astronaut Charlie Duke had taken over as its spokesman – ploughed on regardless with the election of Miss Universe and her physical statistics, ending with the fact that the Animal Trap Company of America had invented a wonderful new mousetrap but housewives (whatever made them think it was the wives who set the mousetraps?) insisted on buying the old-fashioned kind. There was no mention at all of the intriguing mystery of what Luna 15 was doing.

All this began at 6 am Houston time, noon British time, and for us and the public many boring hours followed as Houston and the astronauts checked and re-checked their communications systems, while Armstrong and Aldrin, donning their pressure suits, floated between Columbia and Eagle. Ronnie Bedford, short of a story, went off to see Dr Berry and returned to share with me what the astronauts called Chuck's latest 'Berryism'. He had worked out an 'energy budget' to help the medics monitor Armstrong and Aldrin as they worked on the Moon, so that they could recall them before they reached the limits of human endurance.

Getting in and out of Eagle would consume 2000 energy units (Ells) per hour; hoisting 60 lb of lunar rocks from the surface into Eagle by means of a wire pulley would require 1600 units per hour; picking up lunar samples, never more than 100 ft from Eagle, and inserting them into vacuum-sealed aluminium containers, 1400 per hour; lunar photography, etc. 500. Armstrong had been allotted a total of 3625 EUs for an estimated 160 minutes on the lunar surface; Aldrin, scheduled to be on the surface for only 135 minutes, was allotted 3370 EUs.

Finally they were sealed inside Eagle, and they could relax while Mike Collins, remaining in orbit in Columbia, tidied up after them. He had to put the probe and drogue in place between Columbia and Eagle, set up cameras in the windows to cover the separation – there was to be no TV or photography from Eagle during the descent – purge the fuel cells of excess water, and prepare to vent the air pressure from the docking tunnel between Columbia and Eagle.

Only by studying the Flight Plan could one appreciate how exhaustingly busy the astronauts were – for instance, having to don helmets and gloves to make a pressure check to ensure that the hatches were tightly closed, then doff them ten minutes later in order to copy more data from Mission Control, to start checking their gyroscope torque angles, etc. Finally, the occupants of Eagle and Columbia were satisfied that they could hear one another on Simplex A and Simplex B, had loaded a final 'time hack', and Mission Control had made a final check on the amount of orbital drift being caused by the Moon's mascons.

The crew were given a GO for separation, and Apollo 11 disappeared behind the Moon to start its 13th lunar revolution. Mission Control and the world's TV stations prepared for 30 hours of continuous transmission.

Columbia and Eagle reappeared from behind the Moon as two separate spacecraft, Mike Collins having punched the button to undock them two minutes earlier, 100 hours 12 minutes after lift-off from Cape Kennedy.

Mission Control put out two anxious calls to establish contact, but Armstrong and Aldrin were busy:

> EAGLE: Roger, Eagle. Standby.
> MISSION CONTROL: Roger. How does it look?
> EAGLE (Neil Armstrong): The Eagle has wings.

After that last human touch, I reported, it was all numbers for the computers. While Armstrong was conducting a yaw manoeuvre so that Eagle rolled around like a film star, enabling Columbia's TV cameras to send pictures back to Earth so that Mission Control and the watching world could check that its landing legs were fully extended and everything else looked perfect, Buzz Aldrin was busy receiving and reading back the final numbers for the Powered Descent Initiation (PDI), which included reaction to possible emergencies:

> MISSION CONTROL (Charlie Duke): Roger. PDI pad. PIG 102330436 0950 minus 00021 182287000 plus 56919, PDI aborts less than 10 minutes. 105123000, PDI abort greater than 10 minutes, 103400000 107113000, no PDI plus 12, 102442700, NOUN 81, plus 01223, burn time 046000190 plus 01187 plus all balls plus 01911 NOUN 11 103310700, NOUN 37 105123000. Ready for your readbacks. Over.

Just listening to these numbers and trying my hand at getting them down accurately despite the static noises blurring many of them, made me sweat. I was lost in admiration when Buzz Aldrin, who must have been more than distracted as Eagle and Columbia pirouetted daintily around one another, read both the DOI (Descent Orbit Insertion) and PDI numbers back unhesitatingly at high speed.

> CHARLIE DUKE: Roger. Good readback Buzz. Out . . .
> COLUMBIA (Collins): Is that close enough for you, or do you want it to a couple of decimal places?

EAGLE (Armstrong): No, that's good.
COLUMBIA: I think you've got a fine looking flying machine there, Eagle, despite the fact that you're flying upside down.
EAGLE: Somebody's upside down.
COLUMBIA: Okay Eagle. 1 minute until T. You guys take care.
EAGLE: See you later.

With that, Columbia made an 8-second burn of its forward thrusters, so that they drifted safely 1100 ft apart before DOI. Thirty minutes of checking and re-checking followed and seven minutes before Eagle fired its descent engine for 28.5 seconds for DOI we heard:

MISSION CONTROL: Columbia, Eagle – Houston. 3 minutes LOS. Both looking good going over the hill.
COLUMBIA: Columbia, roger.
EAGLE: Eagle, rog.

Once again the vital manoeuvre was made behind the Moon. Now, I reported, there was no mistaking the tension in Mission Control, packed as it was with more than 10 astronauts standing among the consoles, and the viewing room with what we were told was 'the largest assembly of space officials ever seen in one place'. Russia's Luna 15 was responsible for much of the tension, because it too was making some dramatic manoeuvres in a low lunar orbit – about which the busy Apollo crew were not told. I tried to snatch a cup of coffee and a sandwich in the nearby NASA canteen – but it was shut. America might be landing on the Moon, and there were several thousand parched people on the base – but NASA's canteen workers kept their heads and worked only normal hours, which did not include Sundays. I rushed back to my desk again. As expected, Columbia, emerged first:

COLLINS: Houston, Columbia. Reading you loud and clear. How me?
MISSION CONTROL: Roger. 5 by, Mike. How did it go? Over.
COLLINS: Listen babe, everything's going just swimmingly. Beautiful.

MISSION CONTROL: Great. We're standing by for Eagle.
COLLINS: Okay, he's coming around . . .
EAGLE: Houston, Eagle. How do you read?
MISSION CONTROL: 5 by, Eagle. We're standing by for your burn report. Over.
EAGLE: Roger, the burn was on time. The residuals before knowing: minus 0.1 minus 0.4 minus 0.1, x and z now to zero.

Eagle was now in an egg-shaped orbit and, as Apollo 10's Snoopy had done, was sweeping down to within 50 000 ft of the lunar surface. The difference this time was that when it reached that lowest point the PDI programme would start the final descent. The final minutes before PDI were hectic, with Houston several times losing contact with Eagle, and passing messages through Columbia. It was Mike Collins who finally told Armstrong and Aldrin: 'Eagle, this is Columbia. They just gave you a GO for powered descent.' He had to repeat it a second time, adding: '. . . and they recommend you yaw right 10 degrees and try the high gain [antenna] again.'

EAGLE (Armstrong): Roger, understand. [To Aldrin] Elevate the GO circuit breaker. Second gimbal, AC, closed. Second gimbal AC closed? Circuit breaker and over-ride off. Gimbal enable. 8 scale 45 . . .
EAGLE (Aldrin reading off his check list to Armstrong): Balanced couple on, TA throttle, minimum, throttle, Auto, Cdr, Stop Button, Reset Stop Button, check abort, abort stage reset, att. control, 3 of them to mode control, PGNS mode control is set, AGS is reading 400 plus 1, standing by for arming.
EAGLE (Armstrong): Hit Verb 77. Okay, sequence camera coming on.

The 'Twelve Minutes to Touchdown' which I had described in advance, and illustrated with Teledyne Ryan simulations, were much more nerve-wracking than we expected when they actually took place. Only a handful of people watching their consoles in Mission Control understood exactly what was happening; to the rest of the world – to some extent including even Armstrong and Aldrin – it was all a confusing and frightening muddle, the overall effect being of impending disaster. I sat at my desk, half-buried by documents, lunar maps and Press Kits, with two set of headphones pressed to burning ears, trying NOT to listen to BBC producers and the programme they were putting out, trying to disentangle the meaning of the numbers flying between Mission Control and Eagle, more than half-expecting disaster as the voices tightened. Here it is possible to tell the story of the final touchdown with the help of NASA's official history as well as what we heard at the time.

The 12 minutes, starting when Armstrong punched the button for PDI, or powered descent initiation, began at 102 hours 33 minutes 5.2 seconds after launch, when Armstrong advanced the throttle until the descent engine reached maximum thrust – a process which took 26 seconds. Collins in Columbia was following 136 miles behind, listening intently, and repeating transmissions between Mission Control and Eagle when they failed to hear one another. At first Armstrong could neither hear whether the descent engine had ignited, nor feel the deceleration; but his instrument panel told him that all was well. It was mostly Aldrin that we heard calling out instrument readings, enabling Armstrong, helped by Mission Control as well as Aldrin, to concentrate on manipulating the controls.

EAGLE: AGS on [garble]; 10 per cent.
MISSION CONTROL: Columbia, Houston, we lost 'em. Tell 'em to go aft OMNI.
EAGLE: [Inaudible]
COLUMBIA: Say again, Neil.
EAGLE: I'll leave it in SLEW. See if they've got me. I've got good signal strength in SLEW.
COLUMBIA: See if you're getting them now, Houston.
MC: Eagle, we've got you now. It's looking good here. Over.
EAGLE: Roger, copy.
MC: Eagle, Houston. Aft Yaw around angles S-band Pitch minus 9, Yaw plus 18.

EAGLE: Copy.
EAGLE: AGS and PNGS agree very closely.
MC: Roger.
EAGLE: Rate on. Altitudes are a bit high.
PAO [Public Affairs Officer, informing the media and the public; not audible to the astronauts]: 2 minutes 20 seconds, everything looking good. We show altitude about 47 000 ft.
EAGLE: Houston, I'm getting a little fluctuation in the AC voltage now.
MC: Roger.
EAGLE: Could be our meter, maybe, huh?
MC: Standby. Looking good to us. You're still looking good at 3, coming up to 3 minutes.
EAGLE: [garble]. Looks real good. [garble] Our positions check downrange here seems to be a little long.
MC: Roger, copy.
EAGLE: Altitude rate about 2 feet per second greater than it ought to be. (garble)
EAGLE: Altitude [garble] I think it's gonna drop.
MC: I think it's going to drop.
MC: Eagle, Houston. You are Go. Take it all at 4 minutes. Roger you are go – you are Go to continue power descent. You are Go to continue power descent.
EAGLE: Roger.
PAO: Altitude 40 000.
MC: And Eagle, Houston. We've got data dropout. You're still looking good.
EAGLE: PGNCS, we got good lock on. Altitude light's out. Delta H is minus 2900.
MC: Roger, we copy.
EAGLE: And the Earth right out our front window. [That was because Eagle was rolling on its back, with the astronauts unable to see the Moon.]
EAGLE: Houston, you're looking at our Delta H program alarm?
MC: That's affirmative. It's looking good to us. Over.

EAGLE: 12-02, 12-02.

PAO (Apparently unaware of the developing crisis, but in any case determined to reassure the outside world): Good radar data. Altitude now 33500 ft.

EAGLE: Give us the reading on the 12-02 program alarm.

MC. Roger. We got –

Here Astronaut Charles Duke paused, and in a moment of desperation looked around at the engineers and technicians monitoring their consoles. He had a fraction of a second in which to advise Armstrong and Aldrin whether to continue their descent or instruct them to abort and try to return to a safe lunar orbit. Steve Bales, a 26-year-old engineer charged with monitoring the performance of Eagle's computers (he was known as GUIDO, the Guidance Officer) recognised that the vital one calculating the rate of descent and distance from the surface was overloading. It was for him to judge whether the computer would clear itself; if it would not he must recommend the abort. His calmly assessed 'GO Flight!' in the hushed Center was addressed to Gene Kranz, the Flight Director, who nodded approval to Duke who instantly instantly repeated it:

MC: – we're Go on that alarm.

EAGLE: Roger. P30.

MC: 6 plus 25 throttle down.

EAGLE: Roger, copy. 6 plus 25.

PAO: We're still go. Altitude 27 000 ft.

EAGLE: Alarm. It appears to come up at 16–68 up.

MC: Roger, copy. Eagle, Houston. We'll monitor your Delta-H.

EAGLE: Delta-H is looking good now.

MC: Roger, Delta-H is looking good to us. Right on time.

EAGLE: Throttle down is better than in the simulator.

MC: Rog.

EAGLE: AGS and PGNCS look real close.

PAO: Altitude now 21 000 ft. Still looking very good. Velocity down now to 12 hundred feet per second.

MC: You're looking great to us, Eagle.
EAGLE: Okay, I'm still on slough, so we may tend to lose [? signal] as we gradually pitch over. Let me try auto again now and see what happens.
MC: Roger.
EAGLE: Okay, looks like its holding.
MC: Roger, we got good data.
PAO: Seven minutes 30 seconds into the burn. Altitude 16 300 ft.
MC: Eagle, Houston, it's descent 2 fuel to monitor, over.
EAGLE: 72.
PAO: Altitude 13 thousand 5, velocity 91 hundred feet per second. [They MUST be going to crash, I thought.]
EAGLE: Made it (?) switch over time please, Houston.
MC: Roger, stand by, you're looking great at eight minutes.
PAO: Correction on that velocity, now reading 760 feet per second.
MC: It's the P64. [High Gate – the second part of the descent manoeuvre; a pilot's term indicating that the landing site should be visible.]
EAGLE: Good, roger.
PAO: FIDO [Flight Dynamics Officer] says we're Go; altitude 92 hundred feet.
MC: 8.30 – you're looking great.
PAO: Descent rate 129 feet per second.
MC: Eagle you're looking great, coming up 9 minutes.
PAO: We're now in the approach phase of it, looking good. Altitude 52 hundred feet.
EAGLE: Manual auto attitude control is good.
MC: Roger, copy.
PAO: Altitude 4,200.
MC: Houston, you're go for landing. Over.
EAGLE: Roger, understand. Go for landing, 3,000 ft.
MC: Copy.
EAGLE: 12 alarm. 1201. [Another computer overload warning]
EAGLE: 1201.[They're crashing after all, I thought.]

MC: Roger. 1201 alarm. [In Mission Control all eyes turned to Steve Bales again. 'GUIDO'? queried Kranz. Go' said Bales.]
MC : (Duke to the Astronauts) : We're go. Hang tight. We're go.
EAGLE (Aldrin): 2000 ft. 2000 feet into the AGS. 47 degrees.
MC: Roger.
EAGLE: 47 degrees.
MC: Eagle you're looking great. You're Go.
PAO: Altitude 1,600. 1,400 feet. Still looking very good.
MC: Roger, 1202. We copy it.
EAGLE (Aldrin): 35 degrees. 35 degrees. 750, coming down at 23 [feet per second]. 700 feet, 21 down. 33 degrees, 600 feet, down at 19. 540 feet, down at 30 – down at 15. 400 feet, down at 9. [garble] 8 forward. 350, down at 4.330, 3 and a half down. We're pegged on horizontal velocity. 300 feet, down 3 and a half. 47 forward . . . [garbled] Down 1 a minute, 1 and a half down. 70. Got the shadow [of the spacecraft] out there. 50 down, at 2 and a half. 19 forward. Altitude-velocity lights. 3 and a half down, 220 feet. 13 forward. 11 forward, coming down nicely. 200 feet, 4 and a half down. 5 and a half down. 160, 6 and a half down, 5 and a half down, 9 forward. 5 percent. Quantity light [warning that descent propellant was almost exhausted]. 75 feet, things looking good. Down a half. 6 forward.
MC: 60 seconds [of propellant remaining].
EAGLE (Aldrin): Lights on. Down 2 and a half. Forward. Forward. Good. 40 feet, down 2 and a half. Picking up some dust. 30 feet, 2 and a half down. Faint shadow. 4 forward. 4 forward. Drifting to the right a little. 6 (garbled), down a half.
MC: 30 seconds [of fuel left].
EAGLE (Aldrin): Forward. Drifting right [garbled]. Contact light. Okay [to Armstrong] engine stop. ACA out of detent. Modes control both auto, descent engine command over-ride, off. Engine arm off. 413 is in.
MC: We copy you down, Eagle.
EAGLE (Armstrong): Houston, Tranquillity Base here. The Eagle has landed.

MC: Roger, Tranquillity, we copy you on the ground. You've got a bunch of guys here about to turn blue. We're breathing again. Thanks a lot!

EAGLE: Thank you.

MC: You're looking good here.

EAGLE: I tell you. We're going to be busy for a minute. Master arm on. Take care of the descent [garbled]. Very smooth touchdown. Looks like we're venting the oxidiser now.

MC: Roger, Eagle. And you are Stay for T1. Over. Eagle, you are Stay for T1.

16 First Steps – and Where They Led

Apollo 11 on the Moon

'During simulations we were programmed to abort; on the actual mission we were programmed to land,' Armstrong explained later, when asked whether he had been tempted to abort during four minutes of a bewildering series of 'master alarms'. And few could have resisted punching (and turning) the red button when he was still searching for a safe landing place with less than 30 seconds of fuel remaining.

No wonder that in Mission Control they were holding their breath. A slow-scan, 16 mm black-and-white Maurer camera, set up by Aldrin to film out of a window at six frames per second, and switched on by him 14 minutes before touchdown, provided some poor pictures after they returned of Eagle's shadow rising to meet it; but while the landing was in progress Mission Control had only the mental picture provided by the instruments and the astronauts' commentary to guide them as to what was going on. They did, however, have an exact knowledge of when the descent engine's propellant would be exhausted, and knew that if Eagle had not touched down by then it would fall the rest of the way to the surface, topple over, and be unable to take off again.

Once touchdown had taken place – at 4.17 pm Houston (EDT) time and at 9.17 pm British Summer Time – there was no time for analysis, and little for mutual congratulations. More immediate decisions had to be made: was it safe to stay?

The Flight Plan included stay/no stay decisions at three minutes and nine minutes after touchdown – points at which Armstrong and Aldrin could have fired their ascent engine to return to the correct lunar orbit if staying on the Moon appeared dangerous. At both points Mission Control gave them a 'clear to stay', and they began to relax, and remove their helmets and gloves. After that they would have to wait for

Collins to complete another lunar orbit before they could take off to rendezvous with him. 'What was the atmosphere at Houston, Reg?' I did my required one-minute insert into BBC Radio:

> Enormous relief and satisfaction here at Mission Control at the final touchdown. A mere 37 seconds later than planned, because Armstrong and Aldrin, taking over manual control of their descent engine throttle, had to dodge about a bit to avoid boulders. In Mission Control Dr von Braun was understandably the most excited of the space officials. For this is really HIS day: the fulfilment of his life's ambition. For it was to land a man on the Moon that he's spent 40 years of his life. And within minutes of landing on the Moon, the terror of the unknown is rapidly vanishing. Eagle settled nicely, and should have no trouble taking off again. There's been no big dust storm, which might have taken a long time to clear – proof of that is Armstrong's and Aldrin's vivid descriptions of the lunar surface through the windows.

Eagle had in fact settled with a 4.5° tilt from the vertical, but this was not enough to cause anxiety that it might tip over when the crew moved about; nor would it interfere with the firing of the ascent engine. 'That may have seemed like a very long final phase, 'Armstrong told Mission Control, 'but the auto targeting was taking us right into a football-field sized crater, with a large number of big boulders and rocks for about one or two crater diameters around us, and it required a (garbled) on the 366, and flying manually over the rock field to find a reasonably good area.'

Aldrin followed with the first description of what could be seen through Eagle's triangular windows:

> We'll get to the details of what's around here, but it looks like a collection of just about every variety of shapes, angularities, granularities – every variety of rock you could find. The colours vary pretty much depending on how you're looking relative to the zero phase point. There doesn't appear to be too much of a general colour at all, however. It looks as though some of the rocks and boulders, of

> which there are quite a few in the near area, it looks as though they are going to have some interesting colours to them, over.

Mission Control were politely interested, and urged Armstrong and Aldrin to press on through the simulated countdown for an emergency lift-off – but for once Armstrong wanted to chat, telling them that Eagle 'flew just like an airplane'.

> MC: Rog, Tranquillity, be advised there's lots of smiling faces in this room, and all over the world.
> EAGLE (Aldrin, exchanging grins with Armstrong): There's two of them up here.
> MC: Rog, that was a beautiful job, you guys.
> COLUMBIA (Collins, 60 miles above, not yet out of radio range): And don't forget one in the Command Module . . . It sure sounded great from up here. You guys did a fantastic job.
> EAGLE (Armstrong): Thank you. Just keep that orbiting base ready for us up there now. [garble]
> COLLINS: Will do.

Slowly Armstrong and Aldrin relaxed, despite repeated calls from Mission Control to open and close fuel vents, and Armstrong provided the first simple description of the Earth seen from the lunar surface: 'It's big and bright and beautiful.' His heartbeat, the doctors reported, was 110 per minute when the descent burn began and had risen to 156 at touchdown, when it went down in the 90s. Aldrin was apparently wearing no biomedical sensors.

Eagle had been on the Moon for 116 minutes before the crew and Houston had checked out all the ascent engine systems and were certain that the upper stage was able and ready to take off again if and when required. That was when, despite all the assurances we had been given that the Flight Plan would be strictly adhered to, Armstrong and Aldrin announced that they did not want to take the planned four-hour rest period; they preferred to get on with preparations for their moonwalk and rest after it had taken place. They estimated that after

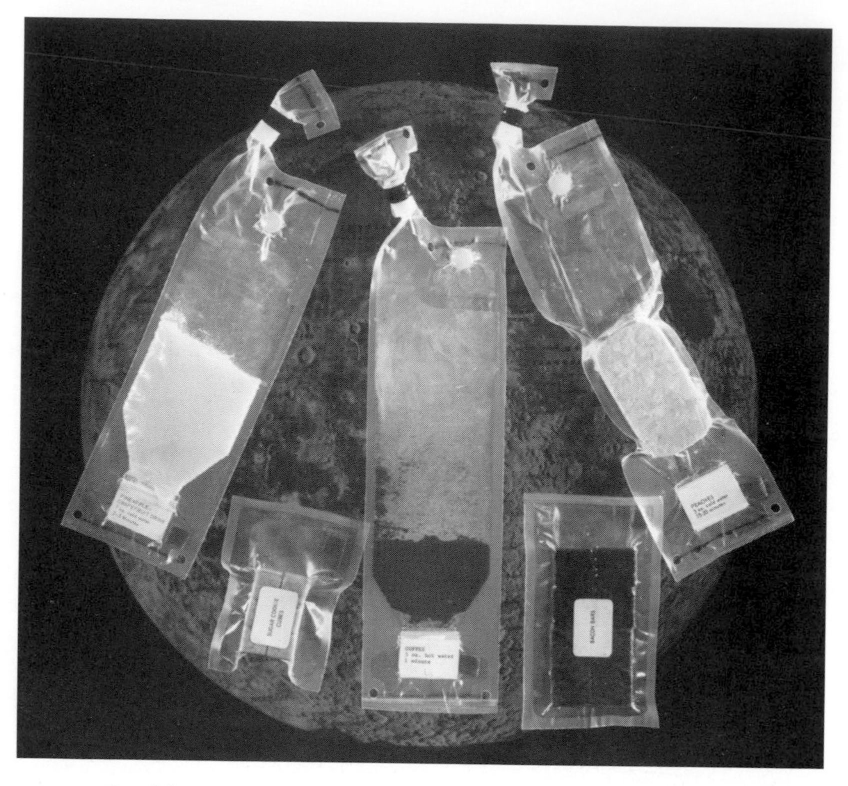

Breakfast at Tranquillity Base: bacon squares, peaches, sugar cookie cubes, pineapple grapefruit drink, and coffee. (NASA)

they had had a meal it would be another two hours before they would be ready to open the hatch. Mission Control had little option but to agree; and as they pointed out, it meant that in the United States the EVA would take place in 'prime time TV' instead of the early hours. Hearts sank in Europe, however; instead of viewers there being able to watch the first moonwalk while they had breakfast, they would have to stay up all night if they wished to watch it.

Mike Collins had just disappeared behind the Moon, still worried about how far Eagle had overshot the landing site, because it would affect his ability to rendezvous and dock with it when it took off again, and Mission Control was still getting Armstrong and Aldrin to check their batteries and temperature controls, when Aldrin called Houston with a note of self-importance. Given the go-ahead, many were surprised to hear him say:

> Roger, this is the LM pilot. I'd like to take this opportunity to ask every person listening in, whoever and wherever they may be, to pause for a moment and contemplate the events of the past few hours, and to give thanks in his or her own way. Over.

'Roger, Tranquillity Base,' acknowledged Mission Control, and Armstrong, somewhat hurriedly, began describing Eagle's footpads, adding that they did not appear to have sunk into the lunar surface at all. It was well known that Aldrin was a religious man, and it was Sunday, after all. But not until Buzz gave me a copy of his book *Men From Earth*, published in 1989, did I learn that this was a coded message for his pastor at Webster Presbyterian Church. The pastor had given him a miniature communion kit with a tiny silver chalice and wine vial, and he was silently taking communion. Using Eagle's note-taking shelf as the altar he ate the Host and swallowed the wine.

While Armstrong and Aldrin went through the laborious process of donning their cumbersome EVA suits and backpacks, Collins, as he passed overhead on his 16th lunar revolution was trying, with the help of Mission Control and much discussion of crater photographs, to pinpoint Eagle on the surface; it was known that it was at least 6 kilometres beyond the target point. But it was still within the planned ellipse, so it was finally agreed that Columbia did not need to make a plane change in order to rendezvous and dock with Eagle when it took off from the Moon.

Armstrong and Aldrin checked and rechecked their pressure suits and communications, and shortly before depressurisation of Eagle was completed Aldrin switched on the slow-scan TV camera set up in the window to watch Armstrong descend the ladder. 'The area around the ladder is in a complete dark shadow', he warned, 'but I'm sure you will get a picture from the lighted part.' It took 45 minutes before the cabin was fully depressurised, with the crew now dependent upon their backpacks for oxygen.

Eagle had been on the lunar surface for 6 hours 21 minutes when the crew were at last ready to open the hatch by pulling it inwards. Armstrong, on hands and knees, needed much guidance from Aldrin as he backed out on to the porch, his antenna and backpack snagging on the door and sides of the hatch. They were worried, with good reason, that they might disturb some of the switch settings on their way out.

It all seemed very clumsy and difficult and by the time Armstrong had got on to the porch he and Aldrin had used up 25 minutes of PLSS (prounounced 'pliss') time – oxygen and other consumables in the personal life support system carried on their backs, which in those early days would last for only 2 hours 40 minutes. Armstrong was also hampered by the fact that he was pulling a lanyard called the LEC (lunar equipment conveyor) which acted as a safety tether for his first few minutes. Aldrin's job was to pay it out as he crouched inside the hatch, and be ready to use it to pull Armstrong back if necessary.

But once Armstrong had got down to the the second step on the ladder everything was transformed. At that point he was able to pull a D-ring which caused the equipment they would need on the lunar surface to swing out from its stowage position low down on Eagle's descent stage. That was called the MESA – an acroynm for modularised equipment stowage assembly – and its contents included a TV camera which dropped into a suitable position to provide 'a knothole view' of Armstrong's first steps on the Moon.

Astronaut Bruce McCandless, now the spokesman for Mission Control, called Aldrin, still inside Eagle, to check and verify that the TV circuit breaker was in, and a moment later exclaimed: 'Man, we're getting a picture on the TV . . . There's a great deal of contrast in it, and currently it's upside down on our monitor, but we can make out a fair amount of detail . . . Okay Neil, we can see you coming down the ladder now!'

To the watching world it was like looking at the negative of a black and white photograph, with the subject so close that it was difficult to identify, but that made it all the more dramatic. Armstrong, intent on descending safely, clearly forgot all about us. He dropped from the last of the nine steps on to Eagle's landing pad:

> ARMSTRONG: Okay, I just checked; getting back up to that first step. Buzz – it's not even collapsed too far, but it's adequate to get back up . . . It takes a pretty good little jump.
> MC (McCandless): Buzz, this is Houston. F2, 1/160th second for shadow photography on the sequence camera.
> ALDRIN: Okay.
> ARMSTRONG: I'm at the foot of the ladder. The LM footpads are only depressed in the surface about one or two inches. Although the surface appears to be very very fine grained, as you get close to it, it's almost like a powder. Now and then it's very fine. I'm going to step off the LM now. That's one small step for man, one giant leap for mankind.

We had speculated for months as to whether Neil Armstrong would have something historic to say at that moment, and on the outward journey he had refused even to tell his fellow astronauts whether he had decided on anything. When the moment came this most taciturn of all the astronauts did not let us down. (He had intended to say 'a' man, but the definite article was inaudible.)

Even while I listened, spellbound, I felt deep anxiety for my newspaper colleagues. It was 9.52 pm in Texas, and 3.52 am in London. How could they ever get their story in the morning papers? But Ronnie Bedford and Angus Macpherson were letting their Editors worry about that; they and scores of other newspapermen gave up trying to follow what came next, and pored over spontaneous descriptions of the moment which led wonderful special editions which are still sold 30 years later as permanent souvenirs. I wished then, as I often did before and after, that I was still working in the printed, instead of the spoken word, for there was nothing for me to do but hope that back in London the BBC producers were having the sense to let the public see and hear what was going on without superfluous explanations from me or from astronomers and geologists packed into the 'Apollo Studio'.

My friends duly got their bylines in large capitals at the top of their stories to show to their grandchildren. What reply would I be able to make when asked by my grandchildren: 'What did you do when Neil Armstrong stepped on to the Moon, Grandpa?' (Actually they never asked, and never cared!)

But behind the hubbub of descriptive journalese around me I could hear on my headphones that Armstrong was continuing with what the flight plan called 'environmental familiarisation'. He had got rid of the lanyard by hooking it on to the ladder, and was trying out life on the Moon:

> ARMSTRONG: The surface is fine and powdery. I can pick it up loosely with my toe. It does adhere in fine layers like powdered charcoal to the sole and sides of my boots. I only go in a small fraction of an inch – maybe an eighth of an inch. But I can see the footprints of my boots and the treads in the fine sandy particles.
> BRUCE MCCANDLESS: Neil, this is Houston. We're copying.
> ARMSTRONG: There seems to be no difficulty in moving around, as we suspected. It's even perhaps easier than the simulations at one-sixth *g* that we performed on the ground. The descent engine did not leave a crater of any size. There's about one foot clearance on the ground. We're essentially on a very level place here. I can see some evidence of rays emanating from the descent engine, but very insignificant amount . . .

There had been much advance and gloomy speculation by some astronomers that on this first visit the astronauts would never be able to move far enough away from Eagle to get clear of the area contaminated by the gases of their own descent engine as it slowed the vehicle for the landing. One had forecast to me that the propellant contamination would extend for several kilometres. In fact it seemed there was no significant contamination – probably as a result of Armstrong skimming the surface before landing instead of touching down vertically. Aldrin's job was to watch Armstrong and remind him what he had to do as he reported on his ability to bend and stretch, reach down, and walk about.

Mission Control reminded Aldrin now and then that it was time to change the camera speed from 12 to 24 frames per second, and back to 12 again, and then to one frame per second. In between he watched and advised Armstrong as the latter took some pictures with a stereo camera he had extracted from the MESA and then scooped up into a plastic bag a 'contingency sample' of lunar soil in case they had no time later to find more carefully selected samples to fill the 'rock boxes', as we had dubbed the special £60 000 containers brought for the purpose. In his bulky suit and helmet Armstrong could not see his own pocket, and Aldrin, peering out of the hatch, had to tell him to push the sample in and pat the pocket shut.

Armstrong took a breather, and looking down at the meter on his chest showing the usage of oxygen from his backpack, reported: 'Oxygen is 80%. I have no flags and I'm in minimum flow.' He looked up and advised Aldrin as he in his turn began backing out of the hatch to feel his way down on to the lunar surface. 'I want to back up and partially close the hatch – making sure not to lock it on my way out,' said Aldrin. 'That's our home for the next couple of hours and I want to take good care of it.' 'A good thought,' replied Armstrong. They had an easy rapport as they worked together, and Armstrong took photographs of his companion descending the ladder. It is his pictures of Aldrin coming down the ladder and working on the Moon that the space books show, for there were to be no pictures of Armstrong himself.

As soon as Aldrin was on the surface Armstrong went back to the MESA attached to Eagle, and pulled a strap to remove a protective thermal blanket. This enabled him to remove a tripod, lift out and mount the TV camera and carry it far enough away from Eagle – 40 to 50 ft – to enable Mission Control and the rest of us to watch them at work. But before walking away from Eagle with it, he unveiled the plaque on the foot of the front landing gear, read it out, and with a fine sense of occasion described how it bore the signatures of the crew members and of the President of the United States. Turning to Aldrin he offered the still camera for a colour photo to be taken of him at the plaque, saying: 'Ready for the camera? I can –'

The only picture of Neil Armstrong on the Moon. In fact this is Buzz Aldrin, taken by Armstrong, but Armstrong himself is reflected in Aldrin's space visor. (NASA)

'No, you take this', interrupted Aldrin, who was busy preparing to pay out the cable for the TV camera. So we got no photograph of the commander unveiling the plaque.

A note in the 184-page Lunar Surface Operations Plan said at this point of ALSCC (Apollo Lunar Surface Close-up Camera) operations:

'Close-up photographs will be taken by either crewman when time is available between or during other tasks. Several times within the EVA are suggested when it may be convenient for the crew to take photos. This is not a requirement to take photos nor does it prohibit them from obtaining photographs at other times which may be feasible.' But there was no mention in the operations plan of either the plaque unveiling or of the flag raising ceremony, which used up a good many minutes of the 2 hours 32 minutes between hatch opening and closing. The omissions were an indication of NASA's wish to distance itself from the nationalistic aspects of the ceremony.

Mission Control commiserated with Collins, passing overhead again, as being 'the only person around without TV coverage' as Armstrong and Aldrin struggled to erect the flag. Of all the jobs he had to do, said Buzz, this was the one he wanted to go the smoothest, since it was being watched by a billion people. They found that below the powdery surface the subsoil was very dense, and they managed to push the pole in only a few inches. But it stood up (until, as Aldrin noticed as they lifted off, the blast from their ascent engine blew it down), and Aldrin in his own words 'snapped off a crisp West Point salute' – photographed again by Armstrong.

No matter that this event was not in the official Flight Plan, the atmosphere at Mission Control, I reported, was one of high emotion. Everyone there stood clapping and cheering. Maddeningly they also did so in the big auditorium around our Press tables, where the lunar pictures were being projected on a big screen, and which had been invaded for the occasion by hundreds of friends and relatives of NASA staff. They much enjoyed demonstrating to the foreign newsmen that this was *their* triumph, oblivious to our need to hear and report what came next. After the salute, Mission Control asked the astronauts to stand together by the flag and in view of the TV camera to receive a call from President Nixon:

> NIXON: Neil and Buzz, I am talking to you by telephone from the Oval Room at the White House. And this certainly has to be the most

historic telephone call ever made. I just can't tell you how proud we all are of what you . . . for every American this has to be the proudest day of our lives. And for people all over the world, I am sure they too join with Americans in recognising what a feat this is. Because of what you have done, the heavens have become a part of man's world. And as you talk to us from the Sea of Tranquillity, it inspires us to double our efforts to bring peace and tranquillity to Earth. For one priceless moment in the whole history of man all the people on this Earth are truly one: one in their pride in what you have done; and one in our prayers that you will return safely to Earth.
ARMSTRONG: Thank you Mr President. It's a great honour and privilege for us to be here representing not only the United States but men of peace of all nations. And with interest and a curiosity and a vision for the future. It's an honour for us to be able to participate here today.
NIXON: And thank you very much and I look forward to – all of us look forward to seeing you on the Hornet on Thursday.
ARMSTRONG: Thank you.
ALDRIN: I look forward to that very much, sir.
MISSION CONTROL: Columbia, Columbia . . . I got a B22 auto-optics pad for you . . . P22 landmark, 1D, LM T1, 110, 26, 56. T2, 11032, 06, 3 miles south. Time of closest approach, 110, 33, 40. Shaft 353.855. Trunnion, 46.495. Roll 0. Pitch 250. Yaw 0. Over . . .

Not even the President could delay Apollo 11 procedures for long. One of Aldrin's jobs was to extract the solar wind experiment from the MESA, deploy it for at least one hour, then collect and roll it up for return to Earth. Looking a bit like a screen for home movies, it had a rod which he had to drive into the surface; then he could pull up a screen of aluminium foil, making sure it was directly facing the Sun – at which, incidentally, the astronauts had to take care not to look.

This was a primitive version of much longer and more sophisticated methods of studying the solar wind in later Apollo and Shuttle flights. It was Armstrong's job to deploy what was to be the most effective of these early experiments – a seismometer package weighing 45 kg, which involved the first major use of nuclear energy on a manned spaceflight. An isotopic heater system had been built into the package to protect the seismometer during the frigid lunar nights. Its nuclear power pack attracted little attention at the time – mainly because few media people were aware of it. He also had to deploy a laser ranging retro-reflecter so that ranging beams from Earth could be reflected back to their point of origin for precise measurements of Earth–Moon distances.

But, peering at the vague, flickering images on the screen, what we most wanted to see was Armstrong filling the rock boxes. The MESA had contained tongs and a scoop to help him pick up what geologists urged should be equal quantities of soil and rocks, and Armstrong quickly scooped up enough dust and small lumps to fill the first. But by the time he had used the tongs to gather 25 more carefully selected rocks and put them in the the second box, they were 15 minutes behind schedule.

Buzz Aldrin did have the camera for a brief period while this was going on, using it in accordance with the operations plan to photograph Eagle and its condition from various angles. Since the descent stage would be left on the Moon and the ascent stage jettisoned in lunar orbit, clear close-up photos would be the only way in which its Grumman designers would be able to judge whether improvements were needed on later missions.

Armstrong had been on the surface for 1 hour 50 minutes when Mission Control gave Aldrin a 10 minute warning to start preparations to return to Eagle. At the time he was collecting core samples – the third type of lunar material called for by the geologists. He asked them to note how hard it was to drive it five inches into the ground. 'It looks almost wet,' he observed – raising hopes, which were not to be fulfilled, that there might be water below the lunar surface.

When the astronauts had been on their life support systems for 2 hours 25 minutes Mission Control began to worry and to chivvy them to collect the close-up camera, parked by Aldrin at the foot of

Eagle, and start back. It was vital to get them on their way up the steps with plenty of oxygen left, in case they got stuck trying to crawl back through the tiny hatch. Before jumping to the first step Aldrin threw on the lunar surface among all the scuffed footmarks a packet carried in a shoulder pocket containing the Apollo 1 mission patch in honour of Grissom, Chaffee and White (the last had been a very close friend), and medals commemorating the dead cosmonauts Yuri Gagarin and Vladimir Komarov. There was also a tiny silicone disk marked 'From Planet Earth' on which were recorded goodwill messages from the leaders of 73 nations.

Once at the top of the ladder Aldrin used the lanyard and a pulley to haul up the rock boxes, later found to contain 48 lb of rocks and soil – less than hoped for, but satisfactory nevertheless. Guiding them from below, Armstrong said dirt was falling all over him, and agreed with Aldrin that it was rather like soot. Finally he too climbed up the ladder and, guided by Aldrin, squeezed inside. Buzz had to tell him to move his foot before he could close the hatch; we assumed that his comment that someone had broken the hinges was a joke, for Mission Control was soon reporting that Eagle was being repressurised and the astronauts were getting their oxygen from the main spacecraft instead of from their backpacks.

Dr Berry informed us that while on the Moon the astronauts had expended only the predicted number of energy units and were 'in great shape'.

After spending two hours tidying up in Eagle and having a meal, the Lunar Module was depressurised and the hatch opened for a second time. They had taken a lot of moondust inside, despite the instructions in the operations plan to clean their spacesuits with their gloved hands, and wipe or kick their boots against the footpad before climbing the ladder – the most likely source of taking possible contamination with them. To lighten Eagle for take-off they threw out their backpacks, boots, armrests and urine bags, and the seismometer recorded the impacts and transmitted them back to Earth. 'You can't get away with anything any more,' commented Eagle when told by Mission Control.

All the hardware left behind, commented one scientist, made the scene look like a bad picnic, and 'set back the anti-litter campaign considerably'.

Mission Control advised Armstrong and Eagle to put their system in the 'caution warning enable' mode, exchanged more congratulations and thanks 'from all of us in all the countries and the entire world'. Aldrin returned thanks and added: 'It has been a long day', and we thought we too would get a couple of hours rest before covering the preparations for taking off and rejoining Columbia. But no, after telling them to 'get some rest and have at it tomorrow', Mission Control started calling them again about their faulty mission timer.

Then, when Armstrong had settled down, lying across the ascent engine with his feet in a makeshift sling, with Aldrin curled up on the floor, ready at last to try to sleep, Mission Control called up with a list of 10 questions which they began to answer. But some were complicated questions from the geologists, and finally Aldrin urged them to wait until after they had rested. Goodnights were said for the third time, and at last, after 40 hours of intense activity, they had five hours of fitful rest. Armstrong was disturbed by Earth-glare, reflected through the telescope on to his face; and both were uncomfortable with moondust smeared over them and the deck. 'It was like gritty charcoal, and smelled like gunpowder from the fireworks I'd launched so many years before on the New Jersey shore' was Aldrin's description in his book.

As soon as they were officially awake a busy three hours followed preparing for the ascent. Aldrin found time to answer some of the geologists' questions about the ground conditions on the Moon; both sides knew he was doing it just in case they did not survive the return journey to Earth. But most of the time was taken up with exchanges of data about the ascent. Mission Control repeated several times how important it was that the rendezvous radar was switched off during the ascent to ensure that they were not bothered with another series of alarm warnings from overloaded computers.

Unknown to them, two hours before they lifted off, Russia's

Lunar 15, which had been making erratic lunar orbits aimed at a soft-landing and soil recovery operation, finally crashed into the Sea of Crises on its 52nd lunar orbit about 500 miles away from Eagle. It probably ranks as the wildest spaceflight mission ever made by Soviet scientists. It posed real danger to the Apollo 11 crew, and reflected the political pressure the Soviets had been under to demonstrate to the world that their many boasts over the years had not been idle. I enjoyed reminding listeners that Cosmonaut Shatalov had said not so long ago that when Americans landed on the Moon Soviet cosmonauts would be there to greet them!

Paul Haney had suggested, before he resigned, that the TV camera, powered from Eagle's descent stage, should be set up so that it could transmit pictures of the ascent stage's lift-off. It was not possible on Apollo 11, but became a highlight of later missions. Fourteen minutes beforehand Mission Control confirmed that everything was Go and Aldrin replied: 'Roger, understand we're number one on the runway!'

Link-up in lunar orbit and flight home

Having witnessed how many thousands of dedicated specialists, plus ships and aircraft, were needed to launch men from Earth, it was hard to believe that two lonely astronauts, a quarter of a million miles away, could launch themselves into lunar orbit unassisted – except of course by Mission Control's monitoring from Earth. In Columbia, coming around on his 25th lunar revolution and the 12th since Eagle had separated, Collins positioned his spacecraft so that its radar transponder would be pointing in the direction of Eagle as it ascended. After hours of exchanges, Eagle's guidance computer was loaded with Program 12, Powered Ascent Guidance, and we heard Buzz Aldrin chanting the countdown:

> Forward 8, 7, 6, 5, abort stage, engine arm ascent – Proceed.

After 21 hours 36 minutes on the lunar surface, Eagle's ascent stage, using the descent stage as a launchpad, lifted off vertically. 'That was beautiful,' continued Aldrin. '26, 36 feet per second up. Be advised of the pitch over. Very smooth . . . A little bit of slow wobbling back and forth . . . We're going right down US 1 . . .' The ascent engine burned for 7 minutes 14 seconds and once again, as it burned, halved Eagle's weight of 10 837 lb. As it rose 25° above the lunar horizon the radars and computers on Eagle and Columbia searched for one another and locked on. At the end of the burn Eagle was in a safe lunar orbit again with a perilune of 10.8 statute miles and apolune of 53.7 miles. The rest of the mission had been rehearsed by the Apollo 10 crew, and went almost according to plan.

But not quite. While things went well, the manoeuvres were made by Eagle, the policy being always to conserve Columbia's propellant for the journey home; and an hour later Eagle circularised its lunar orbit, 15 miles below Columbia, by firing its small RCS thrusters for 45 seconds. A series of small burns combining mid-course corrections and velocity-match manoeuvres, giving Armstrong and Aldrin a busy 3.5 hours, finally closed the gap between them as Eagle chased Columbia behind the Moon on Columbia's 26th revolution. We were all ears as they emerged:

> EAGLE (Armstrong): Okay Mike. I'll get – try to get in position here, then you got it. How does the roll attitude look? I'll stop. Matter of fact, I can stop right here if you like that.
> MISSION CONTROL: Eagle, Houston. Middle gimbal. And you might put out to Columbia, we don't have him yet.
> EAGLE (to Collins): They're tight [presumably the capture latches] . . . I'm not going to do a thing, Mike. I'm just letting her hold in attitude hold.
> COLUMBIA: Okay . . .
> EAGLE: Okay, we're all yours. Roger.
> COLUMBIA: Okay. Okay, I have thrusters D3 and D4 safed . . . I'm pumping up cabin pressures.

We imagined that Armstrong had successfully rolled his tin-can of a spacecraft on to Apollo's nose so that the capture latches had connected, and that Collins had then fired his thrusters to lock the two firmly together. But this exchange followed:

> COLUMBIA: That was a funny one. You know, I didn't feel it strike, and then I thought things were pretty steady. I went to retract there, and that's when all hell broke loose. For you guys, did it appear to you that you were jerking around quite a bit during the retrack cycle?
> EAGLE: Yeah. It seemed to happen at the time I put the contact thrust to it, and apparently it wasn't centred because somehow or other I accidentally got off in attitude, and then the attitude hold system started firing.
> COLUMBIA: Yeah, I was sure busy there for a couple of seconds. Are you hearing me all right, I've got a horrible squeal.

Within minutes I was running stories about the drama of the link-up – so it was quite a shock an hour later when the official transcript of Mission Commentary did not include the phrase 'and that's when all hell broke loose' and there were denials that there had been any drama. We had suspected that unpleasant things were sometimes censored from the transcript. We had to replay our own recordings of the conversation before it was officially admitted that it had happened. We insisted on page 420/1 of the transcript being revised and reissued – and later the astronauts explained that the two spacecraft were twisting against one another as they docked, with a serious risk of jamming the docking probe. Then the only hope of survival for Armstrong and Aldrin would have been to struggle back into their full EVA suits and crawl on the outside from one craft to another – almost certainly abandoning their precious lunar rocks and other trophies.

Everything had gone so well, apart from the delayed landing, that by then the media were hungry for some drama, and not to be denied. If landing men on the Moon was easy, public interest would rapidly evaporate!

No one envied the Apollo 11 crew during the next three hours of drudgery. With a small vacuum brush they did their best to clean the clinging moondust from one another's spacesuits, helmets and gloves, the film magazines and the rock boxes; there was much travelling by Armstrong and Collins between Eagle and Columbia transferring lunar trophies one way, and everything disposable – from the bulky docking probe and drogue to the hated urine and faecal bags – into Eagle.

Now it was Columbia that must be lightened as much as possible for the journey home, ending with the fiery re-entry. While all this was going on, Columbia was pumping precious oxygen into Eagle, to make sure that if there were any moonbugs they would have to travel against the flow to get into the Command Module. With all three crew members back in Columbia – though with Collins in the commander's seat, because he was responsible for firing the main propulsion engine to start them on the way back to Earth – Eagle became the second spacecraft to be jettisoned and abandoned in lunar orbit.

It was a tired and very quiet but still efficient crew that disappeared to start the 31st revolution of the Moon. Well before that Astronaut Charles Duke had called Buzz Aldrin to pass up 'your coming home information, if you are ready to copy':

> ALDRIN: Stand by . . . All right, ready to copy.
> DUKE: Rog, 11. Got two pads for you. TEI 30 [Trans Earth Injection] 30, and then a TEI 31. TEI 30 SPS G & N 36 691 minus 061 plus 066 135 23 41 56 Noun 81 32 – correction plus 32 – 0 11 plus 06818 minus 02 650 181 054 014. Apogee is NA. Perigee plus 00 230 32 86 – correction 32 836. Burntime 2 23 32 628 24 151 1 35 7. Next three lines are NA. Noun 61 plus 11 03 minus 17 237 11 806 36 275 195 04 52. Setstars are Deneb and Vega 242 172 012. We'd like ullage of two jets per 16 seconds, and the horizon is on the 10 degree line at TIG minus 2 minutes, and your sextant star is visible after 134 plus 50. Standby on your readback. I have a TE1 31 if you are ready to copy, over.

Despite static noises which made hearing quite difficult, Aldrin read the numbers back – corrected where Duke himself had made two mistakes – without a pause. A small error, picked up almost at once, was that Aldrin had written 'sextant star 134:10' instead of 134:50. Emerging from behind the Moon for the last time, Apollo 11 had burnt off another 10 000 lb of propellant from its remaining weight of 36 691 lb.

DUKE: How did it go?
ARMSTRONG: Time to open up the LRL [Lunar Receiving Laboratory] doors Charlie.
DUKE: Roger. We got you coming home. It's well stocked.

The flight home took 60 hours, and it was so uneventful that the crew spent much of the time resting, and catching up on other world events. They learned that the people of only four nations had not been told of their achievements – Communist China, North Korea, North Vietnam and Albania. But they were not told about the story which had replaced Apollo 11 as a major source of interest – Senator Edward Kennedy's car accident at Chappaquiddick, which resulted in the death by drowning of a passenger in the car, Mary Jo Kopechne.

There were several question-and-answer periods on the way home in which the astronauts were closely questioned in attempts to identify landmarks which would enable an exact position to be fixed for Eagle's landing; there were also more 'inquests' into what could have caused the dangerous gyrations when Eagle's ascent stage redocked with Columbia, and into possible causes of water-pooling in Armstrong's spacesuit.

The astronauts gave us a colour TV show, which Collins later said was unrehearsed, for they had not acquired any affection for the TV camera and what it represented. But it included the now obligatory philosophising about the fragile Earth and the need to promote peace, with Armstrong throwing in a good look at the sealed rock boxes, and thanking everybody from Congress to the humblest technician for making the mission possible and successful.

About this time we were also reporting that twelve men were being sealed in the LRL to look after the returning crew. They included a chief steward, steward, official photographer, and of course a public information officer, to keep us provided with inside stories. He was John MacLeaish, stout and self-important, preferring to impart his precious information to his US favourites, and the man reputed to have confided to one of the crew that there was a special place in hell for autograph hunters! The astronauts had objected before departure to having a PR man locked in with them during their quarantine, and Deke Slayton had opposed it on their behalf – but had been over-ruled. Meanwhile political interest began to grow in cutting back on the cost of future moonwalks. The scientists helped out by saying that three a year would be too much – they would not have time to sort out all the information gained, and use it to plan the next one.

We were told that NASA had decided to cancel Apollo 20, planned to be the tenth moonwalk and last moonlanding, so that its Saturn 5 rocket would be available to launch Skylab, a manned laboratory, in 1972. Now that the Moon had been conquered, exploiting it already looked less attractive than moving on to something else. We filled in the three-day wait for the splashdown with stories about the huge space station – 'not so much a space workshop as a space hotel', I reported – which was to be assembled in the late 1970s. It would be able to accommodate up to 100 people, with astronauts, scientists and journalists coming and going in a space bus. I looked forward eagerly to being one of them.

A fortuitous thunderstorm in the Pacific splashdown area led to the Apollo 11 crew being given a course correction in the final hours of the return. Collins was not too happy about the change, because he had not had time to rehearse the different sort of entry procedures required if he had to take over and conduct them manually; but it meant that Richard Nixon, newly inaugurated as President, and whose goodwill NASA was going to need, would have a much shorter helicopter flight from Hawaii to the recovery carrier USS *Hornet*, to which he was hurrying to greet the returning heroes.

After the Service Module had been jettisoned, the re-entry routine and splashdown went according to plan; Collins was not required to take over from the computer and fly an unrehearsed re-entry – even though splashdown was 15 miles from the recovery ship, and we got no TV coverage of it. In an 18-knot wind and rough seas Apollo 11 rolled upside down into the 'Stable Two' position; but this crew, which probably remains the only one whose members suffered no space sickness at all throughout the mission, now also resisted sea sickness as they awaited the helicopters and the three frogmen who dropped around them and attached the flotation collars which slowly rolled them upright.

The orderly exchanges to which we had listened for 8 days 3 hours 18 minutes and 18 seconds gave way to a muddled medley as we listened to excited voices from *Hornet* talking to SWIM 1 and SWIM 2, AIR BOSS, and PHOTO 1 all reporting on their activities and advising one another what to do and not to do. It was not done to use names, so they all talked about 'The Big Swimmer'. We were advised that he was Lt Clancey Hatelberg, from Chippewa Falls, Wisconsin. As soon as the three frogmen had made Apollo 11 safe from sinking with the flotation collar, they had to move away upwind of possible moonbugs, and report on the activities of The Big Swimmer. We could just make out a voice, probably Collins', intervening:

> This is Apollo 11. Tell everybody take your sweet time. We're doing just fine in here. It's not as stable as Hornet, but almost. Over.
> HORNET: All right, we copy. Big Swimmer preparing to don suit. One swimmer in raft with full suit [garble] . . .
> PHOTO 1: This is Photo 1. The big swimmer is making adjustments to his garment. He has his helmet on – raised to his shoulders. He's trying to zip it up at this time.
> SWIM 1: This is Swimmer 1 saying Big Swimmer is [garble] at this time . . .

Once Lt Hatelberg, who for some reason did not get into his own BIG until he had been dropped into the inflatable raft, was biologically isolated, the helicopter passed three more BIGs down to him. Then Apollo 11's hatch was opened, the BIGs thrown in to the astronauts, and the hatch closed again. Then Lt Hatelberg's job was to spray the hatch and the surrounding area of the spacecraft with a special decontaminant in case any moonbugs had seized their chance to emerge during the hatch opening.

While all this was going on, the USS *Hornet* had plenty of time to steam the 15 miles to Apollo 11's splashdown point . . .

> HORNET: Swimmer 1, Hornet. Understand Big Swimmer has completed his decontamination of the Command Module, is that correct?
> SWIM 1: [garble] on the flotation collar.
> HORNET: Roger.
> PHOTO 1: Photo 1, Hornet, I passed from Pacific Chief, you are cutting out. You may be releasing your teeth too early on transmission. Over . . .

In due course we heard that Big Swimmer had communicated with the astronauts 'by visual hand signals through the hatchwindow'. I admired the way in which, through all the turmoil as they bounced around in the rolling Pacific, Armstrong, Aldrin and Collins got into their BIGs, and emerged, Armstrong first, into the liferafts. When the frogmen had trouble closing the hatch, first Collins and then the other two crawled back to help close it securely. Despite the the physical discomfort of not yet being readjusted to their normal 1*g* weight, they were determined to ensure that their lunar rocks and photographs neither sank into the Pacific, nor were damaged by sea water. Before being lifted into the helicopter, the astronauts were scrubbed down with an iodine solution by the swimmers, and in their turn the astronauts scrubbed the swimmers.

> SWIM 1: Swimmer 1, decontamination of the Big Swimmer is complete. The Big Swimmer is now decontaminating Swimmer number 1.
> HORNET: Roger . . .
> PHOTO 1: Photo 1, Hornet. We request you reconfirm that they are decontaminating raft number 1.

Apollo 11 Splashdown. Jubilation in Mission Control. From right, Robert Gilruth, MC Director; George Low, Apollo Manager; Christopher Columbus Kraft, Director Flight Operations; Gen. Sam Phillips (looking down) Apollo Program Director; George Mueller (in glasses, looking left), Associate Administrator, Manned Spaceflight. These are the 'backroom boys' who made the moonlanding possible; and in those days smoking was permitted in Mission Control. (NASA)

> HORNET: All right. Photo 1, that is affirmative. Decontaminating raft number 1. The others [garble] . . .

Mission Control in Houston, with an almost superstitious dread of celebrating too soon, waited until Recovery Helicopter One's pilot announced that he had got all three astronauts safely aboard *Hornet*. Then celebrations were unlimited. Every church in Houston was given the GO to ring its bells. Every manager and technician in Mission Control produced a Stars and Stripes, and most still observed the tradition, only just beginning to be questioned, of lighting up a large cigar. A huge replica of the Apollo 11 patch replaced their now dead screens – a patch that was remarkable for its good taste. It remained the only Apollo mission patch not to include the names of the astronauts – a decision made by Collins, its designer, to emphasise the fact that the first moonlanding was everybody's achievement, not that of the astronauts alone.

For the astronauts there were no red-carpeted greetings on the deck of the aircraft carrier. The helicopter and its occupants were lowered by lift directly into the hangar beneath the flightdeck, where the astronauts stumbled out, and through their fogged-up visors followed a white line into the Mobile Quarantine Facility (MQF). A long wait for the millions gazing at their TV sets followed, while Armstrong, Aldrin and Collins took turns to have the best shower of their lives, to shave and change into comfortable blue flying suits inside what the Americans knew as a trailer and what I had to call, so that the British would understand, a caravan.

The wait was eagerly filled by the experts in the Apollo studios, and when they were talked-out I was invited to contribute my usual analysis of the atmosphere at Houston. I included more details of the 'Earth-orbiting workshop' planned for 1972 in which astronauts would spend 52 days at a time.

President Nixon was among those who had to wait until the astronauts had cleaned themselves up, and appeared at last drawing the curtains in the rear window of the caravan, blinking in the glare, for the aircraft hangar had been converted into a vast TV studio. Nixon told 'Neil, Buzz and Mike' that he was the luckiest man in the world – 'not only because I have the honour to be President of the United States, but particularly because I have the privilege of speaking for so many in welcoming you back to Earth.' There were a few jokes and dutiful laughter when he said that when he telephoned them on the Moon the call had been charged 'collect'. He had phoned their wives and invited them to a State dinner in Los Angeles to be attended by the Governors of all 50 States – and would the astronauts come too, right after they came out of quarantine?

Apollo 11 – the happy return. The wives greet the crew – left to right: Armstrong, Aldrin and Collins – who are in quarantine for the next three weeks! (NASA)

'We'll do anything you say, Mr President, just anything, 'replied Armstrong.

More jokes followed. Did they know the result of the All-Star game? Yes, the Capsule Communicators had posted them daily.

> NIXON: Frank Borman says you're a little younger by reason of having gone into space. Is that right? [There had been much talk about Einstein's theories and whether they meant that the astronauts had arrived back at Earth fractionally younger than when they went.]
> COLLINS: We're a lot younger than Frank Borman. [Much laughter]
> NIXON: There he is, over there. Come on over Frank, so they can see you. You going to take that lying down?

> ALDRIN: It looks like he's aged in the last couple of days.
> NIXON: Come on, Frank.
> BORMAN: Mr President, the one thing I wanted – You know we have a poet in Mike Collins and he really gave me a hard time for describing new words of fantastic and beautiful . . . in three minutes up there you used four fantastics and three beautifuls.
> NIXON: Well, just let me close off with this one thing. I was thinking as you came down and we knew it was a success, it had only been eight days, just a week, a long week. But this is the greatest week in the history of the world since the creation. Because as a result of what happened in this week, the world is bigger infinitely . . . the world's never been closer together before. And we just thank you for that. And I only hope that all of us in Government, all of us in America, that as a result of what you've done, we can do our job a little better. We can reach for the stars just as you have reached so far for the stars. We don't want to hold you any longer. Anybody have a last re– How about promotions, do you think we could arrange something? [Laughter]
> ARMSTRONG (Hurriedly): We're just pleased to be back and very honoured that you were so kind as to come out here and welcome us back, and we look forward to getting out of this quarantine –
> COLLINS: Great.
> ARMSTRONG: – and talking without having glass between us.

It was a relief when Nixon called upon Chaplain Pierto of USS *Hornet* to end with a prayer of thanksgiving, and he did so with dignity and nationalistic fervour:

> Lord God, our Heavenly Father. Our minds are staggered and our spirit exalted with the magnitude and precision of this entire Apollo 11 mission. We have spent the past week in communal anxiety and hope as our astronauts sped through the glories and dangers of the Heavens . . . Our reason is overwhelmed with abounding gratitude and joy . . . This magnificent event illustrates anew what man can accomplish when purpose is firm and intent corporate . . . We pour out our thanks-giving . . . May the great effort and commitment seen

> in this Project Apollo inspire our lives to move similarly in other areas of need . . . May our country, afire with inventive leadership . . . blaze new trails into all areas of human cares . . .

It had been a long day – really two days, 20 and 21 July, 1969, that became joined together – during which most of the world's peoples felt they had participated in the first moonlanding and moonwalks. I enjoyed fulfilling the BBC's demand for my own personal assessment:

> The Moon, with its 14.6 million square miles, has become an American island: The first space base, from which man can embark upon the exploration of the solar system. The Moon ranks next to Mars in scientific interest, and that's the next logical step. We shall see NASA getting funds before long for the successor to Apollo – Project Mars. Man will be there by 1980 in my view. Russia, having faltered and failed in the race to the Moon, has already started planning her expeditions to Mars, and there's no doubt that the Seventies will be the decade in which we'll watch the Soviet–American race to the Red Planet. And it's just possible that neither will win. By then, there's certain to be a third competitor – the Chinese.

17 The Moonrocks – and Mars!

The Apollo 11 astronauts did not benefit in the same way as President Nixon from the fact that they had splashed down much nearer to Hawaii. Although the captain of USS *Hornet* had 250 miles less to travel, he did it at a slower pace so that the ship still docked there 55 hours after the recovery. But at least the time all counted against their period of quarantine.

Although their bunks in the trailer were still so small that they could not sit up in them – I had had a chance to try them myself – the astronauts could now relax. In their five-person lounge, the steward served them the strong martinis favoured by the sophisticated American, and they began to think about all the important points they would make during the 'debriefings' that would occupy their quarantine period back in Houston.

They could even check their memories in some respects by revisiting Columbia – and Collins amused himself on one occasion by writing for posterity on the wall of the lower equipment bay: 'Spacecraft 107 – alias Apollo 11 – alias Columbia. The best ship to come down the line. God Bless Her. Michael Collins, CMP.' While the astronauts had been exchanging pleasantries with President Nixon and the outside world through the observation window, the decontaminated Columbia had followed the astronauts down into the hangar, and had been 'docked' with the far-end of the trailer. Before he started work as cook and barman for the heroes, it was Steward Hirasaki's task to re-open Columbia's hatch, check that the propulsion and other systems were safely shut down, and recover the rock boxes and containers of film. After sterilising the containers, he passed them to the outside world through an airlock.

The rock boxes had become more important than the astronauts.

So, while it was in order for the now expendable astronauts to travel together, the rock boxes were flown off the flight deck to Hawaii in separate helicopters so that if one crashed all the samples would not be lost. From Hawaii each was taken in a separate military jumbo jet to Houston – accompanied by a Customs agent with authority to clear them uninspected into the United States. NASA's head, Tom Paine, travelled with the first box, so that he could supervise its opening. The packages of film taken on the Moon were also divided between the aircraft, and on the second, the rock box was accompanied by blood samples from the astronauts, taken during their medical examination in the MQF. We never heard the result of that analysis, presumably because it was so boringly 'nominal'.

For the scientists, and the space correspondents reflecting their activities, the general state of hysteria now actually rose. We were given daily bulletins on the scientists' findings as their studies of the trophies progressed, starting with delighted astonishment at the level of activity on the lunar surface, as sent back by the seismometer left on the surface. The Moon was far more active than anyone had dreamed. But that was all spoilt when a spacecraft technician quietly pointed out how near it was to Eagle's descent stage, abandoned amid the lonely rocks. Like some dying, gasping animal, it was venting its remaining fuel and oxygen, and that was the disappointing source of most of the signals!

The US Federal Aviation Authority had also warned aircraft not to fly near three observatories in California, Texas and Arizona sending up laser beams to be reflected back from the mirror left near Eagle. It was feared that the beams might affect the aircraft instruments and damage the eyes of passengers who happened to look right down them. Initially that too proved a disappointment, because Eagle's position had not been precisely pinpointed – a problem overcome on later missions.

I had filled a cardboard box beneath my feet with discarded Press Kits, and covered my desk with new ones: 'Apollo Spacecraft Cleaning and Housekeeping Procedures Manual – Revision A'; 'The Lunar

Receiving Laboratory Sample Flow Directive', also with revisions; 'Lunar Receiving Laboratory – Facility Description'; and one most valued by the tabloid newspeople, in a beautiful blue cover with a huge crew patch on the front: 'Post Lunar Isolation Menus'. This began:

> The world's first men on the Moon – Apollo 11 astronauts – are eating a variety of appetizing convenience foods during the critical post lunar isolation/quarantine period. The astronauts are in isolation/quarantine in Houston, Texas until August 14 to make certain no hostile organisms are transported to Earth from the Moon. During this period scientists looking for signs of alien organisms have demanded assurance that diet will cause no illness that might mistakenly be attributed to hostile bacteria.
>
> A large percentage of the meals served are frozen prepared foods available in the neighbourhood supermarket. Stouffer Foods, a division of Litton Industries, is providing most of the desserts, soups, and a major portion of the entrees and side dishes placed on the astronauts' dining table during the 21 days of biologically isolated confinement . . . All Stouffer foods meet or exceed NASA specifications for purity based on bacteria plate count, staph tests, salmonella checks and coliform count.

The menus, which were repeated after 11 days, looked fairly routine: fruit juice, cornflakes, eggs and bacon, and tea or coffee for breakfast; soups and sandwiches for lunch; steaks and salads for dinner, with only the occasional 'Cherry Upside Down Cake' to puzzle the un-American eye. The astronauts themselves, we were assured, had chosen the menus during pre-flight 'simulated isolation/quarantine situations'.

It was 2 am Houston time when the astronauts finally arrived at Ellington Air Force base in what Aldrin called their 'fantastic vehicle without wings or wheels', but their wives and eight children were there to greet them through the glass barrier, and take turns to talk to them through a red telephone. From there it was a one-hour road journey to the Lunar Receiving Laboratory (LRL), and only 16 days of their quarantine remained. By then we had watched on closed-circuit TV, with bated breath, the opening of the two rock boxes in the second section of of the LRL, separated by an airlock from the crew section. It was an expected anti-climax when the 48 lb of dirt and rock neither exploded nor burst into flames when exposed to the Earth's atmosphere. Dr Persa Bell gave us a running commentary of the box opening, at first in a near-vacuum, with the scientists' gloved hands reaching through portholes; and of the 20 rock samples in the second box he declared somewhat patronisingly: 'Beautifully selected, but not very well documented.' The fact that Armstrong had been hurried back into Eagle before he could complete the work for fear that his oxygen supply would not last was of no relevance to Persa Bell. In a ritual with religious overtones, crushed lunar dust from the boxes was then mixed with the food of 21 germ-free botanical species, which included oysters, shrimp, two types of fish, cockroaches, quail and 300 white mice. The release of Armstrong, Aldrin and Collins 16 days later would depend upon whether those creatures developed any ill effects, other than constipation, from their moondust meal. The scientists mostly hoped that some of the creatures would react, thus proving once and for all that life does exist outside the Earth; but I found it a sobering thought that if those cockroaches had fallen ill the Apollo 11 crew might have had to continue living in the LRL to this day.

Processing of the film and photographs, for which the TV stations agitated like spoilt children, brought its own dramas. They were subjected to a 47-hours decontamination process, involving the use of an autoclave. Despite trials for over a year, this oven had ruined a batch of test film only a few days before the moonwalk. So one small batch was tried first, and we were told that if that were spoilt the rest would be kept for three weeks until the quarantine period had expired, and then processed in the normal way. The first and only alarm occurred when Teddy Slezak, a photographic technician, opened Magazine 'S', containing, we thought then, pictures taken by Aldrin of the first man on the Moon. Black powdery moondust flowed from the box, and clung around the cassette. Aldrin had dropped the cassette when trying to insert it into a plastic envelope, and had put a note, 'Important', inside

The first moonrocks. On Saturday 26 July 1969, the first box of rocks brought back by Apollo 11 was cautiously opened in Houston's Lunar Receiving Laboratory. The gloved hand, bottom left, indicates their size. They did not explode, catch fire, nor bring back to Earth any germs – good or bad! (NASA)

it. A contamination alarm was sounded, and the medical observers outside ordered Slezak and five others inside, including the PR man, John MacLeaish, to strip at once, and spend five minutes under a shower.

The naked Slezak delayed his own shower while he collected up all the 'contaminated' clothing and sealed it in plastic bags. They were all already inside the airlock, so no further action was deemed necessary. The public was assured that at the worst 120 suspected contaminants could be locked up in isolation, but we heard of no more such dramas until, much later, Collins revealed that an attractive woman technician thought to have been in contact with moondust was installed in the bedroom next to his own. (There were in fact only three separate rooms for the crew, with everybody else accommodated in dormitories, so maybe one of the astronauts gallantly gave up his private room to her.)

Some of the space correspondents, including me, were anxious to move on to NASA's Jet Propulsion Laboratory, in Pasadena, California, to cover the arrival of the first 143 clear pictures of the Martian surface as two Mariner spacecraft, launched five months earlier, travelled past the red planet. But the astronomers and geologists were already quarrelling with one another about what the moonrocks meant, so we successfully clamoured for them to hold a news conference before we departed.

So, four days after the splashdown, we filed into the NASA auditorium to find a dozen of the world's most distinguished scientists lined up on the platform to answer our questions. They proved to be as jealous of one another as competing journalists – or for that matter, any other professional 'fraternity'. Among them were Dr Persa Bell in his role of LRL Manager, with a growing reputation as a bit of a wit at his colleagues' expense; Prof. Clifford Frondel, of Harvard; the venerated Prof. Harold Yuri, old and frail, who had argued for 20 years that the Moon was a dead, cold body; and the formidable Prof. Eugene Shoemaker, of Caltech.

We had just got started when Dr Robert Johnson, also from Caltech, rushed in, sat among us in the audience, fidgeted impatiently, then jumped up to announce that he had just found evidence of 'something like very poor Earth soil' – though so poor, he added, spoiling his own sensation, that it was difficult to imagine that it could have sustained life. His distinguished colleagues sat on the platform shaking their heads in shocked disbelief – whether at his behaviour or his conclusions we could not tell.

Prof. Shoemaker theorised that the Moon was less than 500 million years old, and that the so-called seas, including Tranquillity in which Apollo 11 had landed, had been formed by lava flows. But when he said that the evidence of volcanic activity was 'overwhelming', Persa Bell ruined the theory by commenting that it depended upon *who* was being overwhelmed. Dr Elbert King, Curator of the LRL, producing some welcome hard facts, told us that the first rock sample, after examination with the naked eye and then under a miscroscope, appeared to be granular igneous rock, suggesting a volcanic origin; the moondust varied from dark grey to brownish grey, and there were divisions of opinion among them as to whether it was carbon graphite, which would suggest that there might once have been some form of life on the Moon.

I am afraid that, like naughty schoolboys pitting their masters one against the other, we led them on with pitfall questions, and forced Harold Yuri to admit – in a tense silence that had all his colleagues leaning forward to catch his barely audible retraction – that in the face of all the volcanic activity and lava flows, he 'must consider very carefully whether he shouldn't revise' his long-held views about the Moon being dead.

My summarised broadcast on their collective findings was that the moondust consisted of tiny glass spheres, ranging from dark brown to yellow and yellow-brown. The largest found at that time was only three or four tenths of a millimetre in size, but was described as 'lustrous and reflective'. Their existence, it was pointed out, accounted for the fact that Armstrong and Aldrin had commented that the Moon's surface was 'surprisingly slippery'. Prof. Frondel offered a 'trial hypothesis' to explain the existence of the glass balls: that was that a silicous gas was created by the impact of large bodies crashing on to the Moon; when this gas condensed it turned into glass droplets, which then rained down upon the surface.

What conclusions did the scientists draw about the age-old question of the Moon's origin, now that they had examined 30–35 pieces of it? It was too early – as indeed it remains more than 25 years later – to be certain whether the Moon was a piece of Earth which broke away, or an outside body captured by the Earth's gravity. But all the moonrocks could have been found on Earth; spoiling it all again, Persa Bell said they had 'seen rocks unidentified, but not unearthly!'

One thing they did at last agree upon. They had had an exciting two days in the outer, non-quarantined part of the LRL, not at all diminished when the 90 people there had to don gas masks in a hurry and wear them for two hours after the sensitive monitoring equipment sounded a gas alarm. The gas, it was finally established, was not being generated by the moonrocks. It was the result of an overflowing urinal!

Before I left Houston I did a 'curtain-raiser' for the *Today* programme about the TV pictures of Mars we were so eagerly awaiting from Mariner 6 – 'one more step in America's longer-term plans, culminating, it's hoped, in a manned landing there around 1980'. I accompanied it with an interview with John Hodge, the Englishman from Stevenage, Hertfordshire, who had been placed by NASA in charge of the plans. I asked him how many men he proposed to send on the first, 2–3 years round trip to Mars, and how long they would stay when they got there. Perhaps it is as well that I cannot find the tape recording of his reply.

Richard Francis and his Apollo circus having all gone home, I was rejoined at the Jet Propulsion Laboratory at Pasadena by my favourite camera crew, Keith Skinner and Sim Harris, and we tried to have it both ways in our first report by pointing out that the expected pictures, taken from a 2000 miles flyby, would not determine whether life existed on Mars – but everybody expected that, together with atmospheric measurements being taken, they would show whether life of some sort was possible!

The first two nights during which we sat up in JPL watching fuzzy black and white images building up on the TV screens were rather alarming. Although the scientists expressed satisfaction with the vague monochrome shapes which reluctantly took form on the screens, they were certainly not going to fulfil the expectations I had aroused in my enthusiastic broadcasts: 'For the first time everyone can see for himself that Mars really has a huge ice-cap on its South Pole,' I explained; but on our screens the ice-cap was on the right, instead of dutifully appearing where the south should be. But Mariner 6, though working perfectly, was still half a million miles away. We sat through a second night – in NASA's space programme important developments always occurred at night – and I had to call it 'doubly disappointing'. We saw no Martian 'canals' nor the promised look at one of the two tiny moons, Phobos, in the 17 pictures that arrived; although we did see the famous W-shaped cloud (when it was pointed out!) and evidence began to build that Mars, including the ice-cap, was almost as crater-pitted as the Moon. Critical stories began to appear, and the scientists began to regret allowing us to see the monochrome pictures as they came in; computer processing added colour and much more detail, but took a couple of days, and as usual the media refused to wait, preferring a poor story now rather than a much better story later.

But on the third night we all had our reward. In two dramatic hours, between 2 and 4 am, the closest-encounter pictures flowing back to us through 60 million miles of space, had us clapping with delight as the mysteries of Mars were at last revealed. It was indeed another hostile world, pitted with tens of thousands of craters, differing from the Moon, however, in that there were scouring, sand-filled winds to erode and grind down the craters, so that newer ones superimposed on the old ones were whiter and identifiable. The so-called canals, as everybody feared, were not signs of an advanced civilisation, but appeared when seen more closely as blotchy areas, with no sign of dried-up river beds or any sort of vegetation. A bright spot, known as

Nix Olympica to generations of astronomers and long believed to be a mountain, was redesignated as a crater with a diameter of 300 miles.

For *The Eye Witness* I asked Dr Robert Leighton, in charge of Mariner's cameras for Caltech, to speculate on what it all meant. Like so many, he was a good scientist and slightly resentful interviewee, but could be pushed into making a very prescient assessment:

> TURNILL: Would it be accurate to say that what you've discovered so far is a waterless planet with a poisonous atmosphere, and that if there were any Martians they must all live underground?
> LEIGHTON: Well, that might be one way to put it. Of course it's been known to be a dry planet for a long time, because water vapour is there in very small quantitities, and the carbon dioxide of which Mars' atmosphere mainly is composed is many times more – er, all-present as compared with the Earth. So it really is a rather an inhospitable, uninviting planet.
> TURNILL: Is it a disappointment to find that it's quite such a desolate place, or do you think there may yet be hidden secrets?
> LEIGHTON: Well it obviously has hidden secrets. Whether one is disappointed or not I don't know. That is, a scientist goes to find the answer to his questions, or to ask questions, to obtain data, and then he ponders the meaning of that, and I don't think we're in a position to be elated or disappointed, no matter what we find.
> TURNILL: But its similarity to the Moon seems almost incredible, doesn't it?
> LEIGHTON: It seems quite similar to the Moon, and the more finely we look at it, the more similar it appears. I suppose many people in their fancy hoped that the fact that Mars has polar ice-caps and an atmosphere, and seasons like the Earth, would indicate that some day we would find the little green men and communicate with them, or fight with them, or do whatever is appropriate. But it's becoming more and more apparent that any kind of life on Mars is quite different from the kind of life on Earth, or must be quite rudimentary in its form.

TURNILL: So far as putting men on Mars to explore it like we're doing with the Moon, is that going to be more difficult, or easier than in the case of the Moon?

LEIGHTON: It would clearly be more difficult simply because of the distance and time factors involved. I think the question to ask is, is it sensible? We put men on the Moon, yes, and it's a tremendous feat, we were all thrilled . . . But I think we must ask: What are the functions of men in space? Are they simply to go and make footprints and plant flags, or are they to do something useful? If we had brought back buckets of diamonds I think possibly we could support a continuing programme of man-to-the-Moon. In the case of man-to-Mars, it seems to me that we must exploit the possibilities of unmanned spacecraft there before we would even think of sending a man. It isn't that we shouldn't because it would be unsafe; it's simply that there's no need to, because we can do so much more without man, it's so much less expense.

TURNILL: So your own view might be that the 1980s might be too early to send men to Mars?

LEIGHTON: I can't look that far ahead, but at the rate at which we're progressing in our knowledge, I think that by the middle 70s we will know the answer to the question whether it would be sensible to send a man to Mars. If it turns out to be a sensible thing to do, then it will be a possible thing to do within a ten-year span.

TURNILL: I think the thing that impresses me most about all this is how lucky we are to be on the Earth.

LEIGHTON: In many ways that's right. To me the views of the Earth from the Moon and from the orbital spacecraft show what a beautiful, green garden we live in. It's the only one in the solar system and it must be rare in the Universe.

TURNILL: You must be wanting to push out to have a look at Jupiter now to see if anything there resembles the Earth?

LEIGHTON: My colleague, Dr Murray, is the planetary scientist in our television team. I'm interested, but I'm more interested in the Sun, but we're not going to send a man there very soon!

Home again after six busy weeks, I was required to cover an all-too-familiar British story – the bankruptcy of Handley Page Aircraft, which had unsuccessfully tried to remain independent, in defiance of the British Government's demand that it should merge with other companies. But on Sunday 10 August, the astronauts, unscathed by moonbugs, emerged from quarantine. Dr William Carpentier, quarantined with them – how they must have been relieved that they were not confined with Chuck Berry! – reported that they had suffered 'absolutely nothing' in the way of ill effects, although all three had had direct contact with moondust when Armstrong and Aldrin removed their EVA suits on the homeward journey. Carpentier himself, as well as Hirasaki, had also had bare skin contact while removing the EVA suits from stowage bags, but had also suffered no ill effects. The mice and cockroaches also survived.

Two days later, therefore, the crew held a delayed news conference. Once more the world's TV networks paid out vast sums for the use of the new geostationary satellites, and a world-wide audience was spellbound for two hours as Neil Armstrong, Buzz Aldrin and Michael Collins told their personal story of their expedition. My role, as I watched the picture arriving at the BBC Control Room in TV Centre – where the quality was always far better in those days than the reception on domestic sets – was to act as TV critic.

At last we could see the colour film of the expedition they themselves had taken, and which had all had to brought back for processing. The diffident Neil Armstrong warmed up as the crew ad-libbed their commentary. We learned few new facts, and were grateful when, instead of dwelling pompously on the erection of the Stars and Stripes, they ran thc film fast-forward to roars of laughter to show how they had found that clownlike kangaroo hops were the best form of movement on the Moon's one-sixth gravity. Armstrong said he found it more comfortable than normal Earth-walking. That was when he told us that during all the 'master alarms' sounded by the computers during the touchdown, they were 'springloaded' to land rather than to abort as they would have done during simulations. The test pilot, used to working in seconds, shrugged off questions about having to delay the landing until only 20 seconds of fuel was left. 'It's really more than it sounds.'

No changes in procedures or the spacecraft were needed for the Apollo 12 crew due to repeat the expedition later that year, they declared – though the crew ought to be given nothing to do for their first 15 minutes on the surface except get accustomed to being there. We were all astonished at the amount of film that Armstrong and Aldrin had found time to shoot; and with Collins's help the crew had also taken a thousand still pictures.

Because no one at the time was in critical mood, none of us noticed until much later that there were no pictures of Neil Armstrong on the Moon – only pictures taken by Armstrong of Aldrin. It was not until I was searching for photographs to illustrate a new edition of my pocket-sized *Observer's Book of Manned Spaceflight* that I discovered that the only picture that existed of Armstrong on the lunar surface had been taken by himself – the one in which he was reflected in Aldrin's facepiece.

Suspicion grew that, as a result of resentment at being replaced by Armstrong as the first to step on the Moon, Aldrin had deliberately avoided taking the pictures of Armstrong specified in the Flight Plan. There is some support for this view in the extracts from Mission Commentary quoted above. Armstrong, following the military man's maxim 'Never complain, never explain,' has never publicly discussed it. Buzz Aldrin, however, has always insisted that he never thought about it at the time, being busy with other things – which is also supported by the extracts from Mission Commentary. Aldrin told me himself years later that it was only after they got back and were looking through the photographs together as they were processed that they realised there were 'few' of Armstrong on the surface.

In the subsequent debate about who was responsible for changing the order in which the two men left Eagle to descend the ladder and make history by being the first to step on the Moon, Buzz Aldrin described in his book *Men From Earth* how the tradition had developed from Gemini days that the commander stayed in the spacecraft while 'the number two man performed the EVA. Since Neil was the commander of Apollo 11 and I was the LM pilot, I figured I'd be the first one out.'

After this had been confirmed by quotes from unnamed NASA officials in the media, he then heard rumours that Armstrong would be first out because he was a civilian and Aldrin was an Air Force colonel. When Armstrong avoided discussing the issue with him, Aldrin raised it with their immediate boss – Deke Slayton. Deke, in a discussion with both of them, confirmed that Armstrong would be first – partly because he was the senior man, having come from the second group of astronauts while Aldrin came from the third; but, much more important, because the hatch was on Armstrong's side of Eagle, making it much more practical for Armstrong to leave first and return last, rather than Aldrin, in his bulky pressure suit and awkward backpack, having to manoeuvre past the commander on both occasions. Aldrin said he heard the explanation with 'a flood of relief'.

There seems little doubt, however, that the unpleasantness aroused by these issues played a part in Aldrin's subsequent descent into depression and alcoholism, and the breakdown of his marriage. But he recovered, and 25 years later was still actively supporting America's space programme, autographing his book for me with the words 'With pride in the six lunar flags and great hopes for our future'. He remarried a much younger woman who made him happy and came with him to England for occasional celebrations which I was happy to join. They included one occasion when at a dinner given in his honour I was called upon to describe and illustrate what Apollo 11 was like from the space correspondent's point of view. I touched delicately upon the question of the photographs, and hoped Buzz would interrupt to add his own comments; but he remained silent until the end, when he merely said: 'I felt like interrupting to tell you what really happened!'

The other two have fared better since. Neil Armstrong has never wavered in his preference for shunning the limelight, and disappeared for some years into academic life as Professor of Aerospace Engineering at the University of Cincinatti. In 1986 he played a major part in the inquiry into the Challenger Space Shuttle disaster. Mike Collins made a great impression with an address to Congress in which he demonstrated well-turned answers to difficult questions, such as whether the cost of space exploration was justified in a world in which there was so much proverty: 'We cannot launch our planetary probes from a springboard of poverty, discrimination or unrest; but neither can we wait until each and every terrestrial problem has been solved.'

Had he written the speech himself? When he gave assurances that he had, they appointed him Assistant Secretary of State for Public Affairs, and when he tired of that he became Director of Washington's new Space Museum. As I found when I talked to him there, the new perspective of Earth and its problems as seen from outer space, had unsettled him – as it had done many others – and he made it clear he would be moving on again after a couple of years.

President Nixon, having extracted the maximum political bonus from the moonlanding, added to the restlessness within NASA by cutting the budget, and Moonflights 18 and 19 were among the cuts. The Apollo 11 crew all decided they did not want to fly again; many top executives left the programme, either for more highly-paid jobs in industry or to return to military posts. We were sorry to see John Hodge transfer to the Department of Environment, which, because of the new perspectives provided by spaceflight, had suddenly become much more important. Lower down the scale, people like Poppy Northcutt, who had computed the subtle trajectories that had kept the astronauts safe, decided to pioneer the new generation of women's rights activists, and went off, I believe, to promote those issues in Chicago.

But the enthusiasm of the rocket pioneers like Wernher von Braun for space exploration was undiminished. Although he complained to me about the cancellations, saying that he had built a railway line to the Moon, and that it was to be abandoned after running one train along it, there were in fact still six lunar trains left for him to run.

18 Second Steps on the Moon

Apollo 12

The long-term cuts in NASA's moonlanding programme did not affect the momentum in the short term. While preparations for the second landing in November 1969 were underway, the Apollo 11 crew were required to reap the political benefits of the first by touring the world, meeting heads of state, holding news conference and giving interviews.

They visited 22 countries in 38 days. London was sandwiched between Berlin and Rome, and ill-feeling arose because the US Embassy said they would not have time for ceremonies to receive scrolls giving them life membership of the Royal Aeronautical Society, and specially struck Gold Medals from the British Interplanetary Society and the Air League. There was some outrage among the leaders of these bodies when it was suggested that they should post their awards to Houston!

When the trio did arrive they were suffering from postflight colds and sore throats, and cancelled individual TV and radio interviews. But they fitted in tea with the Queen, dinner with Prime Minister Harold Wilson, and a news conference in which we all joined. It was their 12th in 15 days, and the first in English. It was clear that Neil Armstrong especially hated being asked the same inevitable questions even if they were posed in different languages, but as usual the Russians helped us out by sending up three Soyuz spacecraft containing seven men. In overall command was Vladimir Shatalov, whose docking manoeuvres ten months earlier now appeared to be rehearsals for completing them on this mission. (Shatalov, perhaps the least flamboyant of the cosmonauts, later became a General and head of cosmonaut training.)

Armstrong admitted being 'quite impressed' but all of us were baffled when no dockings took place. The 'Chief Designer' – the Soviet leaders still did not permit publication of Sergei Korolov's name – argued that these 'group manoeuvres' were a necessary command and communications exercise, in preparation for the 12-person space station being planned. The most dramatic event was when the first crew in Soyuz 6 sealed themselves in their re-entry module, depressurised what was called their orbital workshop, and by remote control carried out some molecular cold welding of future space station materials. We had expected the welding to be done in what the Russians called 'raw space', but the same effect was achieved in the depressurised module.

Although I am afraid I called this mission a 'flopnik' on TV, it was welcome to the American space industry. On a brief round-trip to Seattle and Los Angeles as a guest of Boeing in early November 1969 to see their supersonic progress, I visited North American Rockwell (later called just Rockwell) at Los Angeles. With more than enough hardware completed for all the planned the Apollo missions, they could only stay in the space business if money was forthcoming for a space station. Dr Ian Dodds, a British expatriate from Yorkshire who had begun his career as an observer in Buccaneer bombers, gave me a good story for *From Our Own Correspondent* with his designs for space stations which could be clipped together in orbit in 12-person units. They would be dependent upon developing a Space Shuttle to be launched by a piloted, re-usuable rocket, which would break away and return to Earth after sending the Shuttle into orbit. That was the original Shuttle concept, in the long run a much more cost-effective vehicle, which was ultimately abandoned in favour of the present cut-price and expensive-to-operate Shuttle system.

At that time Dodds had America's space future all worked out, and after the success of the first moonlanding one could believe anything. From his Shuttle-built space station, manned expeditions to Mars would set out on journeys lasting only six weeks, because the vehicles would achieve high speeds by ejecting small nuclear bombs, weighing only 2 lb each. They would explode 200 ft behind the spacecraft, thrusting it forward when the blast hit a large 'pusher-plate' at the rear. It would be slowed down by reversing the process when half-way to Mars. The technology was available, but international agreement would be necessary because the system would break the treaty banning nuclear explosions in space.

Thus we all knew, or thought we knew, exactly what were the long-term targets when the countdown began in mid-November for the launch of Apollo 12.

Once again NASA was aiming at what seemed to be the impossible. Despite missing the touchdown point by 4 miles on the first landing, they felt they had learned so much about the unsettling effects on the spacecraft's orbit of the Moon's 'mascons', undocking procedures and urine dumps, that this time they could bring Apollo 12 down so close to the unmanned Surveyor spacecraft which had soft-landed in the Ocean of Storms 31 months earlier, that the astronauts would be within moonwalking distance of it.

I was personally unhappy about this launch because my masters insisted that I cover it from the BBC's New York studio; but I was now so fully briefed, and had been travelling so much on both Apollo, Concorde and other stories, that I had to comply. And on the day I was at least spared the drenching of cold rain endured by the journalists and VIPs who did assemble at the Cape. They included, for the first and only time on an Apollo mission, the President – and it was probably his presence, as well as the knowledge that a postponement might mean a delay for a whole month, that led to the launch going ahead through a rainstorm. At that time the rules permitted launch through rain, but not into a thundercloud. The weather was deteriorating, but appeared to be within the flight rules for an on-schedule launch at 11.22 am, and the Launch Director gave the go ahead. At T+36 seconds two parallel streaks of lightning flashed between the spacecraft and the launchpad, and again 16 seconds later. The Commander, Charles ('Pete') Conrad, called: 'We just lost the [stabilising] platform, gang. I don't know what happened here. We had everything in the world dropout!' Later he explained: 'I knew that we were in the clouds, and although I was

A cheerful Tom Paine, right, newly-appointed NASA Chief, holds an umbrella over President and Mrs Nixon as they await the launch of Apollo 12. Left, Tricia Nixon. (NASA)

watching the gauges I was aware of a white light. The next thing I noted was that I heard the master alarm ringing in my ears and I glanced over to the caution-and-warning panel and it was a sight to behold.'

Had he had time to do so, Conrad would probably have been justified in aborting by firing the escape tower; but the spacecraft automatically switched to a backup power source, and the crew soon restored primary power, checked all the systems, and found that there was no permanent damage. The computer's memory had retained its vital knowledge of the trajectories and course corrections that lay ahead. Whether Apollo 12 had actually been struck by lightning was hotly disputed; but finally it was accepted that it had suffered two 'electrical discharges', that a catastrophe had been narrowly avoided, and that much severer rules must be introduced for the future to guard against launches through electrically charged clouds.

This time, as part of the plan to achieve a pinpoint landing, a 'hybrid trajectory' was flown. Unlike earlier Apollos, the spacecraft was not placed on a course that would mean that if it failed to enter lunar orbit it would skim around the Moon at a distance of 2100 miles and automatically return to Earth. To avoid too much manoeuvering on arrival, the new course would take it within 69 miles, and then, if the rocket engine failed, it would miss the Earth on the return loop by 56000 miles – leaving the crew dependent for survival upon using the Lunar Module's descent engine.

The all-US Navy crew of Conrad, Dick Gordon and Alan Bean switched on the colour TV camera inside Apollo 12 so that for the first time we watched a crew, irretrievably *en route* to the Moon, conducting the major engine firing that took them into this unique trajectory. An error of only 1 foot per second in their speed could mean their touchdown would be almost 2 miles from Surveyor 3 – much too far for them to walk.

As they concentrated on their checklists and countdown to an 8-second firing of the main engine we noticed they were all wearing Navy-type peaked caps of the sort used on aircraft carriers – Conrad's idea because both he and Bean were bald, and caps would protect their scalps as they scrambled in and out of the docking tunnel leading to the Lunar Module. As an all-Navy crew they had decided to use Yankee Clipper and Intrepid as the call-signs for the Command and Service Modules when they separated in lunar orbit.

Once more there was a Soviet space story to enliven their long cruise to the Moon, which took 10 hours longer because of the hybrid trajectory. It was not the sort of story the Russians liked, however, for my friend Robert Hotz, of *Aviation Week*, had pieced together the saga of how Russia's giant booster, bigger even that Saturn 5, had exploded on the launchpad, causing many casualties, and explaining why Soyuz 6, 7 and 8 had done so little when they were all in orbit together. Hotz, as usual, shared the story with me.

Lunar orbit insertion, separation of Yankee Clipper, and the descent of Intrepid, bearing Pete Conrad and Alan Bean, all went so well that it seemed as if Americans had been landing on the Moon for years. Despite the fact that Conrad expressed concern about the landing because the descent engine's blast created a much bigger dust storm than on Apollo 11, he had a safety margin of 58 seconds of hovering time left when he cut the engines and Intrepid settled in the Ocean of Storms, 1300 miles west of Apollo 11's landing site.

As for Surveyor 3, 'when we pitched over just before the landing phase, there it was!' said Conrad. 'I manoeuvred around the crater, landing at a slightly different spot than the one we planned. In my judgment, the place we had prepicked was a little rough. We touched down about 600 ft from Surveyor. They didn't want us to be nearer than 500 ft because of the risk that the descent engine might blow dust over the spacecraft.'

He had made it all look so easy that the media immediately began to get bored with this incredible story! The Ocean of Storms had been picked for this mission because it was thought the surface was covered with debris splashed out when Crater Copernicus, 250 miles away, was created. And it certainly was dusty. This time Conrad and Al Bean stayed on the surface for 31.5 hours, during which they made two 4-hour moonwalks, with a rest in between. Despite the discomfort, they rested in their EVA suits because they feared that if they took them off and put them on again the dust would get into the suits' working parts.

Pete Conrad and Al Bean, whose courage and determination later saved the Skylab missions, enjoyed their moonwalks, but were somewhat boisterous, with the result that Conrad became the first man to fall on the Moon, and when helped up by Bean remarked 'It was no big deal.' Worse was to come, however; when they first erected the colour TV camera it was at an awkward angle, and when Bean was asked to move it he accidentally pointed the lens at the unforgiving Sun – and once again we got no worth-while TV during their spacewalks.

They made up for the loss with 'a stream of chatter' to enable

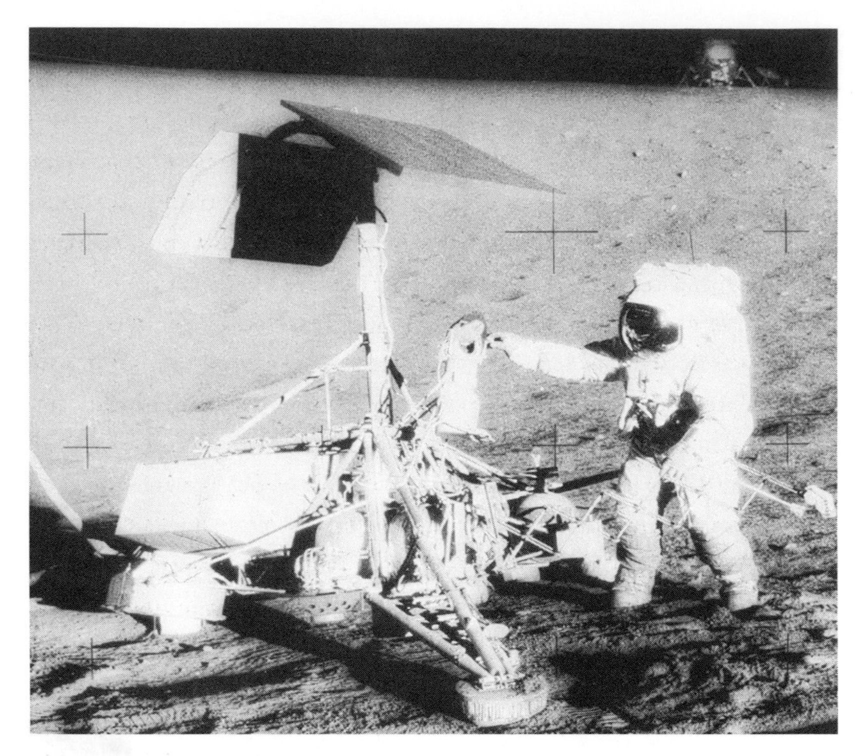

Apollo 12: perhaps the most remarkable, though little known, lunar picture. After a 579 365 km journey, Lunar Module Intrepid, just visible at top, touched down as planned 183 metres from the robot Surveyor 3, soft-landed 31 months earlier. Astronaut Bean photographed Pete Conrad removing the TV camera for return to Earth. (NASA)

Mission Control and the scientists to follow what was going on. A strange phenomenon we would all have liked to see was a group of conical mounds, which Conrad said were like small volcanoes, about 5 ft tall and 15 ft in diameter at the base. They deployed ALSEP, a much more advanced Apollo Lunar Surface Experimental Package than Apollo 11's, with reflectors and seismometers and a nuclear-powered battery, and collected a total of 70 lb of soil and rocks, and what they now recognised as 'glass beads'. On their second excursion they walked over to Surveyor 3, picking up samples on the way and rolling a rock into a crater to see if the seismometer would pick up the noise and vibrations – which it did.

Surveyor 3 had made a rough landing when it touched down in April 1967. First there was a 35 ft leap, and then an 11 ft bounce before it settled, slightly tilted, but still upright on its footpads. After that it transmitted 6000 TV pictures, and dug a trench with an automatic scoop. The task set Conrad and Bean was to photograph evidence of these events, and find out what sort of erosion processes were at work on the Moon. Had time filled in the little trench, despite the lack of wind?

Surveyor, glistening white when it was launched two years earlier, was now looking dowdy with a covering of dust – which it was decided later had been kicked up by Intrepid's descent engine. So that scientists on Earth could have practical evidence of the effects of exposure to the harsh lunar environment, Conrad and Bean cut samples of aluminium tubing and electric cables, broke off a piece of glass, unbolted the TV camera, and broke off the sampling scoop which had dug the small trench after landing – precious samples amounting to 15 lb of Surveyor's hardware.

As they prepared to start back to Earth after an extra day in lunar orbit to take pictures of possible landing sites for Apollo 13, I reported that Conrad and Bean seemed to have been dogged by back luck with their picture-taking activities. Bean, despite his anxious inquiries, was not told that his accidental destruction of the TV camera had turned the moonwalks into a disaster for the advertisement-loaded TV networks and had also been a considerable disappointment for the watching scientists.

> Then, among the litter they discarded on the Moon, they left behind the magazine of black and white pictures they'd taken of their approach to the landing site. And now Conrad has had to confess to Mission Control that they've just knocked a magazine off the camera they've been using to take close-up pictures of Craters Descartes, Lalande and Fra Mauro – highland areas right in the centre of the Moon – for use in choosing a landing place for Apollo 13 in March. That magazine contained 200 pictures, and Conrad thinks they've probably all been ruined. So there was some rush work with

> the computers before Yankee Clipper disappeared behind the Moon on its 43rd orbit so that they could take the pictures all over again as they came round once more.

On their 70-hours' flight back, the crew were shocked into uncharacteristic silence by the news that their three pretty wives had thrilled the media by parading outside Conrad's house near Houston in stunning white panty suits each carrying a sign reading 'Proud', 'Thrilled' and 'Happy'. They passed rapidly on to inquiries as to whether the scientific equipment they had deployed on the Moon was operating satisfactorily.

Like the rest of us, they were intrigued with the result of the decision, when Intrepid had been jettisoned after they had redocked with Yankee Clipper, to crash Intrepid under power on to the lunar surface instead of just leaving it to decay in lunar orbit. The ALSEP seismometers showed that it had set the Moon 'ringing like a bell'; it took seven minutes for the reverberations to build up, and they continued for 55 minutes. Possible explanations advanced by the geologists were that the impact started avalanches and rock slides in many of the craters, and that rocks and soil thrown upwards took a long time to settle in the one-sixth gravity. They decided there and then that on future missions the Saturn 4B upper stage, normally sent off into solar orbit after placing Apollo on course, would also be crashed into the Moon, with an impact 10 times that of Intrepid.

Discussing this between sleep periods, Alan Bean suggested that hollow core tubes should be taken next time to get much deeper core samples:

> You're going to want to up the number of core samples so that you could get down in these areas you're interested in to find out what's going on under there, because it's covered with this layer, and there just ain't no way to figure it out. Before the EVA, during the EVA and afterwards, we've talked about it and thought about it, tried to get the big picture, tried to be more than rock collectors and picture takers, and believe me we worked at it. And I think that by training

> we were pretty doggone good at getting that sort of – not just grabbing a few rocks, but tried to evaluate the picture we want to evaluate; but it just was difficult to do, because the clues just aren't right laying there on the surface. It's got this big blanket of all beat-up soil over every single thing.

After splashdown this crew was again quarantined for 21 days, but precautions against possible moonbug infections were much relaxed. When the hatch was opened after an uncomfortable 45 minutes in a rough sea, the biological isolation garments that gave the Apollo 11 crew so much trouble were replaced with lightweight coveralls and respirators. And plans were announced that, to save £40 million per mission, techiques would be rehearsed so that from Apollo 13 the spacecraft could be re-used. But Apollo 13 was to put an end to all such plans.

19 The Thirteen Story

Apollo 13

Apollo 13 was the perfect media story: three astronauts stranded in a crippled spacecraft.

All the newspapers had long since gone to bed, so it fell to me to break the news to the BBC's listeners on the morning of 13 April 1970. Two hundred thousand miles from Earth and beyond recall, Apollo 13 was being accelerated by lunar gravity towards the Moon. Although uninjured, the three-man crew faced near-certain death.

For the superstitious there had been lots of 'omens' – warning of what was to come on Apollo 13 – although of course they were only recognised afterwards. The first, a story which I was sorry that I missed, indicated that NASA had still not learned all the lessons it should have done from the Apollo 1 disaster about the dangers associated with handling pure oxygen. It was a strange accident worthy of a science-fiction film. The launch vehicle and spacecraft had been erected on the launchpad, and by the end of March 1970 lift-off was less than three weeks ahead. In the early morning the necessary 'chill-down' procedure of the liquid oxygen system was started, and it involved pumping 39 000 litres into a drainage ditch outside the perimeter fence surrounding the launchpad. Normally, sea breezes quickly dispersed it as it vaporised into invisible gas, but on that morning there was no wind.

Three carloads of security and other workers drove to the pad area and parked. The driver of the first car got out and talked to those in the second car until radio instructions came that they should go to the slidewire area. The first driver returned to his car and switched on the ignition. There was a loud pop and it burst into flames. Men in the other cars ran to the rescue, leaving their engines running – and one after another their cars too burst into flames. The guards ran for cover and called the fire brigade, who decided to wait until the oxygen cloud had dispersed – by which time there were three burned-out vehicles and some very shaken workmen. Dumping arrangements were changed, but there were also problems with the fuel lines when loading and unloading the oxygen tanks in Apollo 13's Service Module. These were solved by switching heaters on and off, and it was decided there was no need for further action. And that was the mistake that was to involve NASA in its greatest drama.

There was also an early setback in my personal plans to make this third moonlanding mission rather special. It was to be the occasion of Margaret's first visit to the United States. She had been accompanying me on my European travels for years, because whenever I could we had gone in the car. I had insisted that this time I must once again cover the launch from the Cape and the mission from Houston. We had scraped together the money for Margaret's air fare, got her accredited as Press to accompany me, and we were looking forward to sharing the translatlantic flight to New York and Miami five days before Apollo 13 lifted off.

Then came a Soviet intervention in a new form. After years of haggling, they at last made a 12-mile-wide corridor available for airlines to cross Siberia in a northerly arc to Tokyo – 1500 miles shorter than the North Pole route, in return of course for long-sought improvements in their own Western routes. Japan was to be the first to use it, and I was a guest aboard their inaugural DC8, peering down at hundreds of miles of barely habitable tundra, and at the snow-covered mountain range separating Soviet territory from China's Gobi Desert. It was hard to understand why the Soviet Government made such a fuss about anyone flying over this dreary land, and even more their insistence that passengers must not take photographs through the windows.

The route was still not the shortest possible, for it was carefully planned to keep the passenger jets well away from military areas; all the same the Japanese captain told me when I joined him on the flight deck that Soviet flight controllers were quick to call him if he strayed towards the edge of the permitted corridor. In Japan, I expected to make a quick visit to 'Expo 70' at Osaka, and then return to London just in time to join Margaret on our flight in the opposite direction. But my filmed and telephoned radio reports were overtaken by events. Nine militant Japanese students wielding ceremonial samurai swords took over a packed Japan Air Lines (JAL) Boeing 727 *en route* from Fukuoka to Tokyo and demanded that it be flown to North Korea.

It was one of the first, and still one of the most dramatic, of all hijackings, and for the next four days all my previous plans were abandoned in favour of covering the drama. Foreign Duty Editors kept ordering me to pursue the plane first to Seoul Airport in South Korea, and then to Pyong-yang in North Korea. They were so convinced that I had obeyed these orders that some of my scripts were introduced by newsreaders as coming from Seoul. But knowing I would get bogged down with visa applications and get stranded for days, out of touch with the story, and unable to cover that or get back to London on time, I stayed firmly in my Tokyo hotel room. There I had the benefit of NHK television which was far ahead of the BBC when it came to spontaneous 'outside broadcasts', of regular briefings from JAL's English-speaking female public relations director – and best of all, direct dialling on good-quality phone lines via the new Pacific cable, which was a miraculous improvement upon the telephone lines then available in Britain.

By the time the 99 passengers had been released, and the hijacking students had become heroes in North Korea, I could just get to the Cape in time to start coverage of Apollo 13. But it meant that Margaret had to fly westwards alone on her first transatlantic crossing, while I flew eastwards across the Pacific and the United States. Miraculously we rendezvoused in New York at the hot and rather grubby Bedford Hotel, with an ancient, dust-filled air conditioner protruding out of the bedroom window, ready to separate from its rusting fixtures and fall 14 stories into the street below. I started my Apollo 13 coverage from the BBC New York office, then in the International Press Center.

A postponement was threatened because Jeffrey Lovell, 4-year-old son of the Commander, Jim Lovell, had developed measles, and so had Charles Duke, one of the backup astronauts. Not only Lovell himself, but his two crew members, Ken Mattingly and Fred Haise, were feared to be in danger of developing the disease during the mission. Measles on the Moon? It was unthinkable.

The astronauts' jet trainers were used to ferry blood samples from the crew between the Cape and Houston, where, to the indescribable fury of the crew, our old friend Dr Chuck Berry – who had almost but not quite left NASA – would once again make a 'Go/No Go' pronouncement after carrying out 'hemagglutination inhibition tests for immune antibody to rubella virus'. As usual when faced with a crisis NASA had to break its own rules. We had always been assured that crews would not be split; that if one member fell ill, the backup crew would replace all of them. Instead of that, it was suggested that Lovell and Haise had had measles during childhood, and should therefore be immune to this new infection. But Ken Mattingly, the Command Module pilot, due to remain in lunar orbit while the other two became the fifth and sixth men on the Moon, had never had measles, and thus had no immunity. Chuck Berry's recommendation was that he alone should be replaced by his backup, Jack Swigert. NASA put off making a final decision.

Before leaving New York for the Cape I went to see Dr Kenneth Franklin, at the Hayden Planetarium. He was a watchmaker/astronomer, and had invented a Moon watch which he claimed would solve the problem for future Moondwellers of how to tell the time on a body with a night-and-day cycle of nearly 30 Earth-days. With his help I tried to explain it on our *Today* programme:

> Dr Kenneth Franklin has got it all worked out. What we call a day on Earth – the period between one sunrise and the next, and lasting nearly 30 days on the Moon, is to be called a 'lunation'. That's broken down on the dial of the prototype watch he showed me into 30 'lunes', each equal to about 24 Earth-hours, or one of our days. These lunes are then divided again, not into our old-fashioned minutes and seconds, but into centilunours and millilunours –

which obviously are respectively hundredths and thousandths of a lunour.

The proposed Moon watch uses the conventional three hands to show these readings, plus a little window which, instead of showing the day of the month, shows the moon lune.

So this is how it'll work: our flight plan shows that the Apollo 13 men, German-measle permitting, will land on the Moon at 3.55 am on April 16. Now check your Moon watch: they'll be landing at 45 centilunours past 22 lunours in the 9th lune of the 585th lunation. I ventured to suggest to Dr Franklin that telling the moontime will need a bit of practice. 'Astronomers will soon get used to it,' he said, and then added: 'But quite frankly I'm not used to it myself yet.'

In all the years since, I have never heard the Moon watch mentioned again. Evidently it is taking the astronomers longer to get used to it than Dr Franklin expected.

When Margaret and I arrived at the Cape, we checked in to my favourite Crossway Motel, a safe 3 miles from the Ramada and Holiday Inns, which were already centres of international media hysteria.

Jim Lovell continued fighting a rearguard action, loyally seeking a month's postponement so that he could keep his original crew together. As the first astronaut to make a fourth spaceflight he argued from a position of seniority. But the recovery forces, consisting of 4300 US servicemen, 47 aircraft and five ships, continued to spread themselves around the world.

Lovell assured us that none of his crew was superstitious about being launched on 13 April, which in fact his Italian forebears regarded as a lucky day. And nothing, it seemed, could stop him being launched on his 'lucky day'. Only 24 hours before that his pleas for a postponement were finally over-ruled by the amiable NASA Administrator, Tom Paine – to the great relief of everybody except Mattingly and Lovell. One argument in Lovell's favour had been that Swigert was a large man weighing 60 lb more than Mattingly, and possibly less

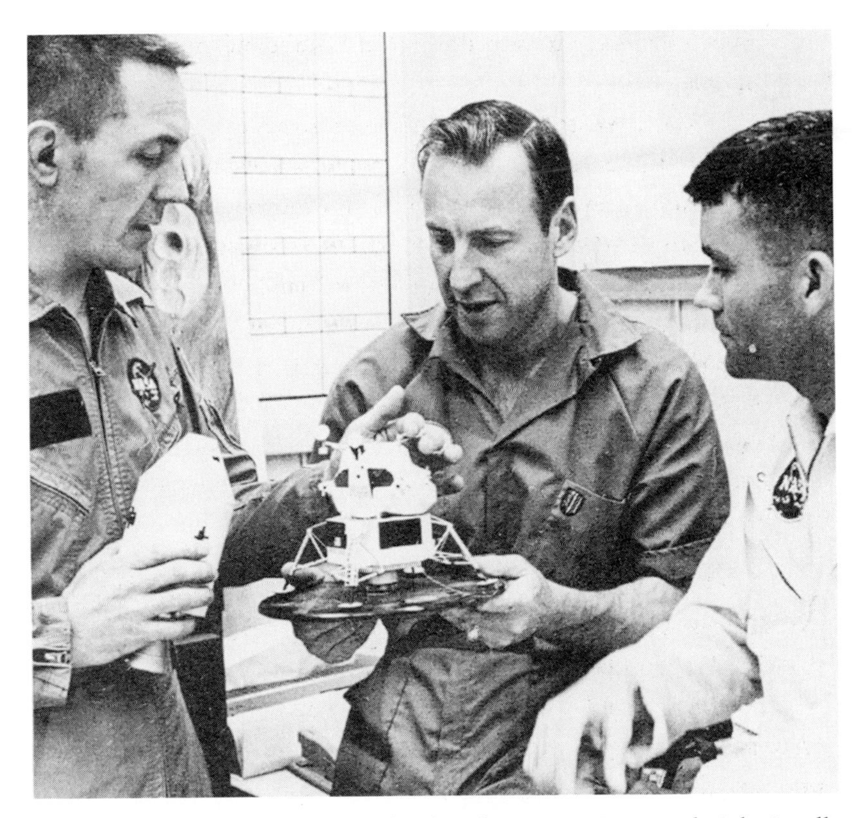

Apollo 13 Command Module pilot John Swigert, Commander John Lovell, and Lunar Module pilot Fred Haise in the Astronaut Training Facility before launch. They are examining a model of the Lunar Module in which Lovell and Haise expected to land on the Moon. They did not expect it would serve as a lifeboat to tow them back to Earth when their Command Module was crippled by an explosion. (NASA)

nimble when twisting and turning in the spacecraft to reach the switches. Other factors that weighed against Lovell were that another delay in this mission would cost NASA about $700 000 at a time when their budget was being cut, and that the Saturn 5 rocket would have stood on the sea-shore launchpad exposed to weather and corrosion for six months, longer than was considered advisable. Deke Slayton, Chief Astronaut, told us in his dry way that everybody was going to feel a lot better if Mattingly did develop measles, and the beleaguered Chuck

Berry added 'Amen' to that. But Mattingly remained obstinately measles-free, and as it turned out was more than rewarded two years later with a place on the much more successful Apollo 16 mission.

The total cost of this mission, I reported, converted from dollars to pounds, would be £155.5 million. Broken down, that was £77 million for the Saturn 5, £23 million for the Command Module Odyssey, £16.5 million for the Lunar Module Aquarius, £10 million for the onboard experiments, and £29 million for operations. I never discovered how many hundreds of millions the impending disaster would add to that.

Apollo 13's launch was by then scheduled for 13.13 hours local time; we did not fail to ask Marilyn Lovell, an established favourite with the TV networks since Lovell had immortalised her on Apollo 8 by naming twin lunar peaks 'Mount Marilyn', whether she was worried. 'I'm not a bit concerned,' she assured us. Launch was on time. At 13.13 hours on 11 April I was in my accustomed front row seat of the Press Stand, A13, to watch the still awe-inspiring blast-off. As on Apollo 11 the dazzling gold flame from Saturn 5's five engines was two-pronged. But for once von Braun's creation behaved less than perfectly, perhaps because of long-exposure to the salt winds. There was dismay in Mission Control when the second stage centre engine shut down two minutes early – an eternity in rocketry. But the onboard computer-brain, responding thousands of times faster than a human could do, kept the other four engines burning 30 seconds longer to depletion, and then kept the Saturn 4B third stage burning an extra 30 seconds. Thus the necessary 17 500 mph needed to enter Earth orbit was still achieved.

There was anxiety that the 4B would not have sufficient fuel left for the later 6-minute burn needed to increase speed to the 24 000 mph required for lunar orbit insertion; but Werner von Braun's generous fuel margins as always proved equal to the occasion, and three Earth orbits later Apollo 13 had started on its 60-hour journey to the Moon.

Marilyn Lovell followed the official post-launch news conference with one of her own – an occasion which has remained unique so far as astronauts' wives are concerned. It took place at the Cape's Press Center, and she was flanked by her two sons and two daughters. Their ages ranged from three to 16, and two were so small that only their eyes showed above the top of the table normally graced by NASA's stop brass. But they all enjoyed answering our questions, and Marilyn declared herself 'selfishly delighted' that her husband had announced that this fourth mission would be his last. I asked her what she thought of the decision that the Apollo 13 crew would again have to undergo three weeks' quarantine when they got back, despite previous findings that there was no lunar life likely to contaminate the Earth. 'I'm sure they have their reasons', she said, 'but personally I think it's ridiculous!' It contrasted pleasantly with the broadcasts I was doing about 'the hazardous landing in the uplands area known as Fra Mauro' – which was named after a fifteenth-century astronomer monk – where Lovell and Haise planned to collect moonrocks five billion years old, to take bore samples 10 feet deep, and climb a rim crater several hundred feet in height. For the next two days everything went so smoothly that media interest began to wane.

Margaret and I transferred ourselves from the Cape to a motel just outside Mission Control near Houston, and for the early evening news bulletins on 13 April, 45 hours into the flight, I reported that Mission Control, listening to soft music being played in the Command Module, commented to Jim Lovell that they thought 'he was running a rest home'. Lovell, comparing the journey with his Apollo 8 flight to the Moon, when the crew did not catch a glimpse of the Moon until they actually arrived, said that this time the spacecraft was rolling smoothly every 10 minutes in 'the barbecue mode' and on each revolution he was getting a good look first at the Earth and then at the Moon. 'A pretty nice merry-go-round' he observed with evident satisfaction.

Having done my late-night pieces for TV News and for the 6pm and 9pm radio news bulletins, Margaret and I drove 10 miles to the River's Edge restaurant at Clear Lake, where as usual Red Adair, famous for fighting fires on oil rigs, was at the bar in a huge Texas hat and high good humour. After enjoying a splendid steak and Californian wine, the routine was to call briefly into the News Centre to make sure that all was well, record a progress report for the BBC's 7 am bulletins, and then go to bed. (By the time one got up again, the 1pm news was hungry for another update.)

The mission was still progressing so routinely that there was only one other reporter in the Press area. The crew was just finishing a 49-minute TV transmission which included views of the Moon through the windows and a demonstration by Fred Haise from the Lunar Module 'Aquarius' of a new mobile drink bag for use on the lunar surface, because the Apollo 12 moonwalkers had got very thirsty. As Haise was seen showing how difficult it was to get into his hammock, Lovell joked that sleeping and eating were Haise's favourite pastimes.

> CHARLIE DUKE, Mission Control Capcom (already recovered from his measles): Okay Jim. It's been a real good TV show. We think we ought to conclude it from here now, but what do you think?
> LOVELL: Roger, sounds good. This is the crew of Apollo 13 wishing everyone there a nice evening. We're just about to close out our inspection of Aquarius and get back for a pleasant evening in Odyssey. Good night.
> DUKE: Thank you 13.

Writing about it later, Lovell commented: 'On the tapes I sound mellow and benign, or some might say fat, dumb and happy. Nine minutes later the roof fell in.'

Those nine minutes were occupied with Duke passing routine instructions about terminating battery charges and checking signal strengths. I cannot imagine now why Margaret and I waited so long; but we were turning to leave when it happened:

> DUKE: 13, we've got one more item for you when you get a chance. We'd like you to stir up your cryo tanks. In addition we have shaft and trunnion for looking at Comet Bennett if you need it.
> HAISE: Okay, stand by.

It was 55 hours 55 minutes into the mission. Apollo 13 was 204 700 statute miles from Earth, free-falling towards the Moon at 2224 mph.

Margaret Turnill and the author broadcasting from Houston. (John Peterson)

HAISE: Okay, Houston – Hey, we've got a problem here.
DUKE: This is Houston, say again, please.
LOVELL: Houston, we've had a problem. We've had a main Bbus interval. [An undervoltage warning on the CSM's main bus B.]
DUKE: Roger. Main B interval. Okay, stand by 13 we're looking at it.
HAISE: Okay, right now Houston the voltage is looking good. And we had a pretty large bang associated with the caution and warning there. And if I recall, Main B was the one that had an amp spike on it once before.

Thankful that I had not left for bed five seconds earlier, I hurried to my desk. The words we heard were calm, but they had an edge to them. I knew at once that this was the start of the sort of drama in space that media newsdesks dreamed about. In accordance with our unwritten agreements, I told Margaret to ring Ronnie Bedford and Angus Macpherson to advise them to dress at once and come to Mission Control. Both refused, saying there was nothing they could do until their next editions, nearly 24 hours later. Bedford had taken two sleeping pills. 'Ring them again,' I commanded, 'and tell them they MUST come.' They did, and arrived grumbling – but thanked me later.

Listening to the rapid exchanges – controlled but urgent – between the crew and Mission Control, I was thankful that London was still asleep, and that I had a couple of hours to listen quietly and try to understand the developing crisis. The problem for the astronauts and Mission Control, of course, was that until they understood what had gone wrong they could not prescribe a remedy – or know whether there was a remedy. Houston had not noticed that as a result of the 'bang' – in fact the explosion of one of the two oxygen tanks housed in the centre of the Service Module – there was a 1.8 second break in the flow of telemetered data.

The explosion had ripped away one of the SM's protective panels, and that damaged the high-gain antenna mounted outside it. But transmissions switched automatically from narrow beam to wide beam width. A confusing factor was the repeated automatic firings of the attitude control thrusters, misleading Mission Control into thinking that perhaps they were malfunctioning. In fact, they were trying to overcome the effects of the explosion on the spacecraft's attitude. Jim Lovell, as he countered a whole series of further master alarms, watched his warning lights blinking and gauges falling relentlessly towards zero, and hoped that the readings were due to instrument malfunctions.

'Houston, are you still reading?' called Haise at one point, clearly fearing that communications would fail with the fuel and oxygen supplies.

'That's affirmative. We're reading you. We're still trying to come up with some good ideas here for you.'

'It looks to me, looking out of the hatch,' interjected Lovell, 'that we are venting something – we are venting something out into space.'

'Roger, we copy you're venting.'

'It's a gas of some sort . . .'

Mission Control was pretty certain that the gas was the all-important oxygen, needed not only by the fuel cells to enable them to provide light and power, but for the Command Module's environmental system to enable the astronauts to breathe. They did not say so, however. Exchanges with the crew became even more difficult to follow as Astronaut Charlie Duke was replaced by Astronaut Jack Lousma at Mission Control; the Apollo 13 crew addressed him as 'Jack' while Lousma frequently also addressed Jack Swigert by his first name. There was fierce concentration on instructing the Apollo crew to reduce their power consumption, to eke out the Service Module's dying supplies long enough to switch to Lunar Module power supplies – though no one had yet mentioned that, either. Twenty minutes had passed since the explosion:

> LOUSMA: Okay 13, this is Houston. We'd like you to go to your G & C [Guidance & Control] checklist, the pink pages 1–5. Do power down until you get a Delta of 10 amps. Over . . .
> SPACECRAFT: (garbled) . . . the pink pages 1–5.
> LOUSMA: Did you copy our power down request?
> SPACECRAFT: Roger, Jack, we're doing it right now. Where did you say that was located, Jack?
> LOUSMA: That's in your systems checklist, pages 1–5 . . . I'd also check for those pages in your Launch Checklist. They're emergency pages. Pink pages 1–5 – and we'd like you to get your power down . . .

We all knew it was an emergency, but to a reporter it was a relief to have it officially admitted at last! One by one Jack Swigert was ordered to shut down the life-giving fuel cells, querying it each time – especially in the case of the third and last, knowing that they could only be restarted by the use of ground power. 'Fuel Cell 1 is closed,' he reported. 'Fuel cell shutdown procedure is complete' – well aware that that could prove to be a death sentence.

Mission Control did its best to be reassuring: 'Okay, 13, we've got lots and lots of people working on this, we'll get you some dope as soon as we have it, and you'll be the first one to know.'

Eighty minutes into the crisis and the Public Affairs Officer announced: 'We now show an altitude of 180521 nautical miles. Here in Mission Control we are now looking towards an alternate mission swinging around the Moon and using the Lunar Module power systems because of the situation that has developed here this evening. We now show a velocity of 3210 feet per second. This is Apollo Control Houston.'

Yet another 10 minutes of exchanges, and at last Mission Control admitted it:

> MC: We're starting to think about the LM lifeboat.
> SPACECRAFT: Yes, that's something we're thinking about too . . .

At last I could start writing my story for the BBC's early morning bulletins. I knew all about the 'Lunar Lifeboat'. The hours spent before every mission – usually on aircraft flights – reading up the procedures for the remotest emergencies now paid off. I had been particularly fascinated by a quite recent theory that, if the Command and Service Modules failed, the Lunar Module's engines for landing on and taking off from the Moon could be used instead to tow a crippled Command Module, the only section heatshielded for re-entry, back to Earth orbit. But a maximum three days' supply of oxygen and water for two men visiting the Moon must be eked out to keep three men alive for longer than that while they continued their trajectory around the Moon in order to get on a return-to-Earth trajectory. The drama spoke for itself, so I kept strictly to the facts:

> Despite the fact that it's admitted here that Apollo 13's power system has suffered a serious accident, there's confidence that it will be possible to bring the three astronauts home. There's no question now of a landing on the Moon. As Chris Kraft, Deputy Head of the Manned Spacecraft Center, has just been telling a news conference, 'They're concentrating on keeping things in Apollo 13 under control' . . . The lunar craft is being used as a lifeboat to bring Lovell, Haise and Swigert back home to an emergency landing in either the Atlantic or Pacific in just over three days' time . . . Trouble struck

> Apollo 13 at about 4 am . . . It's undoubtedly the grimmest situation that's occurred so far in America's manned spaceflight programme. But, though the atmosphere here is undoubtedly very tense, the exchanges between Mission Control, Jim Lovell and his crew have been unwaveringly calm and steady.

I knew as this went out on all BBC news programmes that my personal problems were about to begin. I was still doing my broadcasts on a crude 'lash-up' on a metal table top in the Press Center: two telephone lines were plugged into my Uher tape recorder – one to send, one to receive. The Uher was also linked to Mission Control, so that I could record that and feed highlights to London. But now the system was jammed with incoming calls. As they arrived at their desks, not only the Editors of the many radio and TV news programmes, but of *Today, World at One, PM,* Radio Two, the *World Tonight*, and World Service all sought confidential chats about the crisis, plus priority for major contributions to their personal programmes.

People like William Hardcastle, normally antagonistic to using staff correspondents, were always the most ruthlessly insistent when there was no alternative. They fumed as Margaret came into her own. She took their calls and told them I was much too busy covering the story to talk to them; but she would log their requirements, and if only they would get off the line I would be able to use it to make my broadcasts as their programmes came round!

As the day progressed I was almost continuously on the air, either doing 'straight' reports into TV and radio news bulletins or being interviewed live into successive programmes. Apart from keeping half an ear on Mission Control's exchanges with the crew, I had no time to attend the Press briefings or to call my own friends and contacts for the background on what was happening. But there was generous help from Angus and Ronnie, who found time to hiss the latest into one of my ears while I was fielding questions from London with the other. And Margaret, always a more effective reporter than I was when given the chance, was an invaluable assistant. She went off on a tour of exploration of NASA's Johnson Space Center facilities, visiting the groups of astronauts and technicians, each of which was devising answers to the many problems to be overcome if Apollo 13 was to be brought safely home.

All over America space workers by the thousand were pouring spontaneously into NASA centres and space contractors' facilities, offering their services and expertise. Because Margaret did not look like a reporter, and because anyway everyone was busy doing what mattered, for once they got on with their work and left her to her task. It was her information that provided my first TV satellite report:

> For once, Mission Control ISN'T the most important place. Just now, it's the astronauts' building. The desks, each labelled with a famous name, are empty: they're all working in the simulators, adding their special bits of knowledge to the re-entry programme. Al Shepard, America's first man in space, has suspended his preparations for the Apollo 14 flight next October. Astonishingly relaxed, he went off to one simulator, while Ken Mattingly – who STILL hasn't developed measles – went off to another.
>
> Buzz Aldrin thought the technical recovery from the Apollo 13 disaster was a much greater feat than a third moonlanding would have been. There was a compliment for the contribution Britain's brain-drain mathemeticians are making in this enormous recovery effort.
>
> Deke Slayton, the astronauts' boss, has been telling us that throughout America over 50 000 technicians are now fighting to save these men's lives. That of course is what's raised the morale of Lovell, Haise and Swigert. The little jokes are creeping back: Fred Haise, for instance, complaining that instead of moonrocks his only souvenir will be a Marine Corps foxhole shovel, which he should have used to dig a trench on the Moon.
>
> Behind these jokes, of course, the hidden worries – brushed aside at our news briefings. Was the Command Module's heatshield damaged by the explosion in the Service Module? There's no way of telling till re-entry – a fast one, at 24 600 mph – actually begins. When the Command and Lunar Modules separate, is it likely they

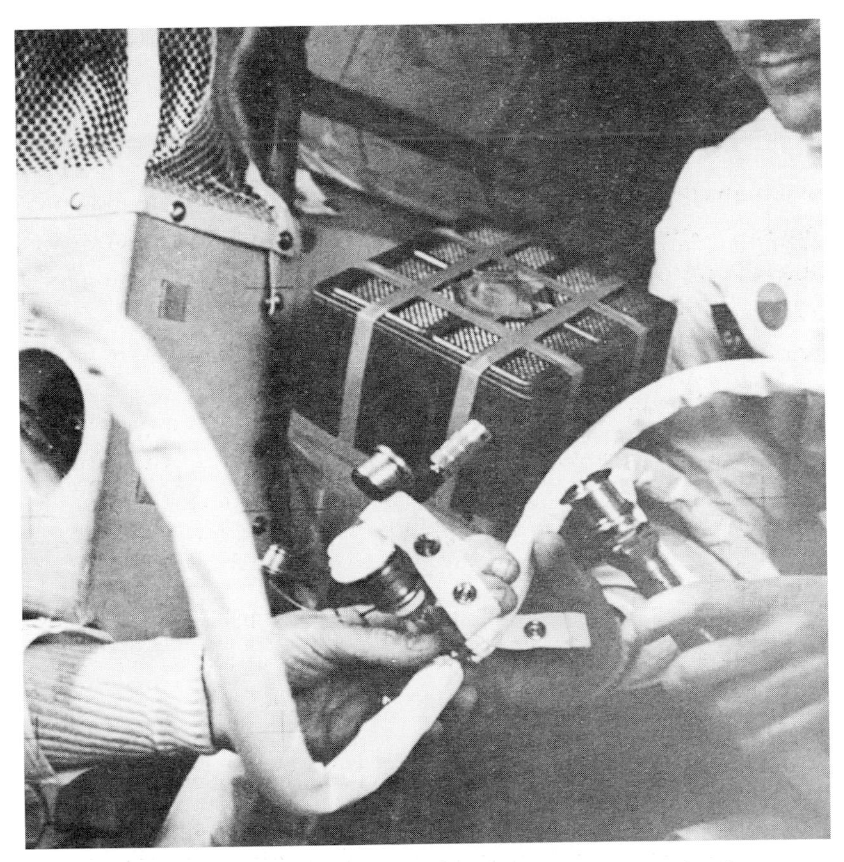

Apollo 13: Inside the Lunar Module on the way back to Earth, John Swigert, left, helps to convert the Portable Life Support system into a makeshift air conditioner. The system was devised and tested on the ground, and instructions were then radioed to the crew. (NASA)

could collide? The 30-ton LEM is packed with all the equipment they DIDN'T leave on the Moon. They're going to try to push the LEM away with a popgun effect by releasing the latches and letting the pressurisation in the docking tunnel do the rest . . .

Deke Slayton summed it up for me: 'We're in good shape right now; but you can't count that something else won't crap out on us before we get them home!'

He had actually summed it up for Margaret – but for once I felt entitled to claim someone else's interview as mine!

My next problem was being overwhelmed with help. Before the mission, with interest falling after two successful moonlandings, the Foreign Editor had rehearsed his usual moans about the cost of continuing to cover Apollo – despite the fact that my free ticket provided by Japan Air Lines to cover the new Tokyo route enabled me to return via the US, so that the BBC only had to pay for the internal US flights. *Now*, so far as cost was concerned, the sky was the limit. Two engineers arrived from New York, to organise live satellite transmissions for TV – then costing about £5000 for ten minutes – and to install improved radio facilities. Charles Wheeler arrived from Washington, and John Osman from New York, plus two TV camera crews from I knew not where – all asking impossible questions like: 'What shall we do?' In desperation I despatched John with one camera crew to join the media mobs gathering outside the homes of the astronauts' families 10 miles west of the Space Center. Charles, the perfect colleague as always, effortlessly took over many of the broadcasts – but resolutely refused to do the 'glamorous' TV transmissions. It was my story, he insisted.

At last I had time to work on the more technical details of the astronauts' fight for survival, and I was pleasurably surprised, twenty years later, to find some of my broadcasts still on discs in the BBC archives:

> . . . still the most serious problem being fought in the spacecraft and on the ground is air conditioning. For hours we've been listening to exchanges between Mission Control and the astronauts giving step by step explanations as to how to rig up an emergency container which can be attached to the Lunar Module's air conditioning. The problem is that the LEM air conditioning can't cope with providing clean air for the whole of the docking tunnel and Command Module as well. One of the teams on the ground has been devising 'do-it-yourself' air conditioners, made entirely of course from materials available to the astronauts: spacesuit hoses, plastic disposable bags, and, most important, lithium-hydroxide (that's charcoal) which can extract the poisonous CO_2 as it builds up in their atmosphere . . . It's

Apollo 13: even while they fought for survival, the crew took some of the most striking lunar pictures. This farside view shows Mare Moscoviense to the left, while the large crater on the horizon is International Astronomical Crater No. 221. (NASA)

an unbelievable story of persistent improvisation and inevitably enormous courage on the part of the astronauts.

Jim Lovell, exhaustedly going over the engine-firing procedures needed as they disappeared behind the Moon, rebuked the indomitable Swigert and Haise when he saw they had their cameras out and were busy taking photographs. 'If we don't make this manoeuvre correctly, you won't get your pictures developed,' he told them. 'You've been here before, and we haven't!' they retorted – and those pictures turned out to be some of the most useful lunar close-ups taken on all the missions.

So, with split-second firings of the LM descent engine, Apollo 13 was nursed around the Moon, and placed in the return-to-Earth trajectory which it had left in preparation for lunar orbit just before the explosion. As Lovell pointed out later 'the biggest heart-stopper' was hardly noticed on Earth. That was because it was difficult for us to follow in the cross-talk – it was called 'P.C. + 2', for pericynthion plus two hours – and even more difficult to describe in limited broadcasting terms. Before engine firings, of course, it was essential that the spacecraft should be perfectly aligned if the firing was to place it in the desired trajectory; incorrect alignment would result in Apollo missing the Earth and becoming stuck for ever in solar orbit.

The inertial guidance platform had been transferred from the CM to the LM, and normally its accuracy was checked with the Alignment Optical Telescope, or sextant, by lining it up with a suitable navigation star. But Apollo was now surrounded by a cloud of debris, travelling with it after the explosion – 'false stars' Lovell called them – and it was impossible to sight a real star. Lovell describes in NASA's *Apollo Expeditions to the Moon* how 'a genius' in Mission Control came up with the idea of using the Sun to check the platform's accuracy. The Sun's large diameter could result in considerable error, but even the debris could not block that out!

On instructions from Houston, Lovell rotated the spacecraft, with Haise tensely looking through the sextant. The air-to-ground transcript, Lovell recalled, read like a song from *My Fair Lady*:

> LOVELL: OK. We got it. I think we got it.
> HAISE: Yes. It's coming back in. Just a second.
> LOVELL: Yes, yaw's coming back in. Just a second.
> HAISE: Yaw is in . . .
> LOVELL: What have you got?
> HAISE: Upper right corner of the Sun.
> LOVELL: We've got it . . .
> HOUSTON: We're kind of glad to hear that.

Flight Director Gerry Griffin logged: 'If we can do that, I know we can make it' – but strangely, in his change-of-shift briefing to us soon afterwards merely commented almost casually: 'That check turned out real well.'

Neil Armstrong and Buzz Aldrin, first men on the Moon on Apollo 11, David Scott and other astronauts, had all been helping from Mission Control with their previous experience. That was at 73 hours 46 minutes into the mission, with another 68 hours 57 minutes to splashdown – but from then it was a mostly a question of endurance. No urine could be dumped overboard, because it would disturb the trajectory; much ingenuity was needed just to find ways of stowing the urine as the quantity increased. Spacecraft temperature was 38°; water was condensing on the walls, the windows were frosted. Lovell and Haise were able to don their lunar boots, but Swigert of course, not being due to take part in the moonwalk, had none. He suffered from wet, cold feet for two days; in Lovell's words: 'We were as cold as frogs in a frozen pool.' Perversely, the crewman on watch had to use stay-awake pills to prevent himself from dozing, while the other two tried in vain to do so.

In the Press Room, Margaret kept us going with endless tea – and the American jokes about our ever-hot teapot faded away and were replaced with thanks for an English cuppa. The ultimate compliment was when the haggard-eyed *Le Figaro* correspondent, the prototype Frenchwoman, tottered over to beg a cup for herself. Even more important, Margaret maintained files of the mission commentary transcript for Ronnie, Angus and myself. In the end it topped 500000 words and about 1200 pages, issued a few pages at a time by teams of NASA typists. As we could not listen every minute, it was invaluable for reference, but when deadlines coincided with crisis points, it was never quick enough. Then a crowd would huddle around our table as one of us played and replayed our recordings of muffled or obscure exchanges. Individual words were triumphantly identified like a crossword puzzle, acronyms and the meaning of programmes – 'VERB 49', 'VERB 24', 'P27 Update', 'POO and ACCEPT', 'That was some P52', 'Your NOUN 63 looks good' – explained by the more knowledgable, or hunted down in the mass of flight plans, procedural programmes or other documents we all accumulated in our large cardboard boxes.

> ... The immediate future of America's manned space programme depends on the astronauts photographing the damaged Service Module as it's jettisoned. How far should Mission Control hazard their safety – they'll be pretty hard-pressed at that stage – by asking for photographic activities as well?
>
> I saw Dr Paine, head of NASA, in the canteen just now, wearing dark glasses. He's the man they're all going to for final decisions. He's had the job too of comforting the wives during their long ordeal. Marilyn Lovell and Mary Haise, with their seven children, live five miles down the road from here. Dr Paine's been to see them, because they're not sleeping much, but listening continuously to their private feed from Mission Control. The astronauts' wives, at a time like this, have developed a way of drawing together in a tight group. Neil Armstrong, a neighbour of Mrs Haise, keeps popping in with reassurance as well.
>
> Watching it all I'm finding it hard to believe: all this brave and incredible improvisation – and of course the incredible courage of the astronauts. It's sad the one big success of the mission has gone almost unnoticed: the extraordinary, 4-hour seismometer readings that followed when the 13-ton, third stage rocket crashed into the face of the Moon.

I only discovered that by reading Mission Commentary – for I heard or read every word spoken during that mission; and despite tabloid headlines – Australian examples were 'Ship of Shame' and 'Haggard Astros in Terror' – there never was any panic or hysteria. The place to find that was as always on the world's newsdesks and in the TV and radio studios far from the action. Members of the international space corps in the Press Centre matched in tense, but calm concentration, the atmosphere in Mission Control. Few of us there seriously doubted that the astronauts would survive. But I was conscious all the time of suppressed resentment back at TV Centre and Broadcasting House, as I answered for the hundredth time the question: 'What's the atmosphere at Houston, Reg?' that I failed to join the media game of outdoing my rivals in 'shock-horror' pieces!

For me, as always, there was nothing to match straight facts when it came to drama. A 23-second burn of the small reaction control thrusters slid the 3-section Apollo 13 right into the centre of the re-entry corridor, thus compensating for the inaccuracy resulting from using the Sun before making the major burn. Then, as they approached Earth, the crew had in turn to jettison the crippled Service Module and then the Lunar Module.

Only four hours remained before splashdown, only 41 116 miles from Earth, when Jack Swigert in the Command Module flicked the pyrotechnic switch that severed the bolts holding the Command and Service Modules together; Jim Lovell and Fred Haise were in the Lunar Module, where Lovell fired the reaction control thrusters to push the Service Module gently away. As they drifted apart, the crew pressed their faces against the windows, now almost free of debris, and we listened to their exclamations:

> There's one whole side of that spacecraft missing . . . Right by the high gain antenna, the whole panel is blown out, almost from the base to the engine . . . A lot of debris is just hanging out the side . . .

'Take pictures – but don't make unnecesary manoeuvres,' instructed Houston urgently – then, striking a lighter note: 'James, if you can't take better care of a spacecraft than that, we might not give you another one.'

Someone started speculation that Apollo 13 might have been struck by a meteorite, and I joined in the speculation in my TV pieces, though few of us believed it.

Preparations to jettison Lunar Module Aquarius followed. The crew packed themselves into Command Module Odyssey, which Lovell described as 'a cold clammy tin can . . . The walls, ceiling, floor, wire harnesses and panels were all covered with droplets of water . . . The chances of short circuits caused apprehension, to say the least. But thanks to the safeguards built into the Command Module after the disastrous fire in January 1967, no arcing took place.'

They drew their last power from Aquarius to charge Odyssey's batteries, persuaded the computer to start up, sealed the docking tunnel and 'rather rudely' blasted loose Aquarius' 30 tons with pressure in the docking tunnel to make sure it cleared cleanly.

Now splashdown was only 75 minutes away, with Odyssey accelerating through 12 900 mph towards the Pacific, a mere 13 200 miles below. 'Farewell, Aquarius, we thank you,' said Mission Control. 'She sure was a good ship,' replied one of the astronauts. A few rare moments of self-indulgence followed, in exchanges between Jack Swigert and Capcom Joe Kerwin, an astronaut and medical doctor:

> SWIGERT: All of us here want to thank all you guys down there for the very fine job you did.
> LOVELL: That's affirm Joe.
> KERWIN: Tell you, we all had a good time doing it . . .
> SWIGERT: You have a good bedside manner, Joe. (Laughter)
> KERWIN: Say again, Jack.
> SWIGERT: You have a good bedside manner.
> KERWIN: That's the nicest thing anybody has ever said.

Seven minutes before re-entry began, Capcom invited the crew to pass down any phone numbers they wanted called – but the astronauts wisely ignored that. Had they done so, the numbers would have been blocked by a million media calls.

NASA's re-entry and splashdown preparations were being matched with much less orderly calm by the world's media, including the BBC. Aquarius was jettisoned at 5.53 pm British time; the spacecraft was due to re-enter the Earth's atmosphere at 6.53 pm, with splashdown at 7.7 pm, followed by all the recovery operations, and hoisting the crew aboard USS *Iwojima* to be greeted by President Nixon. It could not have been better arranged as primetime viewing and listening.

I had to do my TV 'unilaterals' at 5 pm, just before the main events. By now, of course, BBC programmes were littered with'Apollo studios', packed with chairborne experts on astronomy and space theory, and little knowledge of manned spaceflight. The producers of these programmes were determined that, with their own 'feeds' of

Mission Commentary, they could really cover the story much better by themselves, since Charles and I tended to diminish the drama by stating the facts.

But tiresome as it was, they felt they ought to have a Houston dateline in the programme now and then! So I was instructed, via Margaret, to use my 10-minute unilateral to send one minute on the 'final moments', for use in BBC1's 5.50 News and BBC2's 7.30 pm Newsroom, and then follow on with another one-to-two minutes saying 'something likely to be useful' in BBC1's main news, then at 8.50 pm. As always it was: 'Leave the facts to us, Reg. We want the atmosphere at Houston – perhaps something on simulators, and how they worked out the problems of guidance to splashdown!'

Almost as exhausted as the crew, I found myself standing in front of enormous searchlights erected on the balcony of what is normally NASA's exhibition centre, and which was now the international Press Center. Satellite capacity was still so limited that only one TV transmission could be made at a time from the US to Europe – and ALL the European countries wanted to use it at the same time. Engineers and technicians, lights and cameras, had been forcibly pooled on NASA instructions. I was just one of a sweating mob of near-frantic reporters and correspondents queuing to project his face and voice across the Atlantic. Margaret and Charles did their best to keep me up to speed with what was actually happening, while I worried about saying something in my allotted time that would still be relevant whether or not the crew splashed down safely. As my predecessor – a German – was hustled out of the lights, I was thrust into them, while European Broadcasting Union engineers in various switching centres in Houston, New York, Paris, Bonn and London, tried to redirect the picture to BBC TV Centre at White City. As always it seemed everybody succeeded except the BBC engineers. On my earpiece there was an international cacophony, matched only by a rising crescendo of rage and screams emanating from those around me as my 10-minute slot ran out, and others started claiming their time slots. Amid this, I thought I identified someone along the line shouting: 'CUE'.

Too tired to ad-lib, and too dazzled by the searchlights to read my notes, I tackled my 1-minute piece for immediate use:

> When they saw the extent of the damage, Flight Controllers here were astonished that, after the explosion, Apollo 13 and its crew didn't just vanish for ever. As photographs and blow-ups enable the experts to look closer, there's criticism, too. No.2 oxygen tank, which blew up, can't be seen. But it's now known that when it did all the contents of Tank No. 1 were lost as well. There's a cross-feed without a valve, so the astronauts, helplessly watching their oxygen venting into space, couldn't turn it off. The trouble, it's thought, started 9 hours earlier with a pressure surge. That put the warning sensor out of action, so the crew got no indication that, in effect, a time bomb was building up behind them. Jim Lovell originally said this would be his 4th and last flight; tomorrow he'll be telling us whether he's changed his mind, and now wants another chance to go to the Moon.

I suspected London didn't feel that was nearly dramatic enough for the early TV News – but London's comments were submerged by yells around me. I had already been occupying the satellite for more than 10 minutes. Hurriedly I embarked on the longer piece, which had to be usable whether Apollo 13 survived re-entry or not:

> The day started quietly, but steady confidence suddenly changed to somewhat nervous tension when Jim Lovell, who's now clocked up 30 days in space, made what was probably his biggest mistake in all that time. Clearly cold, and somewhat jaded, with only three hours' sleep behind him, he punched up the wrong computer programme for the mid-course correction. It would have fired the main descent engine instead of the little 100 lb thrusters. They spotted it here at once, and no harm was done. But there were five hours to go.
>
> Minutes later the astronauts' boss, Deke Slayton, was at the mike, telling them they must break open their medical kit and take dexedrine tablets – stimulants to sharpen up their reactions . . .

> When the disaster struck, over three days ago, everyone thought it was at the worst possible moment, since the spacecraft was a quarter of a million miles from Earth. As it's turned out, it was that that saved the astronauts' lives. For it gave them the opportunity, by running the simulators non-stop, to work out re-entry techniques they'd barely dreamed of.
>
> One thing that every journalist here is convinced of is this: after the terrific technical feats achieved here on the ground, and the heroic fight put up by the crew, the Apollo 13 men deserved to live.
>
> Investigating the cause of the explosion, and modifying future craft, is likely to be embarrassingly costly just now, in view of the savage cuts recently imposed by the US Government on the NASA budget. What with that, and internal pressure to divert some of the moonlanding rockets and spacecraft to orbiting laboratories, it seems inevitable that at least two more moonlandings will be cancelled. Reginald Turnill, BBC, Houston.

I had barely completed the dateline before, in my turn, I was hustled out of the spotlight. My feeling that this must have been my worst-ever broadcast was confirmed long after when the Director General Charles Curran (to whom, in accordance with BBC traditions, I never actually spoke myself) told Margaret when she boldly sought him out on some social occasion: 'Reg looked very tired on that last TV throw, but I didn't blame him.'

By the time I got back to my table, Lovell, Swigert and Haise, who had elected not to struggle into their bulky spacesuits for re-entry, had hit Earth's atmosphere at a speed of nearly 25 000 mph, and were going through the seemingly endless 'blackout' period – actually only about four minutes – when communications cannot penetrate the ionised air particles created by the friction heat as the vehicle's speed is burnt off. Slight misgivings at Mission Control that the heatshield, to which the Service Module had been attached, might have been damaged by the explosion, were magnified by the media into a final 'shock-horror story'. Deke Slayton had repeatedly said that there was no evidence that the heatshield WAS damaged; but in the BBC's TV studio, I was told, James Burke was on screen with eyes closed and holding up a hand with fingers crossed. We shuddered.

Inside Odyssey the main sensation for the crew was that the deceleration shook free all the droplets of condensation. 'It rained inside the CM,' wrote Lovell.

Once we saw the parachutes blossom as Apollo slowed for the final splashdown – four days after the explosion but only 1.8 km from the target point in the South Pacific – it was inevitably all anti-climax. We still did our best to keep the story going with tales of how up to 100 000 scientists and technicians across America had kept every relevant simulator going while they worked out the recovery techniques. President Nixon had flown out to the USS *Iwojima* on what he alleged was the most exciting day of his life and 'to speak for people around the world' as he welcomed back the astronauts: 'It has been said that adversity introduces the man to himself. Ladies and gentlemen, I present to you three men who have been introduced to themselves as much as anybody in the whole history of man.'

Then he had them flown to Hawaii, where their families had been taken to greet them. One photographer just managed to get a rather fuzzy picture of the passionate kiss exchanged by Jim and Marilyn Lovell.

Lovell himself later described the cause of it all. The thermostatic switches to the heaters in the oxygen tanks had not been modified when the permitted voltage was raised from 28 to 65 volts DC. During tests on the launchpad the wiring had been subjected to very high temperatures which probably welded the switches shut. 'Warning signs went unheeded, and the tank, damaged from eight hours' overheating, was a potential time bomb next time it was filled with oxygen. That bomb exploded on April 13, 1970 – 200 000 miles from Earth.'

There were individual medals for the astronauts, who all declared that they were willing to make another moonflight, but were never given the opportunity. There were collective medals for Mission Control and all the others whose technical skills brought them safely

Mission Control celebrates Apollo 13's safe recovery. On screen is Lovell, aboard USS *Iwojima*. (NASA)

home. Grumman, makers of the Lunar Module, presented North American Rockwell, makers of the Apollo Command and Service Modules, with a salvage bill for over £200 000 for towing the crippled spaceship safely home from the Moon – a joke not wholly appreciated by the recipients.

The BBC recognised Margaret's contribution by allowing me to charge $75 for 'secretarial assistance', and sent me to Washington and New York as a stand-in while Charles Wheeler rushed off to cover revolution in Trinidad. That gave Margaret a chance to do some sightseeing – but I was much too busy to join her, covering China's first satellite, New York police corruption, a Wall Street plunge, and anti-Vietnam student demonstrations around the White House.

For me the most important personal bonus resulting from the Apollo 13 story was that never again did I have any resistance from successive Foreign Editors when I announced that I was booking my flights to cover future manned space missions.

20 Last Men on the Moon

Apollo 14

Within ten days of Apollo 13's recovery China had launched its first satellite, and the Soviets were claiming an unprecedented feat (though the US Air Force said 'Bah, we've already done that four times!') in launching eight small satellites on one rocket. But this mounting international competition did not save NASA from continuing criticism. One moonlanding, maintained many politicians, was quite sufficient for both prestige and scientific purposes; how could the cost of six more be justified? As I had forecast during the Apollo 13 crisis, the Apollo 18 and 19 missions were cancelled, leaving only four more; this reduced to eight the number of astronauts with a chance of getting there, and cost Swigert the promised command of Apollo 19.

But Apollo 14, due to have followed its disastrous predecessor in October 1970, was delayed only four months, during which time a third oxygen tank was installed in the Service Module, and a powerful battery was added to the Command Module to provide electricity if a similar accident occurred again.

Back in Cape Canaveral's warm January sunshine, I found the NASA launch team full of new hang-ups. After the measles scare that preceded the Apollo 13 launch, this crew had been placed in quarantine for three weeks before the flight. To protect them from coughs and colds, their last briefings on abort situations were taking place with plate glass windows between them and engineers; warning bells were rung when they moved from place to place, to warn technicians to get out of the way. To witness the late afternoon launch on the last Sunday of the month, a bigger crowd flocked to Florida's beaches than even for Apollo 11, fearing that this might be the last moonflight. There was also a general feeling that the Cape's whole future was at stake on this mission, and that if things went wrong again the launch site for the Space Shuttle, by then chosen as Apollo's successor, would be moved to California.

Wernher von Braun, with his usual skill in such matters, did his best to cool things down by pointing out that not one cent of the $21 billion cost of Apollo was being spent *on* the Moon; the money was all going into US universities, advanced industries and technicians' pay packets. Alan Shepard, the somewhat acerbic commander, was to become the only one of the original seven Mercury astronauts to get to the Moon. Having been America's first man in space, doing a 15-minute ballistic 'lob', he had had to wait nine years for another mission as a result of persistent ear trouble. Since neither of his companions, Ed Mitchell and Stuart Roosa, had been in space before, they were the least experienced of all the Apollo crews, but it was not to affect their achievements.

There was doubt about that in the early stages, however, for Roosa, as Command Module pilot, had to make six attempts before he was successful in docking with the Lunar Module and extracting it from the Saturn 4B upper stage. There had never been problems with this manoeuvre before, so it was suspected that he and Shepard, working with him, were at fault. Finally, between them they rammed home the Command Module's docking probe by brute force, and it was feared that the probe had been damaged. If that were so, although they were by then committed to going around the Moon, there could be no moonlanding. However, the long outward flight provided plenty of time for the crew to withdraw and dismantle the docking probe to inspect it and the docking latches. They found nothing wrong and were given the go-ahead for a full mission.

They fired themselves into an accurate 313 km × 107 km lunar orbit (figures which were matched almost exactly on every successful mission), and later the Command Module ('Kitty Hawk') took the Lunar Module ('Antares') down to a low point of 15 250 metres before it separated for the final descent. That gave Shepard more fuel for hovering while he found a safe landing site on Fra Mauro – the area in which their predecessors should have landed. It was not an easy descent, for Shepard and Mitchell had to work hard, over-riding lots of master alarms from a faulty computer telling them to abort. Shepard had no intention of doing so, and when he finally stepped on the Moon, 10 years after being selected as an astronaut, he commented 'It's been a long way, but we're here.' He found it all the more fulfilling because many had argued that at 47 he was too old to be sent to the Moon.

Whether or not age was a factor, Shepard and Mitchell did get exhausted during their two spacewalks on the Moon, which totalled 9 h 24 min – even though they had been provided with a two-wheel trolley for towing their equipment and the lunar rocks they collected. Ed Mitchell had a 'thumper' device for firing small explosive charges on the surface to generate readings on the seismometers left behind by Apollos 11 and 12, together with new ones they had deposited. Only 14 of his 21 charges detonated, but a much more effective reaction had been caused by the impact of the 15-ton Saturn 4B upper stage, which had hit the lunar surface the day before the astronauts' landing. Dr Gary Latham, of Colombia University, said that that and previous impacts suggested that the Moon had no crust like Earth; millions of years of bombardment had reduced the lunar surface 'to a jungle of broken rocks, perhaps 30 miles in depth'.

On their second EVA the astronauts, towing their cart, set out to trudge 1.6 km to the rim of Cone Crater, 122 metres above their landing point and 38 metres deep. But after 2 h 10 min they were 50 min behind schedule; Shepard's heart rate had reached 150 and Mitchell's 128. They turned back short of the rim, and one of the 'highlights' of the mission, rolling stones down the inside of the crater, was lost; the geologists had expected to make all sorts of subtle deductions about adhesion and the effects of one-sixth lunar gravity.

Before they re-entered the Lunar Module, Alan Shepard still had energy left to carry out what seemed then a harmless little joke, but which was to have disastrous consequences for NASA. He had concealed two golf balls in his spacesuit, and using the extension handle of the scoop used for obtaining lunar samples without bending down, settled down to a game of golf. We watched the occupants of Mission Control rocking with laughter as his first swing appeared to miss; but with the second he made contact, claiming that in the low lunar gravity, it had travelled 'miles and miles and miles'. Forever afterwards this small incident was to be used by critics of the space programme to jibe at the high cost 'of sending a man to play golf on the Moon'.

As they returned to the LM, Mission Control reminded Shepard to take with him the 100 ft nylon tether they had used on the Moon, and for hauling their harvest of 44 kg of moonrocks and soil from the surface to the interior. The tether would be needed if they failed to redock with the Command Module and had to transfer by spacewalking. However, what would have produced another exciting drama for the space correspondents proved not to be necessary. During the journey back to Earth, Roosa carried out NASA's first experiments into the possibility of using microgravity for processing vaccines and materials for commercial use on Earth. TV coverage of this pioneering work was ignored on America's prime-time Sunday evening shows – and with such experiments still under way on the Space Shuttle more than 20 years later, with little progress apparent, maybe the TV moguls got it right!

That was the last crew to have to endure the quarantine period as a precaution against possible lunar infections after returning to Earth.

Apollo 15

The five months' wait for Apollo 15, the next lunar mission, was enlivened by the launch of the first Soviet space station, Salyut 1, and the successful docking with it of the three-man spacecraft, Soyuz 11. But after spending 23 days inside Salyut, the crew died when Soyuz 11 became depressurised during re-entry.

Once again I disappointed BBC news editors by refusing to give any credibility to speculation by many science correspondents – who included my BBC science colleague – that their deaths probably demonstrated that the human body could not endure such long periods of weightlessness. This was the line that scientists opposed to humans in space hoped would be accepted.

When I arrived at Cape Kennedy (as it was still temporarily called) to cover the Apollo 15 launch preparations, Jim McDivitt, no longer an astronaut but manager of the Lunar Module and lunar exploration programme, told me that NASA had switched 200 technicians on to a two-week study of the Soviet disaster, to see whether there were any implications for America's manned spaceflights. Their main conclusion was the obvious one that it was a mistake for the Soviet cosmonauts not to have worn spacesuits during re-entry. These would have enabled them to survive accidental spacecraft depressurisation, and Apollo astronauts should therefore be suited up at hazardous moments like dockings and undockings as well as for lift-off and re-entry. Jim Lovell, fit and active after Apollo 13, pointed out that during their planned three days on the Moon, Astronauts David Scott, Al Worden and Jim Irwin would have to depressurise their Lunar Module five times.

For the American public, eager for heroes as reference points for their nation's superiority, this was the perfect crew – so it came as all the more of a shock when it was all over and it emerged that their behaviour seemed to have been less than perfect. Colonel Scott, 39, had for years been the prototype all-American boy – blue-eyed, fair-haired, 6 ft tall, open in face and character, having spent his whole life collecting awards, ranging from an MIT thesis on interplanetary navigation to swimming and handball victories.

It was an all-Air Force crew, all successfully married and with a total of eight children. Scott's companions were Lt-Col. Jim Irwin, 41, who as Lunar Module pilot was to accompany Scott on the Moon; and Major Alf Worden, 39, Command Module pilot, whose job was to remain in lunar orbit. They too had logged thousands of hours as military jet pilots, won lots of medals and, we were told, enjoyed activities like skiing and squash, paddleball, camping and fishing.

The countdown went so smoothly for this 12-day flight that I had to find other points of interest; one was that the eve of the launch marked the 21st anniversary of the first rocket to be launched from Cape Canaveral. It was a 60 ft V2, of the type used to bombard London, and called 'Bunter-8'; its significance was that von Braun's team had given it an upper stage – the start of the multi-stage rocket technique which was to enable its future generations to take men to the Moon.

Apollo 15 was the first of three 'J Series' missions, which very successfully fulfilled their aim of exploiting the scientific potential of the Apollo hardware by providing progressively longer stay-times on the Moon. Just before entering lunar orbit the crew donned their spacesuits and this time deliberately blew off a 2.9 m × 1.52 m panel from the Service Module to expose eight scientific experiments. These enabled Worden to spend six days operating mapping and panoramic cameras, as well as spectrometers, two of which were on 6 m-long extendable booms, for assessing the composition of the lunar surface, solar X-ray interaction, and particle emissions. The scientific instrument module thus exposed also contained a 35.4 kg sub-satellite which was successfully ejected after the moonlanding to spend a year studying the Moon's mascons and other phenomena.

Leaving Worden in what was now called 'Endeavour' – spelt the British way by NASA after Captain Cook's ship – Scott and Irwin departed in the Lunar Module ('Falcon'), and cleared the 3960 m Apennine Mountains as they dropped down to land in a basin near Hadley Rille, a 366 m chasm named after an eighteenth-century English mathematician. Falcon came to rest on 30 July 1971 within a few score metres of the aiming point, but with a 10° tilt because one footpad was in a 1.52 m crater. The tilt was within acceptable safety limits, but resulted in a hatch which swung inconveniently against them as the astronauts crawled in and out in their cumbersome spacesuits for three lengthy EVAs.

Because everything went so well my reports for BBC news lacked drama, with the result that they were preceded by descriptions of banner-waving fishermen off Southsea beach protesting that their interests had been ignored in Common Market negotiations; by Professor Barnard's first double transplant operation in Cape Town, and by riots in Belfast and cholera in Spain. I only got back into the lead when, after orbiting the Moon 12 times, Scott and Irwin failed to undock the Lunar Module. Worden in the Command Module reopened his hatch, floated into the docking tunnel and tightened up the connecting power line – and separation was achieved 35 minutes late. There were more alarms on the surface of the Moon, when Scott and Irwin, having pulled the lanyard which caused their new Lunar Rover to fold down, found that one of the two batteries operating the wheels would not work. No front-wheel steering was available, but they strapped themselves in and managed very well with rear-wheel steering.

'It's a real bucking bronco,' declared Scott, as they drove more than 10km around craters, stopping periodically to collect rock samples. When they reached the rim of Hadley Rille, 4km from the Lunar Module, the world shared their view when they turned on the TV camera mounted on the Rover, and deployed the umbrella-like antenna to transmit excellent pictures direct to Earth. On this first EVA, however, energy expended and thus oxygen consumed was 17% greater than expected, and Mission Control warned that activities would have to be curtailed. But before it ended, the third ALSEP (Apollo Lunar Surface Experiment Package) was deployed, and this was an exciting moment for geologists. They were confident that by using the Apollo 12, 14 and 15 ALSEPs they could finally pinpoint and even predict the monthly quakes occurring in 11 areas (mostly in the Copernicus region) each time the Moon passed at its closest point to Earth.

The second EVA was considered the most rewarding. Scott excitedly pronounced that he had discovered what they were looking for: a 10cm piece of anorthositic rock. For non-geologist journalists like me this meant frantic research not only on how to spell it – it still does not appear in ordinary dictionaries – but to explain simply what 'anorthositic' meant: rock believed to be part of the pristine Moon, and thus, perhaps, containing a geological record of its formation. I wish I had been the one to simplify things by dubbing it 'the Genesis rock'. (Later, geologists decided that the rock was 'only' 4150 million years old; although that was 150 million years older than any previous sample, it was *not* as old as the Moon itself!) The third EVA was the shortest – 4 h 50 min – because Scott and Irwin had got overtired on their earlier excursions, and they also had difficulties with the drill provided to obtain 10 ft-deep core samples. But they covered 28 km on the lunar surface and had collected a total of 27.3 kg of rocks and soil by the time they climbed back into the Lunar Module.

Before doing so they carefully positioned the TV camera on the Rover so that at last the lift-off from the Moon could be seen on TV. We watched an astonishing two seconds when our black and white view of the LM on the stark lunar landscape burst into green and red as the ascent stage rushed abruptly upwards to rejoin Worden.

At a TV press conference held during the journey home Scott and Irwin said they had both fallen twice during their lunar activities, but were uninjured and had no difficulty getting back on their feet – although it was important to do it slowly.

The official report that 're-entry was nominal' was less than accurate. The whole world could see on TV that one of the three 25.3 m parachutes had failed to open fully, and the recovery carrier called an urgent warning 'Stand by for a hard landing!' But despite the fact that Apollo 15 hit the Pacific at 36 instead of 22 mph, seven miles off target, the astronauts emerged none the worse from the heavy splashdown.

The unexpected sequel to this mission, which effectively ended the astronaut careers of all three, was kept secret for several years before it was made public. Irwin himself described what happened during their third day on the Moon: 'We played post office, and cancelled the first stamps of a new issue commemorating US achievements in space. With our own cancellation device, which worked in a vacuum, we imprinted August 2, 1971, which was the first day of issue.' The crew had taken 400 unauthorised postal covers or envelopes, in addition to about 250 official ones, as part of their 'personal preference kits'. Some were to be sold in a complicated deal involving a German businessman, and the three crew members expected to use the $8000 they would each get to set up a trust fund for the education of their children.

The story leaked out when the sale of the envelopes at $1500 each began earlier than the astronauts expected. After NASA and Congressional inquiries into the legitimacy of this commercial enterprise they were reprimanded and punished by being removed from their assignment as the backup crew for Apollo 17. Scott, Irwin and Worden always felt they had been harshly treated, and their opportunity to erase the reprimand came in 1983. NASA was short of a payload on Shuttle mission STS-8, and planned to send up two tons of stamped envelopes in a deal with the US Post Office. Worden maintained this was no different from what they had done, and sued NASA for the return of the Apollo 15 envelopes which had been confiscated. NASA was advised that it would lose the action, and in August that year returned the 12-year-old envelopes. By then their value was estimated to be about $500000; whether Scott, Worden and Irwin actually got anything like that for them I have never discovered.

David Scott was back in the headlines again for a brief period in 2001. By then 69, his marriage had failed, and he and Anna Ford, a TV newsreader, were discovered by the media to be together in England. Once again the stamps 'story' was resurrected! It seemed they might marry – but after a few months that relationship was reported to have ended abruptly. Anna Ford complained unsuccessfully that their privacy had been violated by the publication of long-lens pictures taken of them together on a secluded but public beach.

Apollo 16

Just as the Apollo 13 failure had prompted opponents of human space exploration to question the justification for continuing with it, now the success of Apollo 15 was used to support arguments that since everything worth-while had been achieved there was no point in paying for two more Moon missions.

The Soviets, it was pointed out, were very successfully exploring the Sea of Rains with their robot, Lunokhod 1, which was sending back thousands of pictures plus mechanical soil analyses. Small groups of US and Soviet astronautical engineers and technicians were beginning to meet for friendly discussions on the possibility of developing a common docking system for their spacecraft (which led to the Apollo–Soyuz mission in 1975); and the Soviets proudly displayed the 120 grams of moondust returned to Earth by Luna 16, unabashed by the fact that their US competitors had by that time acquired 67 kg.

American astronauts, however, were foremost among those who urged that the success of Apollo 15 justified making their last two missions much more adventurous. It was argued that the target for Apollo 16, – and its crew, astronauts Young, Mattingly and Duke – was another highland area, north of Crater Descartes, which might not be all that different from Fra Mauro. Some geologists wanted a landing much further south, in the rugged Aristarchus area nearer the Lunar Pole, while some of the astronauts favoured a landing on the Moon's little-known farside, which would mean that all communications would have to be passed via the orbiting Command Module. But NASA chiefs, having achieved so much at such little cost in human life, became less and less inclined to take risks – in the long term, probably, a disastrous policy.

Low morale among its thousands of workers had become a major problem, and one of my first stories when I arrived at the Cape for Apollo 16 included an interview about it with the Director, Dr Kurt Debus. With hundreds of layoffs already announced to follow that launch, and with thousands more after the last moonlanding, Apollo 17, managers were having to guard as never before against shoddy workmanship that could jeopardise these missions and even cost the lives of the crews. 'We must and do have a morale problem,' agreed Debus. 'However, we have intensely watched out for any consequences of this problem, for instance in terms of sloppiness of work, or of covering up mistakes. So far we have not found it, so we do not have any derogatory consequences that we can detect.' Debus himself was then 63, and I questioned him about his own job prospects. He was disarmingly frank on the subject. He was legally entitled to stay in his job until he was 70, he said, and had every intention of doing so. It could have been my public disclosure of this that led to his being forced out not long after, along with thousands of his own redundant employees.

As usual there were unexpected stories to be ferreted out by a careful reading of Press Kits and Flight Plans. What was all this about a special potassium-laced diet? Our old friend Dr Chuck Berry, who kept reappearing after it was thought he had left NASA, revealed that during Apollo 15 they had kept secret the fact that Scott and Irwin had suffered from irregular and even double heartbeats during their spacewalks, and that similar irregularities had been noticed with previous astronauts. Potassium loss affected nerve and muscle action, but Chuck admitted it was unpalatable to take. They had considered sending a supply of bananas with the Apollo 16 crew, but abandoned it because bananas did not keep well. So the crew were being ordered to drink potassium-laced coffee, plus lots of orange juice. During their three long spacewalks they would be provided with an orange-based drink to suck through a tube inside their space helmets. Additional drugs had also been added to the mission's medical kit in case of really serious heart attacks.

Another unexpected story was that this crew was to take back to the Moon 4 grams of moondust gathered by Apollo 12. Geologists could not decide whether it had become magnetised while lying on the Moon's surface, or during its quarter-million mile journey back to Earth. So, having been thoroughly demagnetised, it was to be taken back to the Moon then returned to Earth stowed in exactly the same way as on the previous occasion.

Three days before Apollo 16 was launched on 16 April 1972 NASA's cash shortage was emphasised by the announcement that the Apollo 7 launchpad had been sold as 'US Government surplus' for the equivalent of £3000. By then the geologists, thankful to have a launch at all, were reconciled to the landing site in what was now described as 'the mountainous Descartes region, 7000 ft above the mean lunar level'. This area differed from the Appenine Mountains in that they thought it was volcanic. The discovery that this was quite wrong and that what the astronauts brought back were impact brecchias – 'coarse-grained rocks of anorthositic composition that were metamorphosed by impact cratering', to quote Dr Farouk El-Baz of the Geological

Society of America – was just one result which later obscured the memory that this proved to be a most hazardous mission.

At one point Mission Control was seriously considering the necessity for an emergency return to Earth, repeating the Apollo 13 procedure of using the Lunar Module as a lifeboat. Young, Mattingly and Duke were kept troubleshooting with a dozen technical problems on the way to the Moon. Minor problems involved communications, during which exchanges between Apollo and Mission Control were repeatedly interrupted by a Spanish gentleman's affectionate exchanges with his girlfriend, during which he sang to her 'I can remember it well.' There was a so-called 'Shredded Wheat' crisis soon after Apollo had successfully extracted the Lunar Module from the Saturn 4B. Mattingly saw a stream of particles apparently venting from the Lunar Module, and it was feared that its propellant was leaking. 'You'd better get inside and have a look,' ordered Mission Control. Young and Duke opened up the docking tunnel and were inside the Lunar Module in the record time of 17 minutes. Everybody was surprised and relieved when the instruments showed normal pressures and full propellant tanks. Apollo's small thrusters, it was decided, were harmlessly burning off a coat of white paint which had been added to the Lunar Module for the first time on this mission.

Problems mounted steadily. The steerable antenna needed for clear communications remained obstinately stuck in the stowed position – and while Mission Control and the astronauts seemed to communicate quite well with blurred communications, space correspondents found it even harder than usual to fathom what was going on. Command Module Caspar and Lunar Module Orion undocked on lunar orbit 12, with both sweeping low in tandem over the lunar surface in their elliptical orbits. But Young and Duke found that their landing radar would not check out in its 'self-test mode'; it showed spurious readings, probably, thought Young, caused by reflections from the lunar surface.

Then came the major crisis. It was a memorably dramatic moment when Mattingly emerged from behind the Moon much earlier than expected. Everyone knew there was something dramatically wrong except us, clamping our head phones to our burning ears. 'No CIRC' – meaning 'no circularisation burn' – both Caspar and Orion were calling. 'Anticipate a wave-off for this one,' Mission Control was telling Caspar. 'We'll set you up for the next one'. That meant no moonlanding – but why? 'Ken's right in front of us, maybe about 600 feet, so we have a visual on him,' said Young. He and Duke had been able to see and hear, as they travelled around the back of the Moon, what none of us could hear: that Ken Mattingly had decided not to fire his main engine to lift Caspar back to a safe orbit during Orion's moonlanding, because a secondary backup circuit appeared to be faulty.

So Young and Duke already knew that they must postpone their moonlanding, and be ready to attempt a redocking with Orion and an emergency flight back to Earth with the Lunar Module once again acting as a lifeboat. Both craft were ordered to continue orbiting and to reduce the gap between them, ready for re-docking, while Mission Control studied the problem. 'I thought we'd lost the mission,' said Jim McDivitt.

But during a 6-hour wait, oscillations that Mattingly had experienced in one of his thrust vector control loops were reproduced in North American Rockwell's California facilities; that data was compared with readings from the Command Module as Mattingly passed across the face of the Moon; then reports on 1969 tests at North American were dug out and checked, and everyone was satisfied that there was an adequate safety margin for Mattingly to go ahead. So, after three extra, very busy, lunar orbits, Mattingly successfully fired his engine to circularise Caspar's orbit – and Young and Duke started their descent engine and finally touched down on the Cayley Plains of Descartes only 3 m from a crater some 7.6 m deep.

That was at the end of the last ten of 100 hours battling to overcome technical problems, so it was not surprising that even the taciturn John Young took part in some wild outbursts of enthusiasm as he and Duke at last approached the surface and became the first humans to gaze upon the rugged lunar highlands. The drama of the final exchanges, not always clear in our headphones, was duly recorded in Mission Commentary. I have inserted the meaning of abbreviations when I have been able to check them:

> MC: Roger, you're go at 8 [minutes].
> ORION (Young): I can see the landing site from here, Charlie.
> CASPAR (Mattingly, repeating in case MC had not heard): Go at 8, John's got a visual.
> MC: We copy.
> ORION (Duke): 130, we're right on, John . . .
> ORION (? Duke): Enter 360 minus 01 72 Denter 367 is coming up, and I'm starting the clock, I mean the camera.
> MC: Go at 9.
> DUKE: Ehey, we're under 12 000 John. Go at 9 coming down at a 182, a little steep – hey we're going to be right on it, just about right on, maybe 10 feet – 10 000 feet, standby. 64 at 8,200 PRO –
> ORION (? Young): Pitch over –
> ORION (? Duke): Pitch over – hey there it is, Gator, Lone Star, right on.
> YOUNG: Call me the Pings [the acronym for PGNCS, or Primary Guidance Navigation System] Charlie.
> DUKE: Okay, 38 degrees, Palmetto in sight, North Ray, looks like we're going to be able to make it John, there's not too many oblocks up there.
> MC: Rog, you're Go for landing.
> DUKE: Okay, 4000 feet, 42 LPD [Landing Point Designator], 3,900 feet.
> YOUNG: Two to the South, Charlie.
> DUKE: Okay, it's in. 41 LPD, 30,000 feet on profile.
> ORION (? Duke): And we're coming right down – it's going to be a little fast.
> ORION (? Duke): It looks 41 LPD 2,000 feet 60 on profile.
> ORION (? Young): Okay.
> ORION (? Duke): 42 LPD, couple more in, 1,400 feet 44 now looking

good. Out of a 1000 feet, right on profile, 54 LPD dropping out the bottom now, 800 feet, 30 down.
ORION (? Young): Okay Houston, we're going to be just a little long –
MC: Roger.
ORION: – but we're just now abeam of Double Spot.
MC: Copy.
ORION (? Duke): 23, 22 down at 500 feet.
ORION: Okay.
ORION (Duke): The big blocks over here to the left, John. Okay, 300 feet, 15 down.
ORION (Young): Okay, okay, take over Charlie.
ORION (Duke): Okay. Okay, fuel is good – 10%, there comes the shadow. Okay, 200 feet, 11 down, give me a couple of clicks up, 5 down at 130 feet, two forward, no more drifting, looking good. Perfect place over here, John, a couple of big boulders – not too bad, okay 80 feet down at 3 [feet per second] looking super. There's dust, okay 3, 50 feet down, down at 4, give me one quick up, backing slightly, okay, 2 down, stand by for contact, come on let her down, level off, let her on down, okay step 6% plenty fast, contact, stop, Boom Probe, engine arm. Wow! Wild man, look at that! Okay 413 –
ORION (? Young): Well, we don't have to walk far to pick up rocks, we're in front of them, open, close, open close.
ORION (? Duke): Old Orion has finally hit it, Houston.
FANTASTIC! . . .
DUKE: Boy, it sure looks like you could make – let's see, Crown Crater from here. I can see Ray Crater from here. Got it. Boy! I almost had apoplexy, that program alarm – and that's your radar breaker . . .
DUKE: Say, Jim, this ridge in front of us does look like a subdued crater, and it may be the raised rim about 50 metres in frontof us, about – oh, 4 or 5 metres tall. That's 30 or 40% of the surface is covered with boulders that are maybe half a metre in size. Out in front of us, and to the right, where we landed –
YOUNG: Wait a minute, Charlie!

With difficulty, Young persuaded his companion to concentrate on the immediate issues of checking that the Lunar Module was powered down and in a safe condition either to take off or to stay. Young said he still felt he would like to start the first EVA immediately, but discretion was the better part of valour, and they would first take eight hours of overdue rest. Before they settled down, Duke sent a final message to Mission Control:

> DUKE: Hey Jim, my hat's off and a case of beer to FIDO [Flight Dynamics Officer]. I'll tell you, that target was just beautiful. Boy! You guys just burned us right in there.

Before starting their first EVA, Young and Duke, anxious to placate the doctors so that their full programme, involving 15 miles of driving in the mooncar, would not be curtailed, sent reassuring messages that they had slept well, and eaten all of their potassium-loaded breakfast. We visualised them exchanging meaningful looks as they did so!

I described them as 'fairly tumbling out on to the Moon, in their eagerness to explore it', although we had no TV as they did so, and much of their conversation was buried in static noise. Later the transcript of Mission Commentary revealed that Young's first words as he stepped off the bottom rung of the ladder were: 'Hey . . . mysterious and unknown Descartes, Apollo 16 is going to change your image!' One of the first things Duke said was: 'Good Lord, look at that hole we almost landed in!'

They had some difficulty, standing side by side, as they hauled on the lanyards which lowered the Lunar Rover from its stowed and folded position in the side of the descent stage to the surface. They had been on the surface for over an hour before the Rover's antennas were deployed; suddenly Mission Control announced that voice communications had improved '900%' and shortly after we also had excellent colour TV of their activities.

For Young and Duke our ability to observe their every movement was not so advantageous. While Duke was boring holes 9 ft deep – with surprising ease compared with the problems encountered by the previous crew – in which to insert sensors for measuring the heatflow from the Moon's interior, we watched, mesmerised, as Young backed into and broke beyond repair the all-important cable between the heatflow sensor and the electronic box intended to transmit the measurements back to Earth. Young was embarrassed and profusely apologetic, since this was the highest-priority experiment.

High spirits were soon restored when they drove around in the Lunar Rover, collecting core and rock samples; one rock picked up by Duke was, he said, the 'sharpest' thing he had ever seen, and its whiteness was explained as being evidence that it had been thrown up from deep below the surface by a meteor impact. Near the end of the 7 h 11 min EVA, Young was filmed by Duke as he carried out the planned 'Lunar Grand Prix' driving the Rover flat out in circles and skidding it to test wheel grip. On one occasion, as he bounced over the rough surface, all four wheels were clear of the ground, and his maximum speed was estimated at 17 km/h. Duke's film of this episode became one of the most used and familiar scenes in retrospective film and TV programmes on Apollo.

Tired but very talkative, Young and Duke clambered back into Orion, very dirty, especially Duke, because he had fallen down three times and discovered that the best way to get up was to roll into a small hole or crater.

When they had got out of their spacesuits and cleaned themselves up as much as was possible in the equivalent of a telephone box each, there was a long 'debriefing' with Mission Control during which Duke said it had been 'doggone exciting' and the less exuberant Young 'pretty interesting'. After agreeing that all their geology training had paid off, it was expected that they would relax quietly before settling down to eight hours' sleep. Instead we were surprised to hear this:

> YOUNG: I got the farts again. I got 'em again Charlie. I don't know what the hell gives them to me. Certainly not – I think it's acid in the stomach, I really do.
> DUKE: It probably is.

> YOUNG: I mean, I haven't eaten this much citrus fruit in 20 years. And I'll tell you one thing, in another 12 fucking days I ain't never eating any more. And if they offer to serve me with potassium with my breakfast, I'm going to throw up. I like an occasional orange, I really do. But I'll be damned if I'm going to be buried in oranges!

Young was telling Duke that he, Duke, had done most of Young's work for him when Chief Astronaut Deke Slayton came on hurriedly from Mission Control to tell them that they had a 'hotmike' – indicating that the world was listening to their private conversation. Young's 'talk button' had apparently stuck in the transmit position. Those correspondents who had been recording at the time made numerous copies of this exchange for colleagues who had not. It was confirmation, if any were needed, of what the astronauts thought about their potassium-loaded diet.

The second 7 hour moonwalk was also an outstanding success. It could more accurately be described as a drive, since Young and Duke travelled 4.1 km to what had been named Stone Mountain. With the Rover's TV camera being operated completely independently from Mission Control, we had spectacular views of the astronauts working on ridges and crater rims 232 m metres above their landing point. Had the Rover failed, it was estimated that they would face an arduous 1 h 33 min walk back to the Lunar Module – but it did not let them down. While the astronauts were having difficulty turning over large rocks to obtain samples from both top and bottom for evidence of cosmic ray bombardment and of how many millions of years they had been undisturbed, the geologists were panning and zooming the camera for close-ups of boulders and other interesting features – and were embarrassed by lack of any evidence of the volcanic activity they had prophesied. There was some consolation for them in the discovery that this highlands area was much more heavily magnetised than the mare areas visited earlier.

Young and Duke worked tirelessly and in complete harmony, and we saw Duke fall at least six times. On one occasion he had to make three efforts to push himself up again. It was paradoxical that they were a quarter of a million miles away from any possibility of help, and yet their every move was closely scrutinised by millions. Towards the end Mission Control was pleading with them to get back inside the Lunar Module, while they were asking for an extension of their time on the surface: 'We're feeling good . . . Tony [Astronaut Tony England] please, another 10 minutes . . . All we're going to do tonight is to sit around and talk!'

Next day despite the discomfort of climbing once again into their filthy spacesuits, they were ready half an hour early for their final spacewalk, reduced to 5 h 40 min because of the timeline required for lift-off and redocking with Caspar. This time they drove 4.9 km in the opposite direction to explore the 400 ft-deep North Ray Crater and Smoky Mountain. They added to their rock and crystal samples, and advised one another not to get too near the edges of craters. Occasionally even Duke admitted he was feeling the strain. 'The old ticker's really pumping now!' he exclaimed on one occasion.

But to the alarmed dismay of Mission Control and the watching doctors, they still had energy at the end for what they called their 'Lunar Olympics'. This was a contest to see who could jump the highest in one-sixth gravity. Young estimated he had achieved 4 ft; Duke, trying to beat it, fell heavily backwards. 'That's not very smart,' observed Young, helping up an obviously winded Duke, whose life-giving backpack fortunately survived undamaged.

Sealed in Orion for the last time, Young and Duke went through a faultless countdown, and they had set the Lunar Rover's camera so accurately that we were able to watch the ascent stage's abrupt lift-off and follow it until it became a tiny dot, while listening to enthusiastic cries from inside it of 'What a ride, what a ride!' Replaying the video in slow motion, Mission Control saw something coming loose; Young was asked to roll Orion for Mattingly in Caspar to inspect it before they docked. The thermal blanket around Orion was 'pretty badly torn' Mattingly reported, but the docking was unaffected.

Apollo 16's journey back to Earth was enlivened for us many times by Mission Control turning on the Lunar Rover camera again to have another look at the Descartes area. 'It looks like a picnic area the day after a Bank Holiday,' I reported. 'The descent stage, like a dead scorpion, forms the centrepiece as the camera pans around the badly littered moonscape, and zooms in for close-ups of the huge footprints left behind.' The camera also provided the scientists with a dramatic view of a lunar sunset ten days later, before the bitter cold of a lunar night killed the batteries.

Despite the discomforts caused by their potassium-laced diet, John Young, Charles Duke and Ken Mattingly were much the fittest crew to return from space, and Duke, despite 11 days of weightlessness, actually ran across the deck of the recovery ship at the end of the welcome-back speeches. As a result, the doctors gave the go-ahead for the proposed long-duration flights of 56 days and more aboard the Skylab space station being planned for 1973.

Apollo 17

During and immediately after the Apollo 16 mission, NASA's administrators were very obviously concentrating on the future – so much so that one wondered how the thousands of engineers and technicians whose future jobs looked doubtful could possibly concentrate on the final moonlanding.

Happily for both the Apollo 16 and 17 crews, they did. But many still had doubts about whether Apollo 17 was worthwhile, and I discussed this with Dr Wernher von Braun. He caught the mood and thinking of the time so accurately, and was so prescient about the inevitability of anti-pollution regulations, that I give the exchange in full.

Having spent much of his life working to get men to the Moon, I asked him, did he think that Apollo 16 had really fulfilled what he had been trying to accomplish? He was much too skilled in public relations to give a direct answer, and careful too not to link himself closely with future projects which had yet to win funding:

VON BRAUN: It's highly gratifying that things which looked very exotic to us [NASA] earlier in the programme have become a sort of routine. It's also downright amazing to see how fast these guys have learned to walk and work on the Moon and act on the Moon as though they have spent half their lives there.
TURNILL: Did you expect them to learn so quickly in the early days?
VON BRAUN: Well, the doctors had told us all sorts of horror stories about how difficult it would be for man to adapt himself, so we were prepared for more difficulties, quite frankly. It just goes to show that the human body seems to be very adaptable.
TURNILL: It's been so successful that I think you've turned your attention to something even more exciting?
VON BRAUN: Well, the Moon has helped us in finding out how it and the Earth was born – a very similar and probably simultaneous process, but it has also taught us a great deal about new possibilities and brought in a great deal of new technology. This enables us now to do all sorts of new things beyond our reach before we had Apollo.
TURNILL: You are working, I believe, on the possibility of harnessing the Sun's power, and feeding it down to Earth; is that correct?
VON BRAUN: Yes, such a proposal has been submitted to NASA and we are evaluating it. There is obviously a great deal of public as well as Congressional interest into any promising method to meet man's rapidly increasing power needs with non-pollutant sources. Of course the Sun's energy is there in abundant and limitless quantities, and if we could somehow use on Earth solar energy that we intercept in space, we could really meet future needs quite adequately.
TURNILL: It would solve all these quarrels about oil?
VON BRAUN: Precisely. The specific proposal we are studying envisions the establishment of a very large array of solar cells – I'm talking about square miles of solar cells – in synchronous orbit, where the satellite goes in 24 hours once about the Earth, travelling

from West to East; and since the Earth spins within the satellite's orbit, such a satellite would appear to sit stationary over one point on Earth. So it would be a continuous, uninterrupted line of sight, day and night, with many points underneath. The idea now is to convert this collected solar energy, which the solar cells convert into electricity, into microwaves, and beam the power down, with the help of microwaves, and build an antenna array on the ground where this power radio-ed down would be put on the power lines, and fed into the public utility network.

TURNILL: You mean the Sun's energy would be fed down in a sort of invisible pipeline, in a concentrated beam. It sounds terribly dangerous.

VON BRAUN: In fact energy density in this beam would be so low that you could safely fly an airplane through that beam. We are envisaging antenna farms on the ground, again in the order of several square miles, and the density in these farms would be so low that you could safely graze cattle in them.

TURNILL: How much power could you get out of such an operation?

VON BRAUN: Conceivably out of one such plant enough to take care of a whole country, or several countries. In other words in the multi-megawatt region.

TURNILL: You mean the whole consumption of, say, the United States could be met by this means?

VON BRAUN: This has been proposed. Now let me add a word of caution. So far this is nothing but a scheme, and we are investigating the feasibility. So far we have not found a technical reason why it would not be feasible. The question is what would it cost (a) to establish such a system, and (b) what would we have to pay for the kilowatt hour if we operate such a system. Right now we are still pretty badly out of tune with electrical rates that people are ready and used to paying. In fact our present investigations indicate that it would be about one hundred times as expensive. However, this is based on today's cost for the solar cells, and today's transportation costs into space. Innovations such as our re-usable Space Shuttle, and conceivably the manufacture of much cheaper solar electric cells, may knock the price very drastically down. There is also the other consideration that we might as well get used to the fact that electricity will become more expensive if we are to meet the anti-pollution regulations that man undoubtedly will insist upon in the future. So electric rates even for fossil-burning hydro-electric plants will go up.

TURNILL: How long do you think it would take to build a pilot plant?

VON BRAUN: This depends very much upon the funding level. I think we have the technology here today to build a pilot plant if we use existing technology for solar cells and existing means of transportation for going into synchronous orbit. For example, with the Saturn 5 you can carry a very substantial load into synchronous orbit, but you limit the system as a demonstrator to a couple of thousand kilowatts. We could probably establish such a demonstrator in the next six or eight years. This of course would be a very small demonstrator of a principle that later on would require far larger and heavier units.

TURNILL: You mean you could establish one of these pilot plants with just one Saturn 5 launch?

VON BRAUN: If you make it small enough, yes. The question is what do you propose to demonstrate; how much power do you really want to beam down? You see, an even smaller demonstrator could be built if you use one of our existing spacecraft that collect solar energy: put a little microwave converter into the thing and radio your energy down. You don't even bother going into a synchronous orbit; you just want to demonstrate that you can bring the energy down to the Earth. If you settle for that, you could probably do it for a few million dollars in a few years.

TURNILL: Is this project as exciting to you, Dr von Braun, as your efforts to get a man on the Moon?

VON BRAUN: Well, I have always looked on spaceflight as a new area of human activity in broader terms. I always felt it would have its exploratory aspects like visiting the Moon or other planets, but I also felt that, just like aviation, it would have to develop its commercial aspects; it must do things for man that he could understand so that he could continue to be willing to pay the bills. We need a bread-and-butter base for spaceflight. It is similar to aviation; aviation does a lot of interesting things in exploration and so forth, but at the same time has found its bread-and-butter base in the airlines and airfreight and a lot of other things. And that's what spaceflight needs too.

TURNILL: Have you felt at all depressed lately at the way the public has got bored with moonlandings already?

VON BRAUN: I find it a little surprising that the public is getting so blasé about the whole thing – that's all I can really call it. A lot of people in America seem to feel that we sent a man to the Moon in order to show the Soviets that we are still first in technology, and now that we have won the war, let's get the boys home, let's do something else like cleaning up our rivers. Well a thing like a space program cannot be turned on and off like a faucet, and I think it would be a lot better to have a moderately financed space program, with a stable funding source over a number of years than this hot-and-cold blowing that we had in recent years.

TURNILL: And you think a Space Shuttle will provide that?

VON BRAUN: Yes, I hope so. The Space Shuttle will undoubtedly provide an entirely new foundation for spaceflight, both manned and unmanned. It will be the cheapest, most economical space transportation system ever devised by man. We are paying about $1000 dollars today to put one pound of payload in a low orbit and bring it back, and $500 dollars to fly it up one way. With the Shuttle this will go down to about $160, and ultimately to as low as $100 or even $50 a pound. This of course will make spaceflight attractive for a lot of things for which it is simply too expensive today. And being the most economical way of transporting anything into orbit, the Shuttle will also do away with the argument that unmanned

> spaceflight is better than manned spaceflight, because if it is the cheapest way to use a manned vehicle to fly even unmanned payloads into orbit then everybody will use it. It will simply corner the market of transportation into space.

Perhaps, if von Braun had lived, his genius would have ensured that the Space Shuttle did not fail so dismally in getting down the cost of placing payloads in orbit. The Challenger accident of 1986 merely forced NASA to recognise the Shuttle's failure to achieve von Braun's expectations sooner than they would otherwise have done. That was because in those days we were all conditioned to believe that whatever targets were set could and would be achieved. So in the eight months between the Apollo 16 and 17 missions, US and Soviet scientists agreed on the first (and for nearly 20 years the only) joint spaceflight, and in that case the target date of 15 July 1975, set three years before for the Apollo–Soyuz linkup, was achieved. [That was possible because the engineers were not continually frustrated by having their budgets cut.]

Six weeks before Apollo 17 was launched on 6 December 1972, the Soviets demonstrated their superior capability in planetary exploration by obtaining a surface picture of Venus – man's first glimpse of the surface of another planet. Interviewed by Michael Aspel – then still a BBC newsreader – I speculated that the Russians would want to start sending cosmonauts to the Moon a year or so after the Apollo landings ended; and they would want to do the spectacular things that NASA had failed to do, like landing in the rugged lunar mountains, and making sure that the first human on the farside was a Russian. And, while the Americans were thinking in terms of man on Mars by about 1986, the Russians would almost certainly get there by 1982!

Three days before the Apollo 17 launch I was given an hour on Radio 4 (but on a Sunday evening when everyone was watching TV!) to review the whole of Project Apollo. It was really a sort of swansong for BBC radio's distinguished senior producer, Arthur Phillips, who was retiring, and to my dismay he insisted on titling the programme *Out of This World*. Such programmes usually ended up being called either that or *Reaching for the Stars*. However, it gave me a useful chance to stretch my verbal wings, even though, as usual, we ended by cutting out my best quotes because we were over-running.

I much regretted cutting out my old friend John Hodge, British-born NASA flight director: 'I don't know how you can expect to explore the Moon with just a few landings. If you try to compare it with Earth, which we did several times, it's very difficult to know where you'd land ... It leaves more to do next time – and I presume there will be a next time. It's unfortunate we couldn't go a little further; but the priorities had to change, and they did.'

The lunar scientists, who felt even more frustrated that, having mastered the technology of exploring the Moon, they were to be deprived of the most rewarding missions, did have one small success with Apollo 17. They succeeded in having Joe Engle, perhaps NASA's most capable and distinguished test pilot after Neil Armstrong, replaced by Dr Harrison (Jack) Schmitt, who thus became the only scientist–astronaut to reach the Moon.

'On the surface, an expert on breccia and volcanic vents is just the thing,' I argued in a pre-launch *From Our Own Correspondent*: 'But to make a safe landing, stay alive and get home again, you still need a top-quality test pilot at the controls, even though radar and robots have taken out much of the terror.' The safety margins were still so narrow that, in order to give Apollo 17's Commander, Gene Cernan, an extra five seconds to pick out a safe place to land in a rocky valley on the edge of the Sea of Serenity, they had had to reduce the Lunar Module's weight by not carrying the 12 lb camera which usually enabled us to see the first of the two astronauts descending to the Moon. The omission meant that there would be no TV from the surface until the Lunar Rover was deployed 70 minutes later.

I began that broadcast, made just before leaving London for Florida, with a quote from one of my BBC bosses: 'Watching a couple of astronauts plodding around on what's really nothing more than a dreary old slagheap has never been the sort of thing that grabs me! What's the use of it?' I tried to provide an answer as a pay-off:

Apollo 17, Cape Canaveral: the last moonlanding crew. Eugene Cernan, Commander, seated; Ronald Evans standing, right, Command Module pilot; and Harrison (Jack) Schmitt, Lunar Module pilot and the first and only trained geologist to be sent to the Moon. They are pictured on a Lunar Rover trainer, with their Saturn 5 launcher in the background. (NASA)

> When they've finished their 10 years of analyses and arguments over their 800 lb of slag, I hope the geologists will be able to keep their promise and tell us the origin of the Moon. Maybe it's academic whether it's a lump of this Earth that broke off five billion years ago. What isn't so academic is how the Moon got to be such a dreary slagheap. With that knowledge, maybe we can do something to ensure that what Frank Borman, looking down from orbit four Christmases ago, called 'This Good Earth', doesn't also become just another slagheap.

The Apollo 17 launch was almost a wake. Thirty astronauts, some of them no longer active, turned up to watch it because they knew it was the last – and it was also the first night-launch of a Saturn 5. The world's oldest man was also there, NASA having checked that he was genuine. He was 130-year-old Charlie Smith, who had been brought to America at the age of 12 as a slave from Liberia. He had been making public statements that NASA had not really been to the Moon, and that the so-called moonrocks were smuggled aboard before the spacecraft was launched. Unfortunately I was too busy to find out whether being present at the launch convinced him of the truth.

When it came at 12.32 am the launch duly turned night into day; looking at the shatteringly pure gold flame at the base of the rocket was as painful and unwise as looking at the Sun. Jack Schmitt, as the first geologist to look back at Earth, was at first encouraged to make a few observations on the subject, and drew some conclusions about continental drift. But then he settled down to provide a running commentary on Earth's weather, which still went on when Apollo 17 was 180 000 miles away. Mission Control hinted that he should desist by pointing out that the meteorological satellites had already identified a cyclone he mentioned – to which Schmitt retorted: 'We're not in competition.'

It was almost a relief, I reported, when Apollo 17 suffered a master alarm, and Schmitt had to join Commander Gene Cernan and Ron Evans in checking out the systems. The worst that happened on the outward journey was that Gene Cernan kept complaining of stomach gas as a result of the potassium diet, and was told that one way to reduce the gas would be to stop chewing gum! When the crew duly went into their egg-shaped lunar orbit, Schmitt had difficulty focusing his eyes on the strange sights revealed, and confessed: 'My trouble was that I thought all the hills were holes. I had the picture upside down!'

Soon after, Schmitt, looking down from lunar orbit, caused excitement among his fellow Earthbound geologists by announcing that he had seen a flash. Ken Mattingly, who had stayed in orbit during Apollo 16, had been disbelieved when he reported a light flash on the Moon, and told it was almost certainly a cosmic ray flash passing through his eyeballs. Schmitt's report raised hopes that Mattingly might have been right, and that there really was some volcanic activity.

But it was agreed that it was more likely to have been a meteorite strike, a sunflash on a mountain ridge, or even a reflection from the tiny sub-satellite left in lunar orbit by Apollo 15.

As on Apollo 16, the Command and Lunar Modules separated on orbit 12. This time they became respectively America and Challenger and there were no problems. Gene Cernan brought Challenger smoothly down between two 2130 m mountain ranges to land between the Taurus Mountains and Littrow Crater only about 400 yards from the aiming point. The area had been chosen by scientists because Alfred Worden, orbiting in the Apollo 15 Command Module, had seen conical mounds similar to those formed on Earth by volcanic debris accumulating around a vent. When Cernan found he had touched down with 3 minutes of fuel left he mourned that he could have used it 'to hover around and look at the scarp' – to which Mission Control replied with emphasis: 'No thank you, we're happy right where you are.'

When they climbed down to the Moon for the first EVA both astronauts were in an excited and emotional state. Aware that they might be the last men on the Moon in the twentieth century, as indeed they were, Gene Cernan announced that he would like to dedicate his first step to all those who had made it possible. At this stage, with no TV coming back, we could only listen when he followed with the exclamation: 'Oh, my God, it's unbelieveable! We landed in a shallow depression. That's why we've got a slight pitch-up angle . . . It's very, very hummocky. There was no dust on landing, but I can put my foot in ten inches or so where we landed . . . I think I landed 100 metres from Poppy.' [Crater Poppy, named after Poppy Northcutt].

Jack Schmitt found the scenery irresistible, wandered off, and had to be good-humouredly recalled by Gene Cernan: 'What are you doing over there? We're supposed to be working!' But Cernan himself was fascinated: 'All these little craters got glass in 'em!' One told the other not to walk too fast as they were laying out the cables, and Mission Control passed a message from the surgeon that they were to slow down as the water temperature in their suits climbed past the redline.

Schmitt confessed that he had not yet learned to pick up the rocks. 'That's why I fell down; very embarrassing for a geologist. My first experience of getting moon-dirty.' 'Calm down,' advised Mission Control, to which he retorted' 'I've never been calmer in my life!' Repeated warnings were sent that they were burning up too much energy as they deployed the fifth US lunar laboratory, but they were right on time, 70 minutes into the EVA, when the Lunar Rover's TV camera was switched on and suddenly we could see, with perfect clarity, the smooth-topped Taurus mountains rising to heights of one to one and half kilometres behind them. An avalanche slide of light material was visible at one point.

Cernan had some initial difficulty drilling through sub-surface rock, but finally succeeded in inserting the two heat probes to the maximum possible depth of 2.36 metres. (The subsequent science report said that the measured heat-flow averaged half the Earth's heat-flow.) In my early morning reports for 12 December 1972 I reported national nostalgia when Cernan deployed on the Moon the US flag which had hung in Mission Control since Apollo 11, and moments of euphoria when he led Schmitt in singing:

I was rolling on the Moon one day,
In the merry month of May...

Inevitably they used up their oxygen faster than expected, and some of the activities planned during the 7 h EVA had to be cut short. After a rest period, the second EVA started 90 minutes late because during their first drive a mudguard had fallen off the Lunar Rover, and dust had been sprayed over their delicate equipment. Under John Young's direction, they had to construct a make-do substitute by taping together four big lunar maps, to be fastened in place with clips from the spacecraft's emergency lighting system.

Schmitt started off slightly depressed, having been told by Mission Control that some of his tasks had been eliminated. His indignant query as to what he was supposed to do was met with an unusually curt rejoinder from Mission Control: 'You're supposed to help Gene, I guess.'

At the end of a 1-hour drive, and more than 4 miles from the safety of the lunar module, we were regaled with a magnificent half-Earth view. The rover's TV camera, operated remotely from Houston, panned around until it found the Earth above the lunar horizon, then zoomed in to give us a perfect colour view, with clouds and blue sea visible as we gazed at ourselves from a quarter of a million miles away.

The moment when Jack Schmitt kicked up orange soil still ranks as a highlight in the whole Apollo programme. On past missions Schmitt had been asking moonwalkers to kick up the surface as they walked, to expose material below, which had been protected from the fierce variation of temperatures over the lunar month. On the surface, temperatures range from −185 °C to +86 °C (a spread of 271 °C) while at a depth of one metre the variation is only a few thousandths of a degree.

Cernan and Schmitt were already behind time on their 19 km tour, and had had some sharp exchanges with Mission Control. After a dispute and raised voices about the wisdom of carrying a rake on the slopes of the South Massif, Cernan told them: 'It's being done; but let's watch those kind of calls, please'; and we heard him comment to Schmitt: 'They can't appreciate the toughness of going up the slope.' To which Schmitt replied 'We told them, though.' A few moments later one of them said 'Oh, relax' – perhaps Cernan to Schmitt.

Tempers soon cooled, but Mission Control was still pushing them when they arrived at 'Station Four – Crater Shorty' and asking for the TV lens to be brushed and dusted, when we heard this:

SCHMITT: Okay, okay – Oh, hey! There is orange soil.
CERNAN: Well don't move till I see it.
SCHMITT: It's all over orange.
CERNAN: Don't move till I see it.
SCHMITT: I stirred it up with my feet.
CERNAN: Hey, it is, I can see it from here.
SCHMITT: It's orange.
CERNAN: Wait a minute, let me put my visor up. It's still orange.
SCHMITT: Sure it is. Crazy. Orange.

CERNAN: I've got to dig a trench, Houston.

MISSION CONTROL: Copy that. I guess we'd better work fast.

SCHMITT: He's not going out of his wits, it really is! . . .

CERNAN: Well, slap me with a little cold water.

SCHMITT (Adjusting the TV camera): Ok, the stuff has been dusted. I think I gave you 102 or something like that.

CERNAN: Fantastic sports fans. It's trench time. Hey, you can see this in your colour TV, I bet you.

SCHMITT: I didn't think there would be orange soil on the Moon.

CERNAN: Jack, that is really orange. It's been oxidised. Go around and get the lunar sounder over here. It looks just like an oxidised desert soil . . . Man, if ever there was a – I'm not going to say it – if ever there was something that looked like a fumarole, this is it.

Having chosen the Taurus-Littrow area for the last landing site because they thought it would provide evidence of volcanic activity, Schmitt's cry of 'Orange soil' created intense excitement among the geologists and astronomers. It seemed that their hopes and prayers had been answered.

Hurriedly the Rover's TV camera was zoomed in, and we could see the orange soil for ourselves. In retrospect it does not sound very exciting, but in all the moonwalks we had seen no real colour at all on the lunar surface, and had had to wait for colour to be revealed when lunar samples were placed under the miscroscope.

With their oxygen supplies running low, Cernan and Schmitt stayed at Crater Shorty – soon, rather optimistically, re-named Volcano Shorty – for an extra 35 minutes. They dug a trench 8 inches deep and 35 inches long, drove a core tube into the deposit, sampled surrounding rocks, and described and photographed the site in detail. While the astronauts made their slow way back to the Lunar Module for another rest period, there was general agreement among the geologists at Mission Control that the rust-coloured soil meant that a volcanic vent had been found, and that there had been, and perhaps still was, water on the Moon. [But three months later, following analysis of the samples

Apollo 17: Jack Schmitt working on huge boulder at Taurus-Littrow. Lunar Rover and its TV camera visible in foreground. (NASA)

brought back, the US Geological Survey announced that it consisted of tiny, orange-coloured glassbeads formed by the heat generated by an ancient meteor impact. That view was in turn reversed in 1975, when it was decided that the 'beads', unusually rich in lead, zinc and sulphur, *were* of volcanic origin after all, and had been thrown out from perhaps 300 km inside the lunar interior during a period when the lunar mantle had partially melted.]

Proposals that on their third EVA the astronauts should return to Volcano Shorty were rejected, because the pre-planned, 8-mile drive

Apollo 17: before the crew returned to the Lunar Module after spending 22 hours 6 minutes on the lunar surface, they carefully positioned the Rover so that its TV camera could provide Mission Control and millions of viewers with excellent views of their lift-off from the Moon and ascent into lunar orbit. Mission Control reactivated the camera many times after their departure to study panoramic views of the Taurus-Littrow area. (RCA artist's depiction)

included Crater van Serg (named after a twentieth-century mining geologist), which had also been predicted as a likely volcanic area. While there was disappointment that no more red soil, nor evidence of volcanic activity was found, there was plenty of visual drama as we watched Schmitt scaling a boulder the size of a London bus to collect samples; and the Lunar Rover was forced up mountainsides so steep that its wire-mesh wheels were dented.

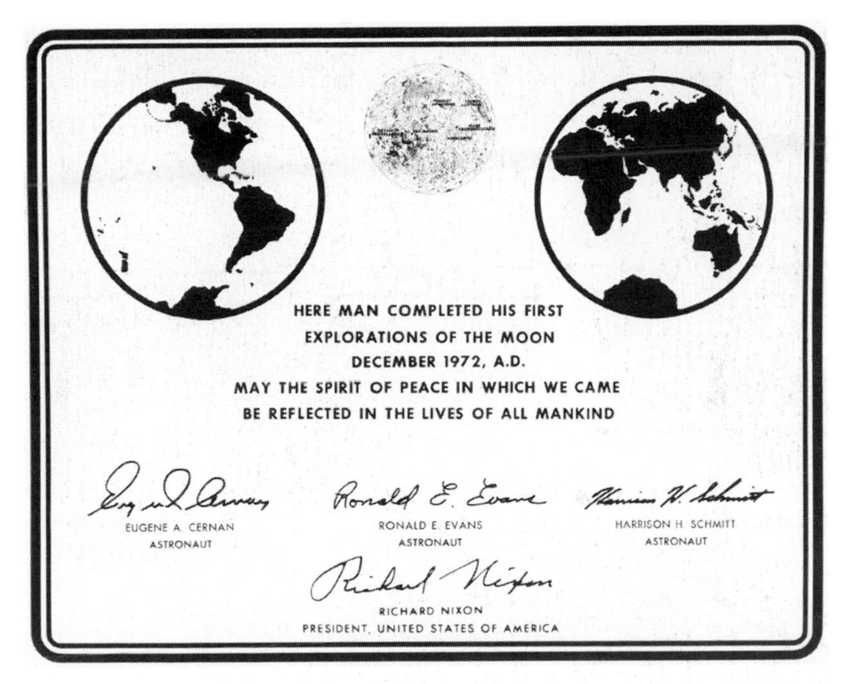

Apollo 17: the farewell plaque attached to a leg of the descent stage of the Lunar Module Challenger at Taurus-Littrow. It is likely to survive there for centuries. (NASA)

Before leaving the surface the astronauts unveiled a plaque on the descent stage of the Lunar Module containing discs of the Earth and of the Moon indicating the position of the six landing sites; like the Apollo 11 plaque, it also bore the signatures of all three crew members plus that of President Nixon, beneath the words:

> HERE MAN COMPLETED HIS FIRST EXPLORATIONS OF THE MOON, DECEMBER 1972 A.D. MAY THE SPIRIT OF PEACE IN WHICH WE CAME BE REFLECTED IN THE LIVES OF ALL MANKIND.

But there was concern at NASA that Nixon had not made time to telephone the crew while they were on the Moon or in space. All he did was to send a message saying 'God speed you safely back to this good Earth' plus an assurance that space exploration would continue – which he then spoilt with the significant prophecy that 'this may be the last time in this century that men will walk on the Moon'. We should have believed that – but we did not because we desperately did not wish to believe it.

21 Apollo's Inconclusive Findings

What did it all amount to? We were being asked to explain that long before the Apollo 17 crew splashed safely down, to be hauled aboard recovery ship USS *Ticonderoga*, amid much flag-waving and speeches about pride in being American that again belied the claims to have carried out lunar exploration 'on behalf of all mankind'.

I went once again to Dr Wernher von Braun for his views on what had been achieved, and what NASA should be doing next, and found him holding court in the lobby of the King's Inn Motel, a few yards from the entrance to the Johnson Space Center and Mission Control. By then, from my point of view, he was the perfect interviewee, with a remarkably clear mind and briefly expressed, well-defined views. Whatever one felt about his past, he struck me then, as I remember him now, a generation later, as an example of the perfect manager.

He was adept at shaping future policy, while making it clear that final decisions were not his – nor his responsibility. With the moon-landings done, he had obtained for himself and his team at the Marshall Space Flight Center a major role in the follow-on, Saturn-launched Skylab space stations, as well as in the development of the Space Shuttle and the space station it was intended to serve. For him future prosperity now lay there. He had the ability to make all these points smoothly while apparently allowing me to direct the course of the interview:

> TURNILL: Dr von Braun, is this the moment you have been waiting for?
>
> VON BRAUN: Well, I would say this discovery of the orange rocks has surely justified every bit of Apollo. It showed that manned spaceflight is not only a very powerful tool to explore the Moon, but

also to verify man's leaders' cosmological theories on how the entire solar system came into being, including our own Earth.

TURNILL: Had you begun to despair of finding anything so exciting as this?

VON BRAUN: The relative uniformity of the material found on the Moon was indeed a little disappointing to me, although of course I am not a geologist, and from the reports I have seen I gather the geologists have read a lot more into the findings even of previous flights than I could. To me rocks look pretty much like rocks, but to them they surely don't.

TURNILL: But this find, just as Apollo is ending – doesn't that make it seem crazy not to go on exploring the Moon?

VON BRAUN: Well I am absolutely convinced that we will keep visiting the Moon. We will just use different equipment. Apollo was, you might say, the sailing ship and dog-sled technology to go to the Pole; next time we go to the Pole we'll use airplanes, and in this particular case I mean that instead of going there with overgrown ballistic rockets, we will use the re-usable Space Shuttle to go the Moon in three phases, at probably one-tenth of the present cost.

TURNILL: Well since that's at least ten years off, won't there be tremendous pressure to reinstate Apollos 18, 19 and 20?

VON BRAUN: I think we've passed the point of no return. Too many people have left the company payrolls that had to support the program. A decision to go on after Apollo 17 would have had to be made about a year ago.

TURNILL: So surely this is the Russians' great opportunity to sweep in and pick up exploration of the Moon at this vital point?

VON BRAUN: It might well be, but I wouldn't be too concerned about that. To come back to my example of the Polar exploration, you have the same pattern too, that there were long gaps between dashes to the South Pole, and every time man came again he was better equipped than the previous time.

TURNILL: So basically your feeling is one of intense satisfaction

The author interviewing Dr Wernher von Braun for a BBC TV programme. He was already fighting cancer, and died soon after in 1977. 'I just envy the youngsters who have a chance of going on where we leave off' he was saying – and would certainly have been astonished that there is still no plan for a manned expedition to Mars. (Margaret Turnill)

rather than disappointment because the find has come so late?

VON BRAUN: Well, I would say there is of course some nostalgic feeling about this being the last Apollo among all of us who were so involved in the program. On the other hand all good things have an ending, and as the saying goes the better is the enemy of the good; and Apollo was very good. What will come in the future is to go to the Moon with re-usable, less expensive equipment; that is most logical. I participated in some of NASA's decisions a year, or a year and a half ago, when this question came up: Should we fly more Apollos after 17? We could have flown maybe two more. And when I looked at the options available to NASA , spending all that money on two more flights to the Moon, and deferring Skylab for another year which in

itself would have cost a lot of money, and taking all that money out of the development of the re-usable Space Shuttle, I was very much in favour of discontinuing myself. Of course I did not make the final decision, but I came out in favour of a discontinuation after 17.
TURNILL: But do you think this discovery should change the emphasis of what's to be done in future: less emphasis on Earth orbital flight and more emphasis on a return to the Moon?
VON BRAUN: No, I don't think so. The Shuttle itself is a very versatile vehicle. [He still pronounced it 'wee-hickle'] You can use it for Earth orbital operations, both for science and for Earth-related applications programs; and you can stick the so-called tug vehicle into the payload bay of the Shuttle to enable you to fly from the low orbit, to which the Shuttle can climb, on to the Moon. It will be probably a two-stage tug so that two Shuttle flights to low Earth orbit will be necessary to go from there to the Moon. There may even be a third vehicle involved that looks pretty much like the LEM [the Lunar Module was originally called the 'Lunar Excursion Module' and most NASA people still called it 'LEM'] and can make a soft-landing on the Moon and then fly back to lunar orbit. The Shuttle itself is the key to everything, and the question is whether you go to the Moon with it, or use it for Earth-related programs, or for orbital-based science programs. It's only a question of mission assignments, and no longer one of major hardware commitments.
TURNILL: Dr von Braun, thank you very much and congratulations on what must be a big day for you.

Apollo 17's Lunar Module had taken off with such a heavy load – 113 kg – of moonrocks that it was technically 18 kg overweight. This was offset by reprogramming the firing of the ascent engine to burn more fuel more quickly. We all held our breath as we watched the last lift-off from the lunar surface, vividly seen and recorded, thanks to the careful positioning by Gene Cernan and Jack Schmitt of the abandoned rover so that Mission Control could focus its TV camera for their ascent.

The total of 385 kg of lunar materials returned by the six successful missions, I reported in 1984 in the first edition of my *Jane's Spaceflight Directory,* should enable techniques to be developed for making manned lunar bases self-supporting. By the end of Apollo 14 it had already been established that both water and oxygen could be produced from the lunar soil, because it contained a high percentage of iron oxide. This, said the scientists, could be achieved by using a solar furnace and introducing hydrogen, which would enable 6.35 kg of water to be produced from 45.4 kg of iron oxide. The water could then be separated into oxygen and hydrogen, enabling life to be supported and fuel provided for space vehicles.

There was much scientific dismay and criticism when, on 1 October 1977, NASA shut down the Houston Control Center, which had been receiving and analysing the stream of information from the five nuclear-powered ALSEPs (Apollo Lunar Scientific Experiment Packages) left on the lunar surface by the crews of Apollo 12, 14, 15, 16 and 17. It was an economic decision, designed to save the $2 million per year cost of the Center. From then on their continous flow of signals, which had been broken only by some occasional mysterious lapses by the Apollo 14 transmitter, went unrecorded. But by then Apollo 12's ALSEP was in its eighth year, and Apollo 17's was in its fifth year. NASA could justify the economy on the ground that 29 years of operating time had been recorded, compared with the cumulative total of six years for which the equipment was designed.

Even so, there was still insufficient data for the geologists to be certain whether the Moon had a molten core. To determine that, they said they needed at least one large 'event' on the farside greater than 10^{19} ergs. It never came before closedown, although the 10000 moonquakes and 2000 meteorite impacts that had been recorded included 18 major quakes on the nearside by 13 May 1972, and on the farside 10^{18} ergs on 19 September 1973.

Many millions of dollars were spent by NASA on developing the space tugs of which von Braun talked, and had they not been cancelled for dubious 'economy' reasons, they would undoubtedly have provided precisely what he intended – a relatively cheap way of returning to the Moon to establish bases there. It would also have provided work for the Space Shuttle more convincing to the public than repetitious 'life-science' missions of doubtful value costing half a billion dollars a time to keep manned spaceflight going while the proposed Space Station was repeatedly re-designed.

Summing up at that time, it was concluded that Project Apollo established that the Moon had no life and no water, apart from the possibility of producing it by a solar furnace. It was not a dead body, as some eminent astronomers had always maintained, but probably had a molten core. (These conclusions of course were dramatically revised in the late 1990s, when the US Navy's Clementine satellite established that there were millions of tons of frozen water distributed over the Moon's surface, with an abundance available at the Poles.)

To use the simile of Dr Farouk el-Baz, the Egyptian-born geologist who trained the astronauts – it was still uncertain whether the Moon was Earth's wife, captured from some other orbit; or Earth's daughter, having broken off, or fissioned directly from Earth; or Earth's sister, having accreted from the same cloud of dust. The Sister theory now appeared most likely.

International geologists are still analysing the rich haul of lunar rock and soil, and making mathematical models based on the behaviour of tiny sub-satellites placed around the Moon by Apollos 15 and 16, and monitored until they decayed and crashed on the surface. New theories are regularly put forward at NASA's Lunar and Planetary Science conference held early every year. At a Royal Society conference in London in January 1994 a joint paper by Professor S. K. Runcorn, of the Physics Department, Imperial College, London, and the University of Alaska, Fairbanks, suggested that at one time the Moon had three satellites; their orbits decayed, they broke up and crashed into the Moon around the equator. Thus it was the debris from these satellites that caused the impact basins, or so-called 'seas', and not colliding asteroids or comets.

At the same conference a rather depressing suggestion was put forward by Dr I. P. Wright and Prof. C. T. Pillinger of the Department of Earth Sciences, Open University, Milton Keynes, England. They had been discussing surface conditions on Mars, as revealed by studies of what they were confident are Martian meteorites found in the Antarctic after being thrown into space by meteor impacts. They suggested that if all the meteorite rocks in that Antarctic area were collected, there would almost certainly be plenty of lunar rocks among them – and they would have made the Apollo missions to the Moon quite unnecessary!

Most of us of course prefer not to believe that. At one time I had intended to publish this book under the title *Applauding Apollo*, ending it with a quote from my last interview with Dr Wernher von Braun in 1977 when he knew he was dying:

> Historians a thousand years hence will say that with Apollo we enabled man to extend his arena of activity beyond his own planet and to make himself at home wherever he pleases. If I were 10 or 15 years old today I would very definitely commit myself and my life to the space programme. There are tremendous opportunities, tremendous challenges out there, and it's a very interesting world ahead. I just envy the kids who have a chance of going on where we leave off.

By 1994, the 25th anniversary of the moonlandings, new judgments were still being made; by 2001, NASA's list of scientific discoveries resulting from Apollo included the statement that 'the Moon and Earth are genetically related; they were formed from different proportions of the same materials' – apparently confirming Farouk el Baz's 'Sister theory'.

So far as von Braun is concerned, my own conclusions were made as a result of a visit to Nordhausen in what had been East Germany, soon after the welcome fall of the Berlin Wall. They resulted in my opening chapter 'The Context – a Twentieth-Century Faust'.

22 Epilogues to Apollo

Skylab – a space station too soon

Skylab proved to be a space station far ahead of its time. It was ahead of NASA's thinking and ahead of the available technology – but no one could know that at the time. Von Braun's big rockets could put huge payloads into orbit, but the returning spacecraft were able to bring little more than their human cargo back to Earth. The Russians were soon to face the same problem with their Salyut and Mir space stations.

Even before the first moonlanding, as reported earlier, planning had been well advanced for 'AAP' – the Apollo Applications Program – intended to provide meaningful work for NASA's manned spaceflight teams between the end of the moonlandings and the time when the Space Shuttle, optimistically billed to provide 'cheap and easy access' to space, became operational. A secondary aim was to make some practical use of left-over Apollo and Saturn equipment. The need for a more convincing title than AAP soon led to 'Skylab' – a scientific laboratory in the sky.

The Soviets had been boasting for years about their plans to create a permanently manned space station, but by the time NASA was ready to launch Skylab in May, 1973, Soviet scientists were facing the low point in their whole space programme. A first attempt had been made in April 1971, with the launch of the 19-ton Salyut 1. Four days after it reached orbit, Soyuz 10, with three cosmonauts aboard, succeeded in docking with it, but the crew either failed, or did not try, to enter it. Two months later the Soyuz 11 crew were more successful. They spent 23 days aboard Salyut 1, but all three cosmonauts were killed when returning to Earth in Soyuz 11. It was a disaster unrelated to the space station; an air valve failed, leading to a rapid decompression, and the cosmonauts died so quickly they were able to give no warning that they were in trouble. Three months later, in October, Salyut 1 was directed to re-enter the atmosphere and burned up. Salyut 2, launched in April 1973, said to be 'similar in design and purpose' to Skylab, became unstable and broke up; and a backup vehicle, disguised as Cosmos 557, and launched rather desperately only three days before Skylab was scheduled, was also a failure.

The countdown for NASA's first double launch was well under way when I arrived at the Cape, thankful to escape from my obligations to report on Concorde's recurrent crises at home. NASA, I found, had not been very successful in convincing the American public that Skylab was anything more than a 'fill-in' programme, as evidenced when, soon after arriving, I mentioned the project to a US Naval Commander. 'Isn't that Russia's space station?' he inquired. America and its media were fully occupied with President Nixon and the long-running Watergate scandal, and Europe was more interested in Skylab than the US at that time. As I have described earlier, I had caused a minor sensation when doing a TV curtain-raiser, by demonstrating inside the mockup how astronauts would at last have a practical toilet: you used an airline seatbelt to strap yourself on to the lavatory pedestal to avoid the possibility of floating away from it at the crucial moment.

Despite the public indifference displayed by the Naval commander, the Skylab space station, weighing what even today can be described as a massive 90 tons (with a docked Apollo it was the same weight as the Santa Maria in which Columbus sailed to America 481 years before) was due to be launched unmanned by the last Saturn 5 rocket ever to be fired. Within 24 hours the first of three astronaut crews was to follow. They were to stay for 28 days, with successive crews occupying the station for 56 and 84 days. Such long-duration flights, I reported, would provide the medical information needed to enable humans to embark on the much longer missions being planned for flights to Mars in ten years' time! The Skylab crews would need only the much smaller Saturn 1 rocket – the first stage of the Saturn 5 – to send them up to the station, since supplies for all the flights,

The three manned launches to the Skylab space station were made from what was dubbed 'a milkstool', because the Saturn 1Bs required to launch the crew were so much smaller than the Saturn 5 required to launch Skylab itself. (NASA)

including one ton of food and three tons of water, were being sent up in advance aboard Skylab.

Space correspondents like myself were full of hope and confidence that Skylab, as big as a three-bedroom house, would provide proof for us, as well as for the man-in-the-street, that space could be used for much more than TV transmissions. 'With the help of a huge section full of telescopes, they're due to take man's first good look at the Sun clear of Earth's dusty atmosphere,' I broadcast. 'They'll be trying to find out how it works: what causes those fearful solar flares? And will it be possible to harness the Sun's energy out there in space, and funnel it back to Earth before man uses up his existing power supplies? They'll be looking Earthwards too, to try to find ways of making more economical use of the resources we've already got; and carrying out welding, crystal-growing and other experiments using the space vacuum – a total of 90 experiments.'

There was plenty of material for pre-launch contributions to radio and TV news programmes, plus my first major piece for *Blue Peter*, the BBC's long-running children's programme. For them I was able to provide pictures of Anita and Arabella, two spiders bred for spaceflight at Houston. Judith Miles, then 16, of Lexington, Massachusetts, had raised the crucial question: could a spider spin its web when weightless? They were to be sent into space hungry, to ensure that they would soon get busy spinning in the hope of catching something for dinner; when they moved, lights would come on in their cage and cameras would photograph them at work. Would they find ways, despite the absence of gravity, of spinning those beautiful, symmetrical webs that we see sparkling with dew in our earthly gardens?

There were more important issues too. During the pre-launch briefings I repeatedly questioned the NASA flight controllers about the ultimate fate of Skylab. I had noted that the Russians made much of the fact that when their Salyut stations failed they posed no threat to populated areas on Earth, because they were provided with re-entry engines so that any surviving debris could be made to fall in the wide Pacific. There were no re-entry engines on Skylab, and my questions about what would happen when its 100-ton bulk finally re-entered the Earth's atmosphere were brushed aside: 'it would be many years before Skylab re-entered, and no doubt ways would be found by then of dealing with it!' Six years later – not the 11 years forecast by NASA – an apprehensive world learned that no such ways had been found.

It was a problem that the mission directors and the thousands of participating scientists around the world did not wish to face as the launch crews worked through two overlapping six-day countdowns. The first crew, led by the indomitable Pete Conrad, was due to follow after it had been established that Skylab had been successfully placed in orbit, its 12-ton telescope mount swung out at right angles, and its windmill-shaped solar wings all deployed to gather power supplies. And we thought it *had* been launched successfully when at last it cleared the pad on a hot and humid Florida day – even though it was still only May. However, as it vanished in the murky overcast I reported, more prophetically that I knew, that the lift-off vibration was terrifying, and that I feared (not for the first time) it would bring the corrugated roof of the Press Stand down upon our heads.

NASA was reluctant to admit, some hours after the flight, that all was not well. This was because DoD, the US Department of Defense, was insistent that they must not disclose to the media that their long-range tracking radars could reconstruct the shape and size of every object in space and were thus able to provide full details of the setback. Slowly our US colleagues extracted from their NORAD contacts in the Cheyenne Mountains the extent of the disaster. The launch vibrations had torn away the pressurised aluminium shield intended both to keep Skylab's main workshop cool and protect the astronauts from micrometeoroid hits. Its jagged edges had snagged one of the main solar panels, preventing it from being deployed, while the one opposite had been torn off altogether. The gyroscopes, needed for turning the spacecraft towards the Sun or wherever else its instruments were required to look, were also damaged. (Many years later I learned that a reconnaissance satellite had been used to photograph Skylab, and that the photos were used to train the astronauts in their repair missions.)

Gloomily we were given conflicting briefings about the possiblity of manning Skylab, but Pete Conrad was insistent that his launch, with his companions Joe Kerwin, a physician, and Paul Weitz, should go ahead. Conrad was a much-respected veteran of three flights, and had commanded the second moonlanding mission; but we were all sceptical about his proposals to take up a 40 ft nylon tether and use it to drag the surviving solar panel into place by brute force. NASA's generous safety margins, it was thought, would enable Skylab to operate on the power from only one panel, and Conrad, faced with scepticism at Houston, rang Dr James Fletcher, the Administrator at Washington, to express his views forcibly. No doubt that helped, for though Fletcher was probably the most cautious of NASA's leaders, one of the Senators was by now scornfully labelling Skylab as 'a billion dollar boondoggle'.

Soon a salvage operation reminiscent of Apollo 13 was under way, and Conrad's crew was among those spending hours in Houston's underwater simulators, practising the deployment of three alternative sunshields. The first requirement when they arrived was to get one in place, for the temperature inside Skylab's unshielded interior had risen to 125 °F.

Ten days late, the first Skylab crew was at last launched in an Apollo Command Module crammed with an additional 180 kg of sunshades and emergency tools. Flight controllers had insisted that all this extra payload must be securely stowed six hours before lift-off; but last-minute modifications to the sunshields meant that they were still being stuffed beneath the centre seat 3 h 40 min before launch, with its occupant, Joe Kerwin, having to wait outside until he could occupy it.

In the end two of three hurriedly devised sunshades were taken. The first, which was to save the whole mission, was the brainwave of Jack Kinzler, a veteran with 32 years' service with NASA. He sent out for three retractable fishing rods, and designed a parasol-like sunshade which could be pushed out of one of Skylab's scientific airlocks and then opened like an umbrella. The second, longer-lasting, twin-pole sunshade, could be rolled back over Skylab's workshop section during an EVA or spacewalk.

Seven hours after launch the Command Module was circling the crippled space station while Pete Conrad was giving us a graphic description of how one of Skylab's 9 m-long solar panels was missing while the other was jammed partially open, trapped by the remains of the torn heatshield. We had known that the surviving solar wing was partially deployed, because it had been providing a trickle of very useful power.

With Conrad at the controls, manoeuvring so close to Skylab that he was in imminent danger of colliding with it, Paul Weitz leaned out of the open hatch with toggle cutters fixed to a long pole and tried in vain to cut the solar panel free. There was much bad language when Weitz caught Conrad's head and the control panel with his end of the pole; so they gave up, closed the hatch, repressurised the spacecraft, and attempted to dock with Skylab. Then came the worst four hours of the whole mission as they tried and failed six times. For the third time that day the crew struggled in cramped conditions into their spacesuits, opened Apollo's docking tunnel and dismantled the docking probe. Nothing wrong was found, so it was replaced and Conrad made a last attempt to dock before accepting that they must make an emergency flight home and finally abandon Skylab. Incredibly, this time the docking, encouraged by brute force, was a success.

After resting, Conrad remained in the Command Module while Weitz and Kerwin, wearing masks and testing for possible dangerous gas, opened five hatches to give them access to the three sections of the space station. Before long we were able to see, via the TV camera set up in Apollo's window, the orange tip of the parasol emerging from the scientific airlock. Conrad and Weitz, taking frequent rests in the blistering heat of the interior, slowly forced through the 7.6 m parasol handle. Then with difficulty the handle was pulled in again, section by section. The parasole's wrinkles and folds were slowly eliminated, until the sunshade, a 6.7 m × 7.3 m rectangle, rested snugly above Skylab's workshop. Protected at last from the direct rays of the Sun, the interior began to cool – though more slowly than it was hoped.

With the windmill-shaped solar panels on the Apollo Telescope

Mount working normally, plus 18 out of 26 operational batteries, the crew had half their expected power supply, but that was enough to enable them to start work on some of the planned scientific experiments – until the sixth day, when another crisis hit them. Two more batteries failed, necessitating the turning off of one of the three air conditioners. The loss of another two would once more mean an emergency return to Earth. Conrad increased his pressure on Mission Control for permission to make a hazardous spacewalk to take out his tether and attempt to free the jammed solar panel, and at the same time deploy the second, 'twin-pole' sunshield – an operation which was still being rehearsed at Houston. With the temperature in Skylab's workshop still obstinately hovering uncomfortably in the mid-80s, deployment of the bigger sunshade was becoming equally urgent.

After three more days of discussion and onboard rehearsals, Mission Control at last gave the go-ahead, and Conrad and Weitz began crawling along Skylab's outside, despite the fact that there were no handholds. Conrad's efforts with the toggle cutters on the end of a 7.6 m pole to cut away the jagged heatshield and thus free the snagged solar wing were unsuccessful. So he crawled on, rather testily insisting to a nervous Mission Control that he was in no danger, and lay across the 1.2 m wide and 9.4 m long solar wing beam, guiding the toggle cutters at close quarters while Kerwin, from the relative safety of the workshop's 'roof', operated them by means of a lanyard. Alas, we could see almost nothing of what was going on, despite Weitz's attempts to send back a TV picture through Skylab's windows.

But it needed little imagination to visualise Conrad fastening his coiled 9.5 m rope to the beam and standing up to obtain leverage over his shoulders. Suddenly the solar panel's beam broke free, swung out and clicked into place. Conrad and Kerwin described how they 'literally took off' as it happened – but were safely held by their tethers. Within minutes electric power was flooding into the eight additional workshop batteries which had been useless since lift-off. Thereafter Skylab had plenty of power for its whole mission despite the loss of the

Skylab: since it was not possible to take a picture at the time, NASA got an artist to depict the drama as Pete Conrad stood on top of Skylab, orbiting at an altitude of 440 km, and with a tether over his shoulder, used brute force to pull out the jammed solar panel, thus making the whole mission possible. As the solar panel sprang out, Conrad said he 'literally took off', but was prevented from departing into space by the tether leading back to Joe Kerwin, science pilot, assisting in the background. Conrad was the prototype 'action man', and it was characteristic that he died in the late 1990s when he failed to take a corner on his motor bike. (NASA)

second solar panel, and deployment of the twin-pole sunshade became less urgent.

This most daring of spacewalks, not really matched even during the more recent hazardous Space Shuttle repair missions, turned the threatened 'billion dollar boondoggle' into a mission so successful that 25 years later it would be looked back upon with nostalgia. On a later spacewalk to recover film from the Apollo Telescope Mount, Conrad even succeeded in bringing one of the two failed batteries back to life by thumping it with a wooden hammer – with the result that 23 of the 24 batteries gave good service until the end of the third mission.

There was a 36-day gap between the time that Conrad's crew returned to Earth – having briefly wrested from the Russians the long-duration record for manned spaceflight – and Skylab's re-occupation by a second crew. That was led by Alan Bean, who was to become an accomplished space artist, accompanied by Owen Garriott and Jack Lousma. With them went spiders Anita and Arabella, who provided me with some sentimental TV pieces for children's TV.

All three members of the second crew suffered badly from space sickness, euphemistically described by Bean as 'stomach awareness'. It delayed for a week the long and difficult spacewalk during which they successfully erected the second sunshade over the top of the now fading parasol. The spiders proved that they could spin webs while weightless, even though the webs appeared to have been spun while they were drunk. Despite the diet of dead flies with which the crew rewarded them, Anita died before the end of the flight, and Arabella was found to be dead after splashdown – a disappointment for Judith Miles, who had been due to open Arabella's cage to give her a chance once more to spin a web in normal gravity conditions.

Six pocket mice and a cloud of vinegar gnats sent up with the spiders were even less lucky; they were all electrocuted by a short-circuit in their environmental system soon after arriving in orbit. But another 75 000 pictures of the Sun were added to those obtained by the Conrad crew, together with observations of six solar flares; and other results included samples of welding and materials processing.

After such a long spell at the Cape and Houston for the Conrad mission, I was grateful that daily 'feeds' of TV pictures via EBU, the now fully-developed European Broadcasting Union, which provided a 'pool' of satellited TV pictures from all over the world, enabled me to cover this second mission from London. This was a period when the see-saw of the BBC's never-ending internecine warfare resulted in my being much in demand by TV News, now far more important than radio. I could cover Skylab without missing other equally important

Skylab: spider Anita demonstrates her ability to spin webs in zero *g* – though they had a somewhat drunken appearance. (NASA)

stories, such as Concorde developments. And it was Concorde's first flight to the US – to Dallas, Texas – which enabled me to reappear at Houston in perfect time for the splashdown of Bean's crew, and to cover the advance planning for the 1975 Apollo–Soyuz mission – which few of us believed would ever survive the suspicion surrounding East–West relationships and their support of opposite sides in yet another Middle East war.

Bean's crew had more than doubled the long-duration record, and that was raised again to 84 days by the third crew – a US record which remained unbroken until America's astronauts began spending months at a time in Russia's Mir space station in the mid-1990s. This time there was a gap of only 23 days before Skylab was reoccupied, and the main story – which again I was thankfully able to cover from England – was to be a spacewalk on Christmas Day to film and observe Comet Kohoutek, rapidly approaching Earth. Its once-in-80 000-years' appearance was expected to be the biggest astronomical event since Halley's Comet in 1910.

Pioneer 10 kept space stories in the news by becoming the first spacecraft to reach the vicinity of Jupiter, sailing safely through its radiation belts in early December 1973. Astronomers at its point of origin – NASA's Ames Space Center in California – announced that our biggest planet was less hostile than expected, and speculated that there might even be plants or other life forms floating in its thick, warm atmosphere. But worries grew about Kohoutek, which British astronomers feared had broken up. A daunting reception had been planned for it. In addition to reorganising the final Skylab mission and programming three spacewalks for filming, NASA was to observe it with four unmanned spacecraft, and send up a series of sounding rockets, balloons, and jet planes carrying telescopes. Dr Kohoutek himself, who had first identified the comet ten months earlier, was to board the liner *Queen Elizabeth II* with a shipload of colleagues to make observations clear of city lights. America's astronomers, whom I rang at observatories like Harvard and in Arizona and Texas, vigorously insisted that Kohoutek was still on its way, but admitted that it was likely to be no brighter than Venus, instead of appearing as large as a half Moon.

We had an early Christmas dinner that year, and I drove to Television Centre apprehensively to cover Skylab's efforts to get pictures of Kohoutek, knowing that the public had been sadly misled by promises that Christmas 1973 would witness a re-enaction of the astronomical events that accompanied Christ's birth.

So far as Skylab was concerned it was pure farce. Astronauts Gerald Carr and William Pogue were late taking up their position on top of Skylab, but having arrived they clipped their camera on a strut ready to get pictures of the comet as its orbit reached its nearest point to Earth – a distance of 13 million miles. Carr at first reported that he could see the comet, and at that precise moment Skylab's computer

Both Skylab sunshades are visible in this last view taken by the third Skylab crew as they departed. NASA's plans to send Shuttle flight No. 5 to boost Skylab into a higher orbit for later re-use before it fell back to Earth were frustrated because the Shuttle was nowhere near operational in 1979. (NASA)

rolled the whole space station far beyond the correct attitude. Inside, the scientist astronaut, Dr Edward Gibson, had to fire the thrusters several times before he could bring Skylab back, and it was a whole orbit – 90 minutes – before the camera could be lined up for another attempt. Then, peering through the view finder, Pogue announced that there was 'No way' they could see Kohoutek. Even on the Earth's dark-side, their own floodlights prevented them seeing into the sky, and

Kohoutek was also too close to the Sun. Mission Control told them to go ahead, and take pictures anyway; the camera was correctly aligned, and something ought to show on the negatives. They were not developed until that last Skylab crew returned to Earth six weeks later; and the fact that we heard no more about those photographs speaks for itself.

But, forgetting Kohoutek, the successful completion of the Skylab missions when Carr and his crew returned to Earth on 8 February 1974, marked the completion of the first era of manned spaceflight. 'Skylab worked better broken than anybody had hoped for if it was perfect' said one astronaut. Apart from the scientific results, man's adaptability to zero *g* was effectively demonstrated by the fact that Carr, Gibson and Pogue, although they had flown much the longest mission, had returned to Earth in the best physical condition.

Before undocking, the last Apollo crew used some of their spare propellant to push Skylab into an orbit several miles higher. By then NASA had indeed begun to worry about the threat its re-entry would pose. There was speculation that the last Apollo spacecraft ever to be launched would revisit it two years later after completing the Apollo–Soyuz linkup.

The first handshake in space

The final Apollo–Saturn launch came 17 months later. In retrospect, this too was a far bigger achievement than we realised at the time. The determination of the astronauts and cosmonauts to make it happen despite repeated political impasses, produced the first fatigue crack in the Iron Curtain which had divided East and West for 30 years after the end of World War II.

For those in the space business there was a more immediate human bonus. It provided a now-or-never opportunity for Deke Slayton, the laconic boss of the NASA astronauts, to make a spaceflight. He was the only one of the original seven Mercury astronauts never to have flown, having been removed from flight status 16 years earlier when doctors found he had a slight heart murmur or 'fibrillation'. Since then he had been Chief Astronaut, organising the selection of his colleagues for each mission as fairly as possible, taking into account their qualifications, medical conditions, and so on. For 16 years he had continued to fight the doctors' vetoes, until finally they had agreed that he was fit to fly.

It was not only the astronauts, every one of whom would have liked his place on this last Apollo flight, who welcomed the decision. It was equally welcome to the small international group of space correspondents. Through all the missions he had attended our briefings and news conferences – never talking much, but always ready with a pithy reply to our questions. We could tell from those replies just how much he would have liked to say if he were not inhibited by all sorts of domestic considerations!

Once again I counted myself lucky to be there to cover the mission. I had been compulsorily 'retired' from the BBC staff on my 60th birthday in May 1975 but the following day had been given a short-term contract to enable me to continue covering the entry into service that year of Concorde, the world's first supersonic airliner, as well as the final Apollo mission.

The latter took place in July 1975, and was called the Apollo–Soyuz Test Project, or ASTP by the Americans, and of course Soyuz–Apollo by the Soviets, as we still called them then. 'Soviets' incidentally, was the equivalent of 'British', insisted upon by Scots, Welsh and Irish, who objected to being called 'English'. It was some years later before I became aware how much, in the cold-war years, Latvians, Estonians, Chechnians and others objected to being lumped in as either 'Russians' or 'Soviets'.

Having recovered from being compulsorily 'retired', Margaret and I set off as usual for the Paris Air Show, where the highlights were my last interview with the terminally-ill Wernher von Braun – described earlier – and witnessing the signatures of 10 nations who agreed to set up the European Space Agency, with Britain's Roy Gibson as its first Director General. And thanks to Margaret sweet-talking some of the Russians, including the formidable and gloomy

Cosmonaut Valery Ryumin – who survived to fly aboard a Space Shuttle in 1998 – I became the first Western journalist to get inside and film a full-scale mockup of their Salyut 4 space station.

After that, Margaret and I set off for the US in good spirits, because I was still able to reclaim most of my expenses from the BBC, and I was being paid a reporter's salary on top of my pension. We went non-stop to San Francisco, where I left Margaret on her own to explore the Fishmarket, the Chinese Quarter ('You shouldn't have gone there on your own!' she was later rebuked, though not by me) and the Golden Gate Bridge, riding the famous trams. I had an invitation to visit Boeing, in Seattle, who lent me their own camera crew to film a piece for TV News about the maiden flight of their 'Junior Jumbo' – a long-range version of the 747, able to carry 300 passengers 6000 miles non-stop, which they unconvincingly denied was an effort to compete with Concorde.

Equally unconvincing was a radio piece I offered to the *Today* programme about a Boeing plan to assemble solar satellites, covering 20 square miles. No doubt this project had grown out of the ideas on the subject thrown out by von Braun towards the end of the moonlandings, and summarised in the interview he gave me at the end of Apollo 16. Now impetus had been provided by the US Government's Energy Research and Development Administration's decision not to build any more nuclear power stations. One suspected that John Hodge, having moved across from NASA, had had something to do with the conclusion that 20 such Powersats, as they were called, would be able to provide all of America's electricity requirements, and Boeing had been given a contract to develop a pilot scheme.

Even 25 years later the proposed solar satellite project seems crazily ambitious: the first one would have taken 100 astronauts, spacewalking for eight hours a day, a year to assemble it 250 miles above the Earth. They would have had to begin by building a spaceport for their own living accommodation; and then, after assembling the huge array of mirrors on an aluminium frame with radiator panels and a microwave power transmission system, supervise its transfer to geostationary orbit 22 000 miles high – a slow job, to avoid the whole contraption collapsing if subjected to any pressure. It was going to cost around $50 billion, twice as much as Apollo. Everyone was perfectly serious about it at that time, for US citizens were worried about potential energy shortages at a time when British citizens were worrying themselves sick about galloping inflation.

Having rejoined Margaret in San Francisco, so that we could recross the US together to Cape Canaveral to start our coverage of ASTP, I didn't give Powersat another thought until I found the fading script when preparing to write this chapter. Like the proposed lunar watch and other such projects which seemed exciting at the time, it must have been financially strangled slowly and quietly so that no one noticed!

For us it was good to be back at the Cape and in the relaxed sunshine of Cocoa Beach. It was 8 July 1975, one week before the ASTP launches were due, and we knew it would be America's last manned flight for four years. No one expected that it would actually be six years before the Space Shuttle, already over one year late, made its first orbital flight.

With a week still to go, pre-launch nerves were very evident at NASA's Launch Control. The countdown for the flight of the last Apollo spacecraft was going well, but violent thunderstorms had hit the Cape every day for a week at launchtime. A day before our arrival a tornado had flattened homes and cars less than 50 miles away from the launchpad. The US Air Force was using 12 Phantoms to rehearse dropping chaff into the thunderclouds to make them discharge their electricity harmlessly, so that America would not face the humiliation of holding up the Russians conducting their simultaneous countdown. Interfering with such natural forces was a dangerous activity, and one of the Phantoms crashed – though the crew ejected and the pilot escaped with a broken leg.

Soviet scientists were even more worried, for it was the first time in their 14-year-old space programme that they had ever publicly committed themselves not only to a launch day, but to an exact minute on that day. At Kazakstan they were counting down two Soyuz rockets and spacecraft, to make sure that they could launch one of them on time.

Tom Stafford, the Apollo commander, had been promoted to Brigadier-General so that his rank matched that of Aleksey Leonov, the Soviet commander, and the two of them, having established an easy personal rapport, were having mutually reassuring chats every day over 'dedicated hotlines' linking the two space centres. They were both said to have spent 1000 hours learning each other's language so that when docking in space Stafford could speak Russian and Leonov English.

Arrangements for newspeople, however, were much less co-ordinated. The day after our arrival we got up at 3 am expecting to be hooked into a joint news conference with Moscow. Absolutely nothing happened, and we all went back to bed. Throughout America's space programmes it was always my belief that the prime function of press officers was to ensure that we never got more than three successive hours in bed.

Media coverage had been a major stumbling block during the laborious East–West negotiations before and after President Nixon and Prime Minister Kosygin had signed the ASTP agreement on 24 May 1972. NASA had insisted that, in accordance with their constitution, details of the advance planning, the launch and the mission, must all be available to the media. The Soviets, reluctant to give up the luxury of cloaking their activities in secrecy, were bitterly opposed to such public scrutiny. They had never announced the name of a cosmonaut before he was in orbit, nor disclosed any details of a mission's objectives until it was either underway or completed. They would follow their usual policy, they said, and America could follow theirs. NASA pointed out that this was clearly impractical, and the Soviets finally gave way. That was probably why the Soviets made a major, last-minute change in the whole project. Originally the plan had been for both the Soyuz and Apollo spacecraft to dock with the Salyut 4 space station but, when Nixon came to sign the agreement, they found it proposed a simpler, direct docking between the two spacecraft, with no involvement of the Salyut station.

This was a big political disappointment for America, undermining one of their main objectives, which was to get some first-hand knowledge of the quality of the Soviet space and missile systems. But by that stage it was too late for either side to withdraw. For NASA, ASTP had other important political benefits. Funding for manned spaceflight was under threat because the moonlandings had been successfully concluded, and ASTP would provide some much needed astronaut activity during the long gap between the end of Skylab and the start of Space Shuttle flights. Making good use of leftover Apollo hardware in this way would also provide 4000 jobs for the next three years, thus holding the technical teams together.

For the Soviets, ASTP was even more desirable. Their space programme had suffered a whole series of failures and setbacks; most worrying of these, from the American point of view, was that in August 1974 the Soyuz 15 crew had had to make an emergency return after failing to dock with Salyut 3. The Soviets were at first very reluctant to disclose what had gone wrong. They also badly needed to improve managerial areas, like quality control, about which they could learn much by working with the Americans. Altogether, the mission was more important to both countries than the active pursuit of the Cold War!

For NASA it was normal but for the Soviets sensational, when the two countries announced not only the proposed date for the launch, but the names of the crews, two years in advance. That was when we had the satisfaction of learning that Donald 'Deke' Slayton, denied by the doctors' decision the chance to become the second American in orbit 16 years earlier, was now to become the first human over 50 to make a spaceflight. America's third astronaut, Vance Brand, had lost his chance to fly earlier when the last three moonlanding missions were cancelled; his qualifcations now were said to include good linguistics. Russia's Soyuz, of course, could accommodate only two people, and their men would both be veterans – Leonov, the world's first spacewalker, and Valeri Kubasov, who had also flown before.

Leonov had been chosen, the Russians told us, because 'he is such a sociable person'; while Kubasov 'is a man of firm energy and great knowledge, though he hides this behind a quiet front'.

Much more difficult than the actual flight, of course, was planning it. Joint working groups were set up, and they had 16 meetings – eight in each country. At first, progress was slow with language proving a major barrier: translations could not be trusted to reflect their respective space jargon accurately. And the Americans were for a long time baffled by Soviet insistence that Soyuz must be launched first. To the Americans this was completely illogical, since Apollo carried so much more manoeuvering fuel (1290 kg compared with about 227 kg on Soyuz) that it could wait in orbit for up to two weeks compared with a limit of 4–5 days for Soyuz.

Only when one of the American negotiators managed to obtain a copy of Russia's 'Project Technical Proposal', took it back to his office, and got an interpreter to spend the whole night translating it aloud to him, was the mystery solved. The Soviets, he discovered, intended to allocate no fewer than three Soyuz spacecraft to the mission, and to countdown two of them for the flight. But they had never mentioned this in discussions! That being the case, there was obviously no NASA objection to Soyuz being launched first. Why had they never mentioned it? The NASA men never found out; but it seems likely that the Soviets were so accustomed to withholding information that it was natural for them not to disclose such details until it was impossible to avoid doing so.

Later the whole project was almost cancelled when Stafford, during training sessions in Russia, insisted that his crew must be allowed to visit Baikonur to familiarise themselves first-hand with the Soyuz spacecraft with which they must dock. His NASA bosses were alarmed when they heard that, when the Russians continued to oppose such a visit, Stafford had announced that, unless the visit was allowed, the whole project was off. Next morning, it was Aleksey Leonov who broke the deadlock. 'What's the problem?' he asked, throwing an arm around Stafford. 'Of course you can go to Baikonur!'

For the correspondents the run up to the launch was even more exhausting than usual. Having been forced to do things the Western way, the Soviets decided to do it even better, and flooded us with Press Kits, Flight Plans, briefings and news conferences. It was a surprise to find that the Soviet Press Kits were in fact longer and in some ways more informative than NASA's; there were, however, the usual omissions which only the well-informed would notice. Examples were that there was no mention that the cosmonauts on Soyuz 1 and 11 had been killed, nor that both the two planned dockings would be performed by the more versatile Apollo spacecraft. (NASA's task force had had to be very insistent to make the Soviets disclose what had caused the Soyuz 11 fatalities, and what had been done to ensure that it could not happen again.)

The most effective correspondents were those who gave most time to studying all these documents – and especially the Press Kits issued by the US Navy and Air Force – on their rescue preparations if an abort situation occurred. Even though an abort never happened, such knowledge provided much information on what could go wrong, and therefore some useful pre-launch 'horror stories' – especially useful during 'holds' in the countdown.

Despite this plethora of paper, we competed to acquire more – copies of documents not issued to us. One of these, obtained by Angus Macpherson of the *Daily Mail*, and shared with us, was a 'Glossary of Conversational Expressions between Cosmonauts and Astronauts During ASTP (Preliminary)'. This crib sheet provided Russian and American versions of commands which might be forgotten under stress, such as 'Stop thrusting', 'Back away', 'Proceed with emergency undock', 'We cannot capture your spacecraft' and 'I will go after you.'

I was glad I had resisted BBC proposals that I should cover the mission from Russia when, on the eve of launch, I heard Moscow announcing that there would be no Soviet newsmen at the Cape, even though nine had applied for accreditation. Apparently they had been forbidden to come because US newsmen in Moscow were insisting on their right to cover the launch from Baikonur and had been firmly refused. Soviet newsmen, a modest 28 in number compared with our hundreds, travelled directly to Mission Control at Houston, matching the fact that Western newsmen were allowed to go to Soviet Mission Control at Kaliningrad near Moscow. The Soviets had been very reluctant in the early planning stages even to disclose the exact whereabouts of their Mission Control.

When the time came, the years of intense preparations were rewarded with on-time lift-offs from launchpads 16 000 km apart. The Soviets confounded the sceptics (and I was among them) by giving us live TV of their launch, preceded by a countdown commentary and pictures of Leonov and Kubasov going aboard. Launch Control at the Cape had to provide a 'Go' so far as they were concerned 20 seconds before Soyuz lift-off. The Apollo astronauts, not due to lift off for another 7.5 hours, were still asleep.

There were no weather problems to delay Apollo; heavy stormclouds rolling around the Atlantic kept well away; and although the USAF had a Phantom ready and eager to drop chaff into the nearest clouds to reduce lightning risks, the launch director wisely decided that this was likely to increase rather than diminish the risk. It was good to hear Deke Slayton, in space at last – though not yet in orbit – savouring the view as he reported, at T+9 minutes: 'Man, I tell you, this is worth waiting 16 years for!'

Our hopes for a quiet 44 hours, while Apollo, in an orbit 55 km below Soyuz so that it travelled faster and quietly caught up with it for the docking, were not to be fulfilled. But the Apollo crew were even worse off. Vance Brand started the mission as he was to end it by incorrectly operating some of the switches. The result was that Stafford had to operate all 16 reaction control thrusters simultaneously to ensure that no air bubbles had formed in them.

Two hours into the flight, Stafford successfully undocked from the 'top' of the Saturn 4B upper stage and then redocked at the other end. This was the same transposition manoeuvre required on flights to the Moon; the difference being that instead of drawing out the Lunar Module for the moonlanding, this time he had to position the 2-ton docking tunnel – really a separate spacecraft – on Apollo's nose so that later on Soyuz could be docked at the other end. But after that had been accomplished, the Apollo crew found they were unable to dismantle the docking probe they had used to acquire the docking tunnel. Unless that was removed, of course, the astronauts and cosmonauts would not be able to pass through it, to and fro between the the US and Soviet spacecraft.

Exhausted by their efforts, the astronauts decided to sleep on it – and when they tried again, Brand cleared the trouble. An electric cable had been blocking the hole used to release the latches. We were astonished to discover at a news conference next day that the troublesome docking probe was the one that gave trouble on Apollo 14, and was brought back from lunar orbit instead of being jettisoned as all others had been. When nothing wrong with it was found, it became the first docking probe to be used for a second time – a rather dubious economy.

The docking tunnel in itself was a major advance in orbital activities; the initial contact ring came from some advanced Apollo designs, while the structure latches at the Soyuz end were provided by the Russians from the Salyut space station design. It made docking operations easier and more reliable, and meant that when at last, 20 years later, more East–West dockings took place, there were no technical, only political, problems to be overcome.

Another thing we learned, as a result of monitoring the exchanges between the cosmonauts and their Mission Control, was that they had been given carefully prepared scripts for all their planned TV transmissions – so we knew there was no chance that either we or NASA would benefit from the accidental disclosure of information. The Soviets, however, were probably more worried about political gaffes than technical disclosures.

When it did come, the first East–West handshake in space was a confused affair. We heard Stafford's voice, saying in Russian: 'I'm approaching Soyuz' and then much laughter at Houston when a female Russian voice, speaking English, presumably from Soviet mission control, pleaded: 'Oh please don't forget about your engine.' We all

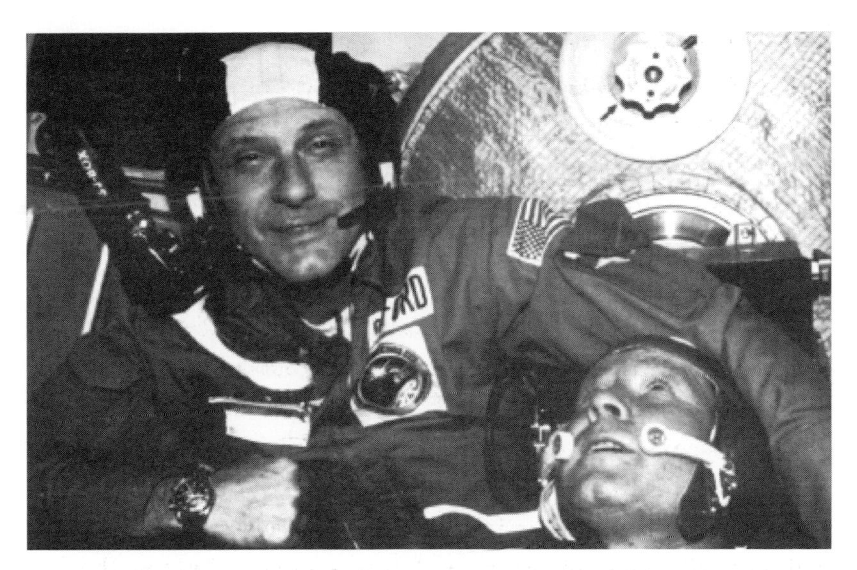

The Apollo–Soyuz Commanders, Tom Stafford and Aleksey Leonov enjoying the first US–Soviet handshake in space in July 1975. They remained lifelong friends. Leonov, artist, traveller and anglophile, came to England in July 1999, to help to celebrate the 30th anniversary of the first moonlanding. He took part with the author and others in a nightlong re-enactment of the drama on TV's Channel 4. (NASA)

knew how worried the Russians were that Stafford would drive too hard into the much lighter Soyuz. Their worries had started when they were shown Apollo video of Stafford's vigorous docking procedures. In fact the new docking system, with its petals instead of probes, worked perfectly first time, as Leonov acknowledged in English: 'Well done Tom, it was a good show. We're looking forward now to shaking hands with you on board Soyuz.'

But before that the crews faced three hours of strenuous work, opening up the docking tunnel, equalising atmospheres, and opening their hatches. That took place amid much painful high squealing on the radio circuits before direct communications with Soyuz were established. The visual effect was also slightly marred by the scripts which the spacemen were carrying around with them. For five minutes after all the hatches had been opened, with Leonov's happy grinning face and reaching hand just discernible through the porthole, the most dramatic thing we heard was Tom Stafford telling Leonov and Kubasov: 'Come over here and do it on camera . . . This is where the audience is . . . Very happy to see you . . .'

The big problem then was feeding Apollo's TV camera and its cable through into Soyuz, because American communications were being used to carry the greetings between President Ford and Mr Brehznev, and the pictures of the astronauts and cosmonauts presenting one another with flags and mementoes, and signing certificates. Ford told Slayton he was very pleased to see him there as the oldest space 'rookie' and asked if he had any advice for future astronauts. 'Decide what you really want to do, and never give up till you've done it!' responded Deke promptly.

During the 44 hours during which Apollo and Soyuz remained docked, the spacemen exchanged visits and hospitality, the cosmonauts producing borshch and spiced veal, and the astronauts beef steak and cheese spread. Meanwhile we in Houston and newsmen in Moscow were preparing separate lists of questions for a joint press conference. The two commanders sat together in Soyuz, with Slayton, Brand and Kubasov in the roomier Apollo, to give their predictable answers as our questions were read up to them. They hoped to make a joint mission to the planets, they said. Leonov in a disarmingly human moment, said he would like to fly with someone who was not 'dull'. Brand was practical and predicted that such a flight would not happen for 20 or 30 years. ('50 years' would have been more accurate!)

Leonov made the most memorable contribution, displaying what he called 'a cosmic portrait gallery'. It consisted of lightning sketches he had made of Stafford (first as he already was, without hair, then 'looking younger' with hair), Slayton and Brand. They showed up well on our TV screens. Leonov, the first and still one of the most accomplished space artists, told us he would like to fly longer and higher than the low-Earth orbit used for the docking, so that his artist's eye could have a better look at Earth. For him and all the other astronauts of that era, that was never to happen.

Correspondents like myself had already shown impatience

because, during the initial approach and docking, the world had watched marvellous pictures of Soyuz provided by Apollo's TV cameras, but not a single view of Apollo – twice the size and much more impressive – from Soyuz. With continuous pictures of Soyuz, the world audience was getting the impression that it was largely a Soviet operation, when in fact not only the TV but all the manoeuvring was the work of the Americans. There seemed little doubt that the Soviet camera 'failures' were not accidental.

There were also some unlikely looking journalists among us (they included Al Shepard, America's first man in space, working for American TV) and a wag had posted a notice in the News Center announcing a football game between representatives of the CIA and KGB who had been accredited as 'media'. NASA urged us to be patient, and to wait for the exciting things planned after the undocking. But from lift-off both crews had had little sleep due to technical problems, and now Moscow casually announced that some of the planned TV would have to be cut.

As the two craft undocked, Soyuz TV was supposed to provide an exciting 15 minutes showing a unique 'solar eclipse experiment' as Apollo was manoeuvred between Soyuz and the Sun, thus 'occulting' it, and enabling Soyuz to provide unique pictures of the solar corona. Slayton carried out the complicated manoeuvres faultlessly, apart from using far more thruster fuel than planned. But there was no Soviet TV. Moscow said the fixed camera was not working, and that the cosmonauts were too busy to accede to Houston's repeated requests that they should use a hand-held colour camera pointing out of a window.

The second docking was accompanied by a frightening medley of voices; astronauts, cosmonauts, both mission controls, and interpreters were all talking at once. But finally it took place, though Moscow was to complain that it was 'heavy' and started hinting that it had caused a pressure leak in Soyuz. They also got worried about Leonov's heartbeats, and urged him to sit still and take medication – which he finally did under protest.

Finally we forced a NASA spokesman at Houston into making a

comment: 'We don't understand why the broadcasts did not take place. Either they couldn't get the portable camera in the right position at the window, or they just didn't want to do it!'

But Soyuz duly made a routine touchdown in Kazakhstan after a flight lasting nearly six days, and Moscow had no technical troubles when it came to providing us with their first ever live TV coverage of a Soyuz descent and touchdown. There was such a cloud of dust as it did so that we thought it had crashed – but it was only the braking rockets raising the dust as it approached touchdown.

Apollo stayed in orbit for another three days carrying out some complicated experiments, which included iceberg studies and checking the then novel theory that earthquakes were caused by 'tectonic plates'. The crew held the first genuine 'press conference' from space, during which we were allowed to talk directly to them. Having given me a moment's personal pleasure as they passed over London by sending greetings 'to the BBC and all our good friends in England', Apollo made what I described as an 'an extremely unpleasant splashdown in the Pacific'. The parachutes failed to jettison and dragged the spacecraft upside down – the 'stable two' position as the Press Kit describes it with unconscious humour – and it took the crew five minutes to deploy the big inflatable balloons carried for just such an emergency to right the spacecraft. They were fighting for their lives, but we did not know it then. The frogmen were dropped in the wrong spot, and had a bad time fighting their way through surging seas, and the crew remained inside the pitching vehicle until it was winched aboard the USS *New Orleans*.

Stafford, Slayton and Brand clambered out on the flight deck to a brass band welcome and the usual 'A-great-job' and We're-proud-to-be-Americans' exchanges with President Ford, and everything seemed to be in order. Dr James Fletcher, NASA's very dull administrator, held the usual post-splashdown news conference, and must have regretted to his dying day opening it with the words: 'Needless to say, we have just witnessed another flawless Apollo splashdown!'

By then it was midnight in Britain, and with my reports for the early morning bulletins done, Margaret and I went off to Houston's traditional splashdown party in complete ignorance of the drama unfolding in Hawaii. It was a nostalgic occasion, with much drinking and cigar-smoking in order, for the party marked the last of 14 Apollo flights, as well as America's last manned mission for some years – until the Space Shuttle began its orbital flights.

When news of a Reuter message spread around the party saying that the Apollo crew had been taken to hospital, we were slow to believe it – and NASA was even slower to confirm it. It took them nearly 24 hours to admit that the astronauts were in intensive care in the US Army Hospital in Honolulu because they had inhaled poisonous gas during the descent. And it was four days before we pieced the whole story together. During the descent Brand had apparently omitted to operate two Earth landing switches, and Stafford had failed to check the omission. The result was that a pressure relief valve, supposed to open to allow outside air to enter the spacecraft as it neared Earth, had opened too soon and allowed nitrogen tetroxide oxidiser, one of the most poisonous gases known, and which was being vented from the thrusters, to enter the spacecraft. Brand became unconscious, but the quick-witted Stafford working desperately as they tossed in rough seas, succeeded in getting out their gasmasks. He and Slayton had put theirs on, and together they fitted Brand's and brought him round, undoubtedly saving his life. Their struggles and coughing went unheeded, for the recovery team was jamming the airwaves.

The recorded exchanges after the splashdown were not usually transcribed, but this time we demanded a transcript and ended with three different versions as the babble of voices were gradually identified by repeated replays. It became possible to pick out the voices of the astronauts amid the calls between the USS *New Orleans* and the recovery team.

The transcript reveals the first sign of trouble five minutes after splashdown. Amid the voices of swimmers saying 'I swam half a mile!' is an urgent appeal from Tom Stafford: 'Get this fucking hatch open.' A few seconds later he is heard again: 'Put (garble) masks on your faces.'

Nobody it seems, is listening to anyone else. The Public Affairs Officer's voice intervenes: 'Five minutes after splash. The spacecraft right at Apex 1 condition'. Stafford again: 'Yeah, well I'm sorry. I'm sorry. You gotta do something to (garble) ends released (garble).'

A swimmer is credited with asking 'Okay, Vance?' but it was probably an astronaut. Not surprisingly they had left their microphones open, and there were repeated calls from the USS *New Orleans* pointing this out, since the astronauts' exchanges were adding to the confusion in the water. Vance Brand replied: 'Yeah, I'm fine', and Slayton and Stafford could be heard checking and querying the procedures for operating the circuit breakers and landing vents.

More confused voices, until it was estalished that all the swimmers were OK, and then Stafford's voice again: 'How d'you feel, okay? I think we passed out for about a minute there.' Brand: 'Huh?' Stafford: 'We passed out for about a minute.' Brand: 'Yeah, I know.'

Between reports from the recovery team on progress in placing flotation collars around the spacecraft there were more snatches of exchanges between Stafford and Slayton while they were fitting oxygen masks and reviving Brand. Stafford referred to propellant, and asked: 'Why the hell did it come in here?'

I was lost in admiration as I studied all this and recalled how impeccably all three had performed during the greetings ceremony while the slow-acting gas was blistering their lungs. After denials that it would be necessary, the wives and children were flown out to Honolulu to visit the invalids, and it was two weeks instead of two days before they went home.

As always, the newspeople benefited from other people's misfortunes, and Margaret and I enjoyed staying on at the Cape to report on the astronauts' health as well as waiting for the launch of two Viking spacecraft on an 11-month flight to land on Mars.

Until it was surpassed by the two Voyager spacecrafts' tour of the solar system, Viking was America's most ambitious unmanned mission. The craft carried some water so that after landing they could scoop up quarter-ounce samples of soil, and mix them with the water.

The theory was that any tiny organisms in the mixture would emit gases to indicate to the 70 scientists monitoring the results whether there were any primitive life forms on the red planet.

Dr Carl Sagan, of Cornell University, was much in evidence, holding forth at news conferences and gathering young people around him for 'workshops' in the garden of the Holiday Inn on Cocoa Beach. He was expounding his theory that Mars might be passing through a 12 000-year ice age, from which it could soon awake and return to life. I interviewed him for the BBC's *World Tonight*, with some trepidation, for he was much addicted to giving newspeople a verbal thumping – very effective in keeping them at bay. Sure enough, when I asked him to explain his strident campaign to send robots instead of men into space, he leaned forward to punch his reply into my microphone: 'Because Men Bugger Things UP!' In those days, alas, such sentiments were not broadcast.

Over the years, Carl Sagan's attitude to manned spaceflight changed, and in 1996 (the year in which he died, much too soon) he was among those calling for a manned expedition to Mars. That was after the announcement that microfossil lifeforms had been detected in a Martian meteorite. My own scepticism about this particular 'discovery' was expressed in a letter used by *The Daily Telegraph*:

> I am doing my best to believe that 16 million years ago a Martian snail crawled into a potato-sized rock, that it was then hurled into solar orbit by an asteroid collision, that 13 000 years ago it landed in our Antarctic, and after three years of study our scientists found the snail's remains. Years ago I asked Bernard Shaw for his reaction to a somewhat similar new theory: 'The magnitude of the lie appals me!' he snapped.
>
> But thank goodness the little Martian potato has been politicised, and space travel is back on the human agenda. We can now hope to see the proposal for *Sprints to Mars* – one-year round trips by six astronauts, with two weeks spent on the surface – proposed 10 years ago by Dr Sally Ride, America's first woman in space, starting in 2010.

Incredibly, the debate over this dubious discovery revived public support for Martian exploration, which had earlier taken a dive following the failure of Russia's Mars 96 to leave Earth orbit in November. Optimism returned in July 1997 with the sensational success of NASA's Mars Pathfinder mission. A tiny robot crawled about the dusty Martian surface and sent back pictures and analyses of its barren landscape which were made instantly available to all on the Internet. The second of NASA's two 1996 launches, Mars Global Surveyor, was only a few months behind. They marked the start of a new era of Martian exploration by robots – but there was little evidence that renewed talk of men on Mars would be backed with funding.

The rather forlorn hope then was that Professor Colin Pillinger's tiny 'Beagle Two' robot lander, which was to renew the search for evidence of life on Mars if and when it touched down in 2004, would open a new chapter!

Skylab's dramatic demise

Skylab was back in the news in mid-1979. At that time NASA was still hoping that it would survive in orbit for another five years. By 1984 there was a reasonable chance that the Space Shuttle would be ready for a dramatic rescue mission. Dollars by the hundred million were being spent on developing the Teleoperator Retrieval System (TRS), to be carried into orbit by the Shuttle, docked with Skylab by remote control, and then fired to place Skylab in a much higher orbit. There it could be parked for later re-use – or at worst placed on a safe re-entry trajectory, as the Soviets did with their Salyuts.

That scheme was finally abandoned as TRS costs rose, and as it became clear that Skylab would fall out of orbit long before the Shuttle was operational. Higher than expected solar activity was steadily eroding Skylab's 50° orbit; and as Britain's Royal Aerospace Establishment began to issue estimated dates for its final fiery plunge into the atmosphere, concern grew, especially over heavily populated areas in Europe, about the bomblike effects which might result from many tons falling on towns. The furore was increased by Russia's loss of control in January 1978 of Cosmos 954, which finally fell, with its nuclear reactor, in northern Canada.

NASA embarked on some rather desperate efforts to reactivate Skylab and achieve some directional control by using its attitude control thrusters, and these may have had some minimal effect. But with every country over which it passed closely monitoring it, Skylab finally fell out of orbit on 11 July 1979.

Disintegration began at an altitude of 19 km, lower than forecast, resulting in a smaller than expected distribution of the debris which failed to burn up. That fell in western Australia, in an area 74 km wide and 7400 km long. No one was hurt, but recovered debris included two large oxygen tanks, a pair of titanium spheres which held nitrogen, and an 80 kg piece of aluminium from the door of the film vault in Skylab's workshop.

It was NASA's biggest technical failure. If Skylab's life had been prolonged, and even more if the backup Skylab had been put to use instead of ending up as an exhibit in the Washington Space Museum, America would have had a facility available for long-duration manned flight and materials processing which was far more versatile than Russia's Mir space station, which was launched 12 years after Skylab was abandoned. The Shuttle would have been able to ferry substantial quantities of microgravity experiments back to Earth, perhaps even in commercial quantities. And it seems likely that manned expeditions to the planets would then have been preferred to the increasingly expensive and laborious construction of the International Space Station.

So far as that is concernered, it will be the end of the decade – 2010 – before anyone can be certain that it will be a commercial success. But in any case it should be able to fulfil Wernher von Braun's expectations by providing a launch platform for a manned expedition to Mars in the decade after that.

23 John Glenn's Apollo Postscript

In January 1996 Senator John Glenn, 76, grey, bald and bespectacled, held a news conference in Washington in front of an enormous blown-up photo of himself as a space-suited young astronaut of 36. He had earlier said he would be retiring from the Senate later in the year; now he announced that, as a member of the American Association of Retired Persons – there were 35 million over the age of 65 – he would be returning to space aboard Shuttle Discovery the following October.

Glenn had been lobbying Dan Goldin, NASA's Administrator, for several years for a flight aboard the Shuttle – a 5-star hotel when compared with the cramped confines of his original Mercury spacecraft. Inevitably there was controversy about the flight. Glenn said Goldin had told him that nobody was getting a free ride into space; his mission would have to have a good basis in science, be peer-reviewed and pass through all the hoops that any NASA science experiment had to survive.

Now it seemed Glenn had satisfied NASA that he could conduct 10 days of experiments that would help scientists to find ways to combat the problems of old age – such as weakened muscles and bones, disturbed sleep and blood flow problems. Goldin announced that, like Glenn's first flight, it would be one of America's defining moments: 'Children will look at their grandparents differently; older Americans will benefit from Glenn's experiments; and Americans of all ages will be reminded that we still have heroes – that this is a nation where we take on bold tasks and risks. This is a place where dreams come true.'

There were plenty of critics who did not accept that. David Anderson, policy director for the Space Frontier Foundation, which advocated opening space to people other than Government employees, denounced it as 'a stunt' which would achieve only 'minor medical knowledge'. Others said it was a political payback for work done by the Senator to distract attention from President Clinton's personal problems.

Whichever it was, it was a mission that I had to cover. It was 21 years since the last Apollo spacecraft had returned from orbit, but I had been kept busy covering for the BBC's *Newsround* and occasionally *Blue Peter* , as well as recording in minute detail in my latest creation *Jane's Spaceflight Directory,* the steady progress made by the Soviets with their Salyut and Mir space stations until at last the Space Shuttle became operational.

The last active Apollo astronauts carried out the drop tests and piloted the first four orbital missions before the Shuttle became operational on its fifth flight in November 1982. But John Young, who had made two pioneering Gemini flights followed by two Apollo flights (he was the ninth man on the Moon), and Vance Brand, blooded on the ASTP mission, were the only Apollo men to fly operational Shuttle missions. Young, having successfully flight-tested Shuttle Columbia with Robert Crippen on STS-1 (the clumsy Shuttle Transportation System acronym is still in use 20 years later!) enjoyed his final command on STS-9, the six-man, 10-day mission which carried ESA's Spacelab laboratory into orbit in November 1983. Until then Margaret and I were at the Cape for each mission, but after that they became too frequent to cover them all.

Space Shuttle developments are only part of this story in so far as they lead up to Glenn's historic flight aiming to prove that there was a place in space for elderly pioneers as well as for those in the prime of life. Despite my own advancing years – I was six years older than Glenn – there had been plenty to keep me too mentally and physically active. On occasions I found myself making the news as well as reporting it. The most unexpected occasion began with the 11th Shuttle mission on April 1984, when Columbia placed in orbit the world's first recoverable satellite, the Long Duration Exposure Facility, which we called 'Eldef'. An octagonal cylinder, it had 57 experiments attached to its sides, ranging from various materials to millions of seeds.

The idea was to recover it 10 months later to see what effect prolonged exposure to the space environment had upon the experiments. My interest was in the seeds, for before the flight started I obtained a promise from NASA that they would let me have some tomato seeds to plant in the BBC's *Blue Peter* garden. The hope was that space exposure would cause them to mutate, in the same way that pink grapefruit had been created by irradiating white grapefruit.

Plans to recover LDEF were shattered when Shuttle Challenger exploded during launch, killing the seven crew members, in January 1986. I had not gone to the Cape on that occasion, but happened to be in a train travelling from Folkestone to London to attend a space conference. When I arrived at Charing Cross station there were TV crews from ITN, CNN and the BBC, plus a BBC radio reporter, holding up placards for 'Mr Turnill' seeking interviews with me. I persuaded them all to come with me to the space conference so that I could find out what had happened before saying anything. ITN set up an impromptu studio for a live discussion, first showing us a replay of the launch disaster. That enabled me to point out that the Solid Rocket Boosters were the weakest link during the launch, for if one of them failed there was no backup and disaster was inevitable.

More than two years later, when Shuttle flights had still not been resumed, I commented in my last edition of *Jane's Spaceflight Directory* that NASA had apparently 'lost the will to fly'. The comment was much quoted in the US, and next time I was there I was peremptorily summoned by the Director of the Marshall Spacecraft Center to justify myself! Before that I was in Washington DC, to cover the report of the Presidental Commission into the disaster. That was issued on 9 June 1986, and Margaret and I were week-ending with Robert Hotz, a member of the Commission, who had by then retired as Editor-in-Chief of *Aviation Week*. A copy of the report, embargoed for 2pm that Monday, lay tantalisingly on his desk, but I resisted the temptation to look at it. Had I still been reporting for BBC News the temptation would have been overwhelming!

LDEF and its precious cargo of space-exposed seeds and other experiments was finally brought back to Earth by STS-32 in January, 1990 – after nearly six years, instead of 10 months in orbit. I successfully reminded NASA of its promise to let me have some tomato seeds, helped to sow them in *Blue Peter's* garden and grew some in my own as well. The seeds, *Californian Rutger's Supreme* germinated splendidly, and that autumn my family as well as *Blue Peter's* audience of over five million, watched us sampling them. They were large and fleshy, very sweet-tasting, but alas very normal. There were no mutations.

We saved some seeds and planted them for a second year. This time perhaps something dramatic would happen. And it did! But it was not the tomatoes that mutated, but MAFF – the Ministry of Agriculture and Fisheries. They had not seen or heard about the TV progress reports, but someone running a space school who disapproved of my activities reported them to MAFF. The Whitehall bureaucrats decided I had failed to obtain an import licence for the seeds – though they could hardly allege that I had acted furtively! – and despatched an inspector to visit my home and greenhouse at Sandgate in Kent.

Happily he made an appointment first, and when he arrived I had arranged an ambush. A TV crew from Meridian was waiting in our lounge. I warned the inspector that if he came in he would be filmed – and half expected him to retreat in a huff. Not of bit of it. Andy Monro, from the Ministry's research establishment at Wye, enjoyed every minute of it. We were filmed going down the garden into my greenhouse, where he formally warned me that I would be served with an order allowing me to grow the fruit, but asking me not to eat it, and to burn the plants, the pots and the earth in which they were grown at the end of the season. It was hoped that would ensure that no possible contamination survived from outer space.

Andy did an interview for the TV crew – for which he was rebuked by his HQ for not having first obtained their approval – and Margaret served tea on the patio while further discussions took place. It made a good TV story, and we duly got our notice from MAFF – as did half a dozen schools which had been given some of my seeds by the *Blue Peter* team. Newspapers around the world carried the story. Headlines ranged from *The Times'* 'Tomatoes from outer space give ministry the pip', to the *Daily Express's* 'Reg Grilled Over "Killer" Tomatoes' to *Florida Today's* 'Britain out to halt attack of U.S. space tomatoes', and the *China Daily's* front page 'Extraterrestrial tomatoes land in Britain'!

One result was that the schools got worried, especially when parents started asking questions, and duly destroyed their plants. We, I am afraid, took no notice. Andy invited me to talk to his branch of Rotary at Ashford, Kent, where I formally presented him with a beautiful descendant of the space tomatoes – still unmutated. By then he had taken early retirement, so he was able to accept it and report later that it made good eating!

Soon after, with my 80th birthday approaching in 1995, I suggested to the European Space Agency that it would make a good story if they allotted me one of the places for journalists on their second parabolic flight – the first stage in an astronaut's training, when his or her reaction to weightlessness can be observed. ESA had started annual leasings of NASA's specially modified KC-135 aircraft so that European students with ideas for space experiments could be given an opportunity to try them out, with three or four journalists being taken along to report the occasion and experience what an aspiring astronaut has to face.

To my delight ESA agreed – so long as I took the required medical tests certifying that I was fit for flight in high-performance aircraft. In 1958, just after being appointed BBC Aerospace correspondent I had initially been refused such a certificate by the RAF's doctors because of an ear perforation dating back to the war. They had hurriedly changed their minds after the USAF had said there was no problem, and provided me with a big story by enabling me to become one of the world's first air correspondents to go through the sound barrier.

When I called the RAF a second time nearly 40 years later I again encountered the familiar British negative-response. There were sharp intakes of breath. 'We've never tested anyone of your age before,' they said doubtfully and finally agreed to consider it and let me know. The decision, I suspected, would be conveniently delayed until after the flight took place. So at the suggestion of a retired Air Marshal who was a former head of the RAF School of Aviation Medicine at Farnborough, I rang the medical head of the Royal Netherlands Air Force. 'Delighted!' they said. 'We spend all our time testing healthy young people and it gets very boring. We've never put an 80-year-old through our tests before, and we'd find it most interesting.'

I had to present myself twice at the Netherland's Naval Air Base, Valkenburg near Leiden – first for straightforward medical tests, and secondly for pressure-chamber tests. The first occasion was a Saturday, and some of the doctors were so interested they came in specially. All went well and I got my certificate. Then of course, when it was almost too late, the RAF came back to me. They were prepared, as a special favour, not to be taken as a precedent, to give me the required tests. But they would charge me, if I remember correctly, £800 – nearly three times the Netherlands' charge. It was a pleasure to turn them down! My Air Marshal friend told me he was ashamed of the RAF's reaction.

On a third enjoyable visit to the Netherlands I joined a group of excited young scientists getting a chance to try out their experiments in space. Our aircraft, a version of the original Boeing 707, had only about 30 seats, for use on take-off and landing. The remainder of the interior, padded like an asylum cell, provided ample space for about 16 experiments per flight with their young scientist operators.

Our two-hour flight gave us 31 parabolas, which meant 31 periods of 20–30 seconds of weightlessness. It was a great experience – the drawback for me being that the laws of gravity ruled that when one returned to Earth one had averaged 1 *g* throughout the flight. So after each 30-second dive, during which the young scientists demonstrated their varied experiments such as the behaviour in microgravity of fish, flames and even a pile of marbles, the pilot pulled straight up into a climb. I found myself instantly converted from weightlessly floating on the ceiling to twice my own weight, and slammed back on the floor. It was necessary to make sure that my aged arms and legs were in an appropriate configuration for the impact and I was more than grateful to Wubbo Ockels, veteran ESA astronaut, for keeping a sharp eye on me.

The author at 80 becoming the world's oldest human to experience weightlessness. Left: veteran ESA astronaut, Wubbo Ockels, makes sure he comes to no harm. (ESA/ESTEC)

ESA had arranged to have a still photographer and a video camera operator on board, and when the latter came my way I waited until I was floating and then addressed his lens: 'It's a wonderful feeling, and I recommend it for all 80-year-olds.' There was just time for that before the parabola ended, and Wubbo broke my fall as I fell to the floor. It would have been nice to have had some photographs too – but the camera operator was one of six among us who became sick and retired to the 30 seats. I kept my lifetime's sick-free record while flying in both military and civil aircraft; all the same, I was glad that I was just a reporter, and not a young scientist trying to justify ideas for space experiments in such conditions.

Once again ITV's Meridian news did an amusing report of the occasion, with the help of the video provided by ESA. My own former employers, as usual, were underwhelmed when offered the story – in accordance with their tradition of washing their hands of former correspondents!

In 1996 Hugh Harris, Kennedy's Director of Public Affairs, and a good friend for many years, visited Sandgate to present me with The Chronicler's certificate mentioned at the start of this book. And later that year I had renewed acquaintance with Glenn, albeit briefly, when he came to the Cape just after his flight was announced. I had reminded him that he kept me, and a lot of others, hanging around for six weeks the first time he flew; and he promised to do better on his second flight.

So at last, having become 'an award-winning journalist' and the world's oldest human to have experienced weightlessness, the time came to cover the last postscript to Apollo. Margaret and I enjoyed a carnival atmosphere when we returned to the Cape. This time we were not on Cocoa Beach, but staying at Titusville with David Henson, a senior Shuttle engineer and a friend ever since he had cabled me a quarter of a century earlier to praise my pocketbook *Manned Spaceflight*. We were also spared the hassle of getting a hired car, for his wife Sandra was away piloting Virgin aircraft in Europe and we had the use of her car. It was good to be there, for an estimated 250 000 visitors arrived to witness the launch and enjoy 80 degree temperatures; over

4000 'journalists' had been accredited compared with 2700 for Apollo 11. While I enjoyed having no deadlines to meet, I felt uncomfortable having nothing specific to do!

On the Cocoa Beach area the hotels were full and there were cheerful crowds spending money on souvenirs. Alan Shepard had died suddenly of a brain tumour in July 1998, and Deke Slayton, with whom Margaret and I had been most friendly, had died the year before that. But the other three survivors of the Mercury Seven, Wally Schirra, 75, Scott Carpenter, 73, and Gordon Cooper, 71, were all there, employed as commentators for the TV networks, ABC, NBC and CBS respectively.

Walter Cronkite, who had become famous during the moonlandings as the CBS anchorman, was there for CNN. Just as the BBC no longer wanted to know me, CBS no longer wanted to know him. During the mission, several BBC reporters, from *Tomorrow's World* and elsewhere, approached me tentatively about interviewing me or taking part in their programmes, but all finally backed off. I sympathised with their natural feeling that this was *their* day; I had, after all, had mine!

Schirra, Carpenter and Cooper held a pre-launch news conference at the Astronaut Hall of Fame. This was a fairly new facility, a moneymaker dreamed up by Alan Shepard and financed by retired astronauts and journalists. It was cunningly sited on Route 405, the main road from Titusville into the Cape, with a full-size mockup of the Space Shuttle advertising space courses and other attractions aimed at enticing the millions of visitors to think it was the main Visitors' Center – which was in fact several miles further on! Glenn's flight, said his former colleagues, would revitalise America's interest in the space programme – which it certainly did – but it was clear that not one of them really wished he was making the flight instead of Glenn; and the truth was, none of them looked fit enough! Schirra looked rather heavy and the other two rather frail.

Glenn, with the help of NASA, kept his promise not to keep us waiting a second time. Discovery lifted off on Wednesday 29 October 1998 only 10 minutes late – and that was due to private aircraft straying into the forbidden zone. Glenn's wife of 55 years, Annie, was with their physician son David, daughter Lynn and two grandchildren on the roof of the Launch Control Center to watch.

John Glenn was ranked as Payload Specialist No. 2 – technically the least important of the seven crew members – veterans with a total of ten previous flights behind them. But inevitably his joyful good humour was to dominate the mission before, during and after it.

Monitoring the countdown and launch from the Cape's Press Mound, we had a dramatic reminder of the fact that lift-off is still the most dangerous moment in spaceflight. We saw, as it happened, an object fall from just above the bell-shaped orbiter engines at the moment of lift-off. After many replays and questions to the launch director, it turned out to be the 11-pound cover holding in place the drag parachute used for landing. If it had struck and damaged one of the gimballing engines' bells, vital for directional control, Discovery would have crashed into the sea. Happily, it did not. This sequence was not included in the subsequent NASA video of the mission.

As it cleared the tower, Lisa Malone, Kennedy's News Chief, ended her commentary by saying: 'Lift off of Discovery with a crew of six astronaut heroes and one American legend.' Lisa, petite and very pretty, had come to the Cape as a teenage trainee press officer in the early days of Shuttle flights. The space veterans were all so impressed at the speed with which she learned her job and passed on her inside knowledge that our praise led to a permanent appointment. And once there she rose steadily through the ranks to the top!

There were mixed reactions to the launch. Chuck Yeager, first man through 'the sound barrier', and one of the test pilots who had spoken scathingly in the early days of the Mercury astronauts as 'spam in a can', dismissed it as a publicity stunt. Cosmonaut Valery Polyakov, 56, with a record 678 days in space, said age increased a person's value in space. Bernice Steadman, a survivor of the 13 women pilots trained as astronauts in 1961 and then dropped, had still not forgiven Glenn for allegedly helping to ditch their chances at that time by calling them '90 pounds of recreational equipment'! Later I talked to Jerrie Cobb, pioneer woman aviator who was another of the disappointed 13. Signatures were being gathered on her behalf urging NASA to give her a similar flight to study ageing in women. But 15 years on, NASA is at last honouring its promise to provide a flight for Barbara Morgan, Christa McAuliffe's backup on the disastrous 'Teacher in Space Mission' of 1986. Barbara has been selected to fly as an 'Educator Mission Specialist' on a Shuttle flight to the Space Station in 2004.

The retired US Marine colonel did not allow the critics to spoil his enjoyment of the nine-day flight. But he got no window seat on the flight deck for take-off and landing; it was no coincidence that he occupied the middle seat on the windowless deck below, between Professor Chiaki Mukai, a highly-qualified Japanese heart surgeon, and Dr Scott Parazynski, a NASA heart specialist. Emergency medical attention could not have been closer. One of Parazynski's tasks was to take regular blood samples from Glenn during the flight, with the result that he became known as 'Dracula' and wore plastic fangs on one occasion.

Safely in orbit, Commander Brown summoned Glenn to the flight deck. His Presbyterian faith was as firm as it was 38 years before, when we filmed him going to church. 'I don't think you could be up here, look down on the Earth at this kind of creation, and not believe in God,' he said. 'It just strengthens my faith . . . I wish there were words to describe what it's like to look out the window and see about a 4000 miles swathe of Earth go by.'

He endured without complaint four nights wearing a sort of hairnet, wired with 23 sensors measuring his breathing, snoring, eye and chin muscle movements and brain waves. He had had to swallow a large pill containing a radio transmitter and thermometer, the signals from which were recorded in a belt worn around his waist. And having done his best to get some sleep, as soon as he woke he had to spend 20 minutes filling in a form about it, and analysing his urine.

The media took little interest in the mission's prime activities. They included the release and recapture, for the fifth time, of the Spartan spacecraft, used to study solar winds and coronas, part of the

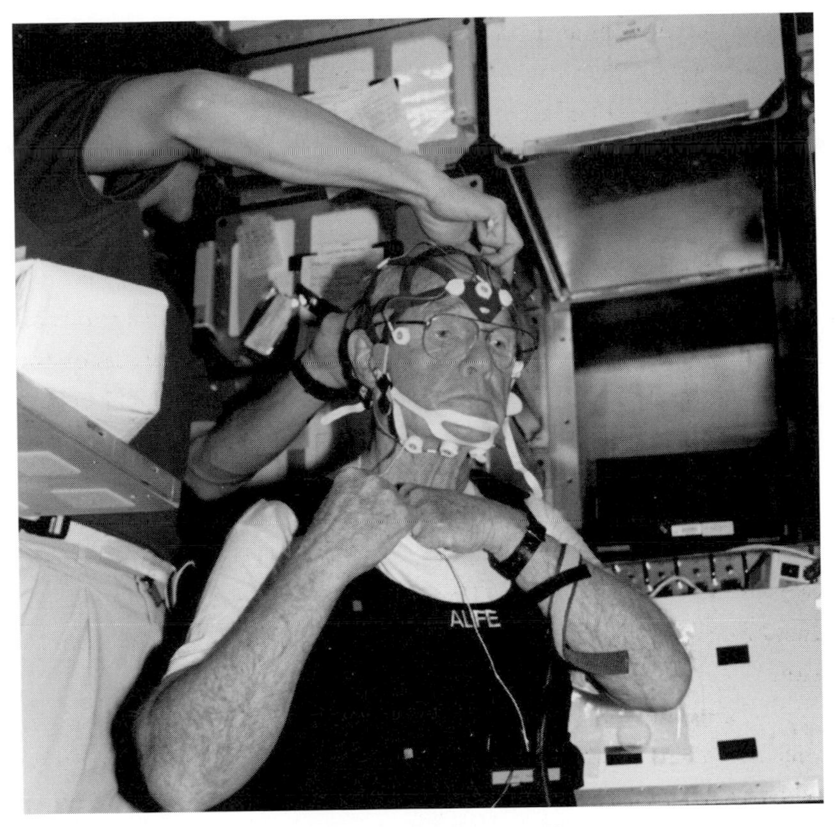

John Glenn, at 77, being fitted with sensors to measure the effect of spaceflight on the aged. His breathing, snoring, muscle movements and brain waves were recorded during the 1998 Space Shuttle mission.

ongoing studies of the solar winds and coronas, to build up a picture of the 11-year solar cycle; rehearsals of the materials and equipment needed for the third Hubble Space Telescope repair mission; and 80 experiments covering studies ranging from the inner universe to the human body. Not that the inquiring space correspondent would ever discover exactly how successful all that work had been; and even in the case of John Glenn's studies into ageing one's inquiries were largely fobbed off with the excuse that they were covered by medical confidentiality!

Glenn was a bit shaky when they came back, as we saw as the

crew emerged and did their usual walk around the outside of Discovery. So the eagerly awaited postflight press conference, usually held a few hours after landing, was postponed until Sunday morning. In the familiar auditorium on the Press site, which for years had often provided me with a place of quiet retreat when not in use, there was just room on the platform for the seven crew plus Lisa. John Glenn was at the far end, as befitted his 'junior' status.

I had to wait a long time to get my question in. Lisa, as I well knew, was obliged to call first on all the TV networks and radio stations, and then the important newspapers like the *New York Times, Washington Post,* and *Florida Today.* There had been a time when my BBC status, plus my seniority, ensured that I would be among the first to be called. Now I began to worry that time would run out before I had my chance. But at last Lisa did call 'Reg Turnill':

'I'm aged 83, and the world's oldest working space correspondent' – and I was slightly thrown by friendly laughter from my colleagues: 'Do you regard yourself at 77 as the cut-off point, or do you think there's still hope for my generation?' That brought applause as well as laughter in which John Glenn joined.

'Still a lot of hope', he said. 'I admire you for still working at 83, and I hope I'm still working at 83. I think too often folk set their lives by the calendar. They think, well, I'm 60 or 70 or whatever it is, and I'm expected to sit on the couch and do nothing for a while. That takes the fun out of life so far as I'm concerned. I believe people, the longer they feel active, the longer they feel good, whatever they can do physically, they should run their lives not by the calendar but what they feel and by their ambitions.

'Old folk have ambitions and dreams too, like everybody else. So why don't they work for them? Don't sit on the couch; go for it, that's my attitude.'

Discovery's light-hearted commander, Lt-Colonel Curt Brown, summed it all up at what was the conclusion of his fifth and probably last flight: 'Don't let this be the final chapter. Instead let it be the first chapter in a new adventure – the International Space Station.'

Bibliography

Listed below are some of the books and documents which the author has used with gratitude since 1956. NASA's activities must be the most recorded in thc history of man.

1956–86 Reginald Turnill's BBC scripts

1956– *Aviation Week* and *Space Technology*

1958– NASA, Rockwell, etc Press Kits and Flight Plans (Mercury, Gemini, Apollo, Viking, Voyager, Skylab, ASTP, Space Shuttle Missions, etc.)

1959– TRW Space Logs, TRW Group

1963 *Project Mercury, A Chronology*, James M. Grimwood (NASA)

1963 Mercury Project Summary, NASA.

1966 *History of Rocketry and Space Travel*, Wernher von Braun *et al.* (Thomas Y. Crowell Co.)

1966 *This New Ocean – A History of Project Mercury*, Loyd S. Swenson *et al.* (NASA Historical Series)

1966 *Observing Earth Satellites*, Desmond King-Hele (Macmillan)

1967 *Lunar Science and Exploration – Summer Study*, NASA

1968 *The Promise of Space*, Arthur C. Clarke (Hodder & Stoughton)

1968 *The Book of Mars*, Samuel Glasstone (NASA)

1969 *Moonslaught*, Reginald Turnill (Purnell)

1969 *A-Z The Soviet Encyclopaedia of Spaceflight* (Mir publishers)

1970 *Inside the Third Reich*, Albert Speer (Warner Books)

1971 *The Russian Space Bluff*, Leonid Vladimirov (Tom Stacey)

1971 *The Language of Space*, Reginald Turnill (Cassell/JohnDay)

1972 *Kennedy Space Center Story*, NASA

1973 *13: The Flight that Failed*, Henry S. F. Cooper (The Dial Press)

1973 *Entering Space*, Joseph P. Allen (Orbis)

1974 European Space Agency Quarterly Bulletins (ESA Publications)

1974 *Carrying the Fire*, Michael Collins (Farrar, Straus & Giraux)

1975 Future Space Programs 1975, Committee on Science & Technology, US House of Representatives

1975– European Space Agency Bulletins

1976	*A Forecast of Space Technology 1980–2000*, NASA
1976	*Outlook for Space*, NASA
1976	Soviet Space Programs 1971–75 Congressional Research Service (US Government Printing Office)
1977	*On the Shoulders of Titan*, Barton C. Hacker and James M. Grimwood (NASA Scientific & Technical Information Office)
1977	*The High Fronter*, Gerard K. O'Neill (Corgi Books)
1978	*Observer's Spaceflight Directory*, Reginald Turnill (Warne)
1979	*Chariots for Apollo*, Courtney G. Brooks *et al.* (NASA History Series)
1981	NASA Program Plan 1981–1985
1981	*Encyclopaedia of Space Technology*, Kenneth Gatland (Salamander)
1981	United States Civilian Space Programs 1958–1978 Congressional Research Service (US Govt Printing Office)
1982	Soviet Space Programs 1976–80, Science Committee, US Senate
1983	Salyut – Soviet Steps Towards Permanent Human Presence in Space, Office of Technology Assessment, US Congress
1983	US Civilian Space Programs, Science Committee, US House of Representatives
1983	Astronauts and Cosmonauts, US Congressional Research Service.
1983	*Diary of a Cosmonaut*, Valentin Lebedev (Texas)
1984	NASA Program Plan 1984–1988
1984	*Spacelab*, David Shapland and Michael Rycroft (Cambridge University Press
1984–89	*Jane's Spaceflight Directories*, Reginald Turnill (Jane's)
1985	Space Activities of the US, etc., US Congressional Research Service
1985	*Kosmonautica Encyclopaedia*, Moscow
1986	Investigation of Challenger Accident, Committee on Science and Technology. House of Representatives
1987.	*Challenger – A Major Malfunction*, Malcolm McConnell (Doubleday)
1987	*Soviet Military Strategy in Space*, Nicholas L. Johnson (Jane's)
1987	Kosmonautica CCCP, Moscow
1989	Astronauts and Cosmonauts, Congressional Research Service.
1989	*Men From Earth*, Buzz Aldrin (Bantam Press)
1998	Soviet Space Programs 1981–87, Science Committee US Senate
1993	*Seize the Moment*, Helen Sharman (Victor Gollancz)
1994	*Moonshot*, Alan Shepard and Deke Slayton (Turner Publishing)
1995	*Exploring the Unknown*, Vol. 1., ed. John M. Logsdon (NASA History Series)
1995	*Spaceflight Revolution*, James R. Hansen (NASA History Series)

1996 *Stages to Saturn*, Roger E. Bilstein (NASA History Series)

1996 *Exploring the Unknown*, Vol. 2, ed. John M. Logsdon *et al.* (NASA History Office)

1996 *To See the Unseen*, Andrew J. Butrica (NASA History Office)

1996 *Aiming at Targets*, Robert C. Seamans (NASA History Series)

1997 *Wallops Station*, Harold D. Wallace (NASA History Office)

1997 *Beyond the Ionosphere*, ed. Andrew J. Butrica (NASA History Office)

1997 *Way Station to Space*, Mark Herring (NASA History Series)

1997 *Space, The Dormant Frontier*, Joan Johnson-Freese and Roger Handberg (Praeger)

1998 *Exploring the Unknown*, Vol. 3, ed. John M. Logsdon *et al.* (NASA History Division)

1998 *The Space Shuttle*, David M. Harland (Praxis Publishing)

1998 *From Engineering Science to Big Science*, ed. Pamela E. Mack (NASA History Office)

1998 *History of the European Space Agency*, ESA

1999 *Exploring the Unknown*, Vol. 4, ed. John M. Logsdon *et al.* (NASA History Division)

1999 *Before This Decade is Out . . .*, ed. Glen E. Swanson (NASA History Office)

2000 *Reconsidering Sputnik*, ed. Roger D. Launius *et al.* (Harwood Academic)

2000 *Challenge to Apollo: The Soviet Union and the Space Race, 1945–1974*, Asif A. Siddiqi (NASA History Series)

2000 *Disasters and Accidents in Manned Spaceflight*, David J. Shayler (Praxis Publishing)

Appendix 1
NASA and 400 Years of Failure

To demonstrate the 'openness' of the United States' space programme compared with the paranoid secrecy practised by the Soviets, NASA swamped the media, and especially the accredited space correspondents, with information. The result was that few had the time or patience to master it all. The Apollo News Reference book consisted of 330 illustrated pages, accompanied by similar-sized books detailing the Saturn 1, Saturn 1B and Saturn 5 rocket launchers. [NASA uses roman numerals for the Saturn launchers, but for clarity arabic numerals have been used throughout this book.]

These were in addition to the Press Kits for each mission, provided not only by NASA but by the many contractors like Rockwell and Boeing and the US Navy and Air Force. The Flight Plans provided another bulky addition.

Two extracts are offered here as relevant to this story. The first page of the Apollo book reads as follow:

> It took 400 years of trial and failure, from da Vinci to the Wrights, to bring about the first flying machine, and each increment of progress thereafter became progressively more difficult.
>
> But nature allowed one advantage: air. The air provides lift for the airplane, oxygen for engine combustion, heating and cooling, and the pressurized atmosphere needed to sustain life at high altitude. Take away the air and the problems of building the man-carrying flying machine mount several orders of magnitude. The craft that ventures beyond the atmosphere demands new methods of controlling flight, new types of propulsion and guidance, a new way of descending to a landing, and large supplies of air substitutes.
>
> Now add another requirement: distance. All of the design and

> construction problems are re-compounded. The myriad tasks of long-distance flight call for a larger crew, hence a greater supply of expendables. Advanced systems of communications are needed. A superior structure is required. The environment of deep space imposes new considerations of protection for the crew and the all-important array of electronic systems. The much higher speed of entry dictates an entirely new approach to descent and landing. Everything adds up to weight and mass, increasing the need for propulsive energy.
>
> There is one constantly recurring theme: everything must be more reliable than any previous aerospace equipment, because the vehicle becomes in effect a world in miniature, operating with minimal assistance from Earth.

An extract from the Saturn 5 News Reference Book effectively summarises the role of the launcher:

> The jumping off place for a trip to the Moon is NASA's Launch Complex 30 at the Kennedy Space Center. After the propellants are loaded, the three astronauts will enter the spacecraft and check out their equipment.
>
> While the astronauts tick off the last minutes of the countdown in the command module, a large crew in the launch control center handles the complicated launch operations. For the last two minutes, the countdown is fully automatic.
>
> At the end of the countdown, the five F1 engines in the first stage ignite, producing 7.5 million pounds of thrust. The holddown arms release the vehicle, and three astronauts begin their ride to the Moon.
>
> Turbopumps, working together with the strength of 30 diesel locomotives, force almost 15 tons of fuel per second into the five engines. Steadily increasing acceleration pushes the astronauts back into their couches, as the rocket generates 4½ times the force of earth gravity.
>
> After 2.5 minutes, the first stage has burned its 4 578 000

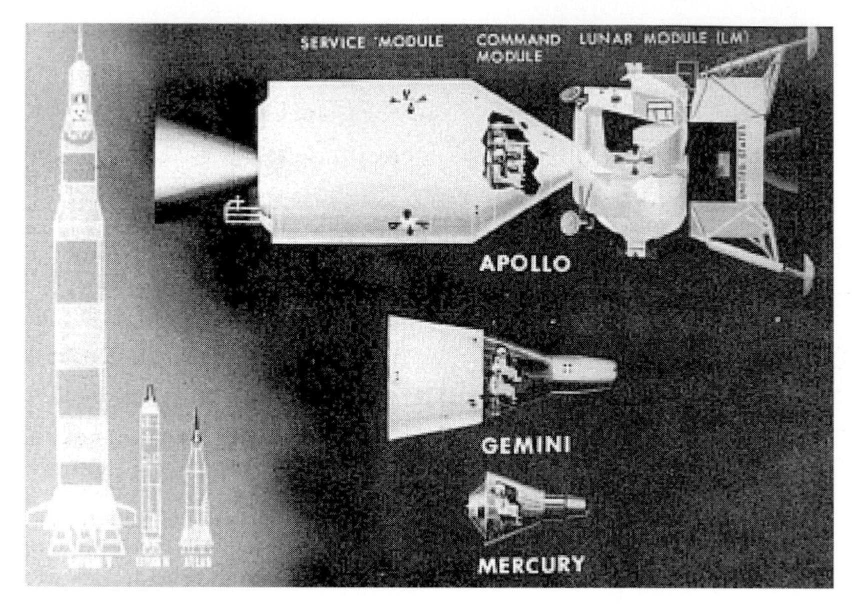

How NASA reached the Moon in three stages: Comparison of three generations of launchers and spacecraft. (NASA)

pounds of propellants, and is discarded at about 38 miles altitude. The second stage's five J2 engines are ignited. Speed at this moment is about 6000 mph. The second stage's five J2 engines burn for about 6 minutes, pushing the Apollo spacecraft to an altitude of about 115 miles, and a velocity of about 15 300 mph. After burnout the second stage drops away and the retrorockets slow it for its fall in the Atlantic Ocean west of Africa.

The single J2 engine in the third stage now ignites and burns for 2.75 minutes. This brief burn boosts the spacecraft to orbital velocity, about 17 500 mph. The spacecraft, with the third stage still attached, goes into orbit about 12 minutes after lift-off. Propellants in the third stage are not depleted when the engine is shut down. This stage stays with the spacecraft in earth orbit, for its engine will be needed again . . .

The astronauts are now in a weightless condition as they circle the earth in a 'parking orbit' until the timing is right for the next step to the Moon . . .

During the one to three times the spacecraft circles the earth, the astronauts make a complete check of the third stage and the spacecraft. When the precise moment comes for injection into a translunar trajectory, the third stage J2 engine is re-ignited. Burning slightly over 5 minutes, it accelerates the spacecraft from its earth orbital speed of 17 500 mph to about 24 500 mph in a trajectory which would carry the astronauts around the Moon. Without further thrust, the spacecraft would return to earth for re-entry.

If everthing is operating on schedule, the astronauts will turn their spacecraft around and dock with the lunar landing module [housed in the far end of the third stage]. After the docking maneuver has been completed, the lunar module will be pulled out of the forward end of the third stage, which will be abandoned. This completes the Saturn 5's work on the lunar mission.

Appendix 2
Von Braun's Master Plan for Mars

Two weeks after Apollo 11 had landed on the Moon, Dr Wernher von Braun delivered to the US Space Task Group a detailed, fully costed plan for landing men on Mars in 1982 and for establishing permanent manned bases on both the Moon and Mars. He had given me a copy of the plan on the day of the moonlanding.

His Integrated Program 1970–1990, summarised here and with some of his illustrative charts over the following pages, was backed with detailed proposals for an initial Mars excursion lasting two years, to include a fly-by of Venus on the return journey. The idea was to make maximum use of both the equipment and technology developed for Apollo. It would have included a Space Shuttle and two Earth-orbiting Space Stations, one in geosynchronous orbit and another in the low orbit chosen for the existing Space Station. von Braun estimated an annual cost peaking at $7 billion in 1974 and running at $6 billion thereafter.

'The next frontier is manned exploration of the planets,' he said. 'Perhaps the most significant scientific question is the possibility of extraterrestrial life in our solar system. Manned planetary flight provides the opportunity to resolve this universal question, thus capturing international interest and co-operation . . . A 1982 manned Mars landing is a logical focus for the programs of the next decade. It will be a great national challenge, but no greater than the commitment made in 1961 to land a man on the Moon.'

His Saturn 5 launchers, which had performed so faultlessly, would have remained in production indefinitely, and instead of ending with Apollo 17, moonlandings would have continued until Apollo 33 in 1978. In the meantime – and von Braun laid little emphasis on this most controversial feature of his plan – a nuclear-powered rocket would be developed. This would be parked in Earth orbit at a safe distance, and used to ferry six-astronaut, 800-tonne spaceships from there to Mars and back. The conventionally powered Space Shuttle now in use would have been capable of operating to the geosynchronous orbit (35 680 km from Earth) instead of being confined to altitudes of around 400 km.

Even now, his proposals, which seemed simple to him, appear breath-taking: 24 men (at that time he was not required to talk of 'humans') in lunar orbit by 1984, and 48 on the lunar surface by 1985; 100 men in low Earth orbit by 1989; 12 men on a temporary Mars base by 1985, with 48 men on the surface and 24 in Martian orbit by the end of the 1980s.

Had it been implemented, his plan would now be costing the US at least half as much again as the current $14 billion per year – but, assuming success, the returns in technological advances and the use of manpower on earth alone would have been many times greater than that. Although costly to build and operate, the geosynchronous space station alone would be highly profitable by providing the ability to service the hundreds of communication and intelligence satellites now operating in that orbit. By contrast, the current International Space Station, supplied by the US Space Shuttle and Russia's Soyuz spacecraft, consumes most of the world's space expenditure, with little indication that it will ever yield a commercial return.

But for the first time in his colourful career, Dr Wernher von Braun's plans fell upon deaf ears. Politicians and the media had had a surfeit of space; the timing was also wrong economically. Few ever heard of the plan and it seems doubtful whether it was even filed!

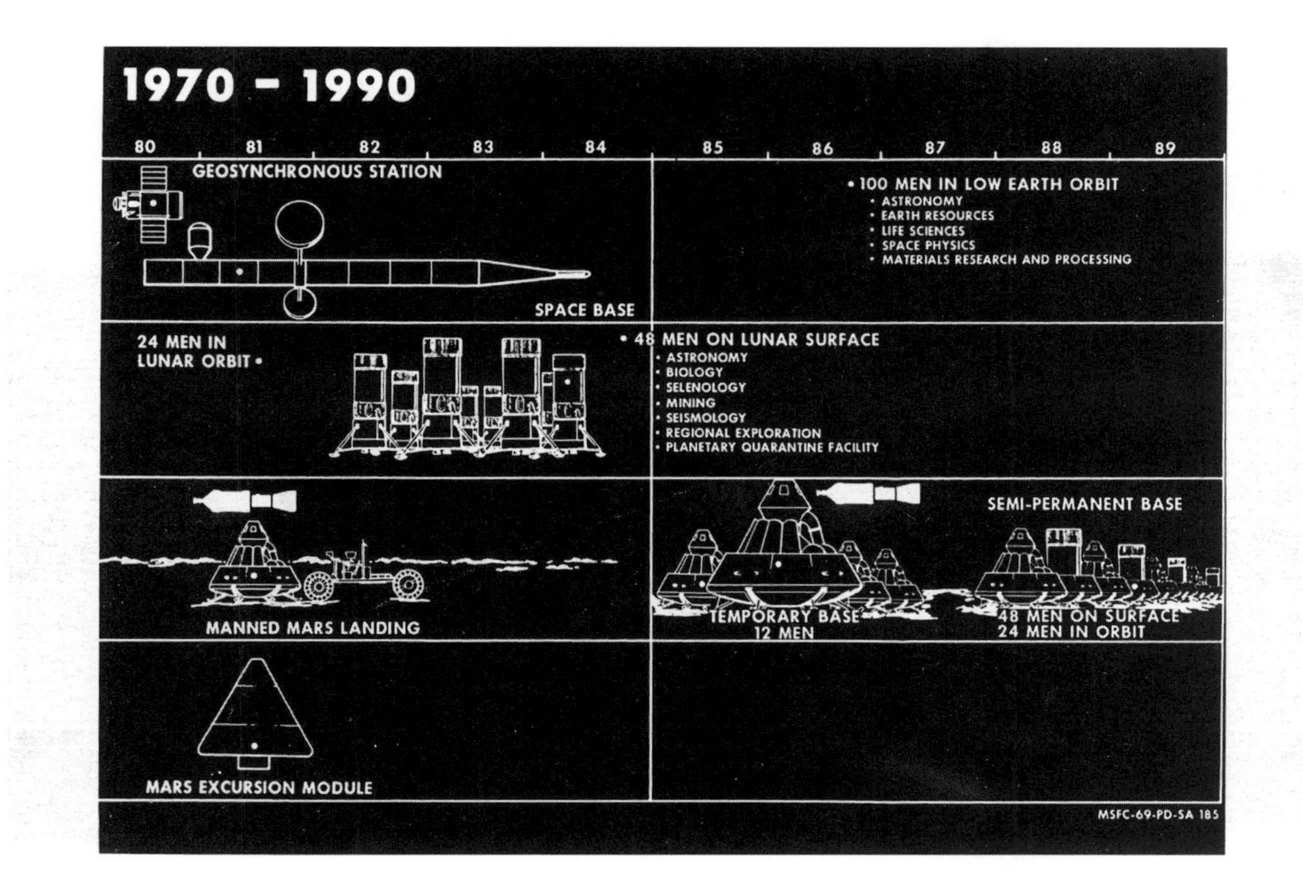

1970 - 1990
80
81
82
83
84
85
86
87
88
89
GEOSYNCHRONOUS STATION
SPACE BASE
• 100 MEN IN LOW EARTH ORBIT
• ASTRONOMY
• EARTH RESOURCES
• LIFE SCIENCES
• SPACE PHYSICS
• MATERIALS RESEARCH AND PROCESSING
24 MEN IN
LUNAR ORBIT •
• 48 MEN ON LUNAR SURFACE
• ASTRONOMY
• BIOLOGY
• SELENOLOGY
• MINING
• SEISMOLOGY
• REGIONAL EXPLORATION
• PLANETARY QUARANTINE FACILITY
MANNED MARS LANDING
TEMPORARY BASE
12 MEN
SEMI-PERMANENT BASE
48 MEN ON SURFACE
24 MEN IN ORBIT
MARS EXCURSION MODULE
MSFC-69-PD-SA 185

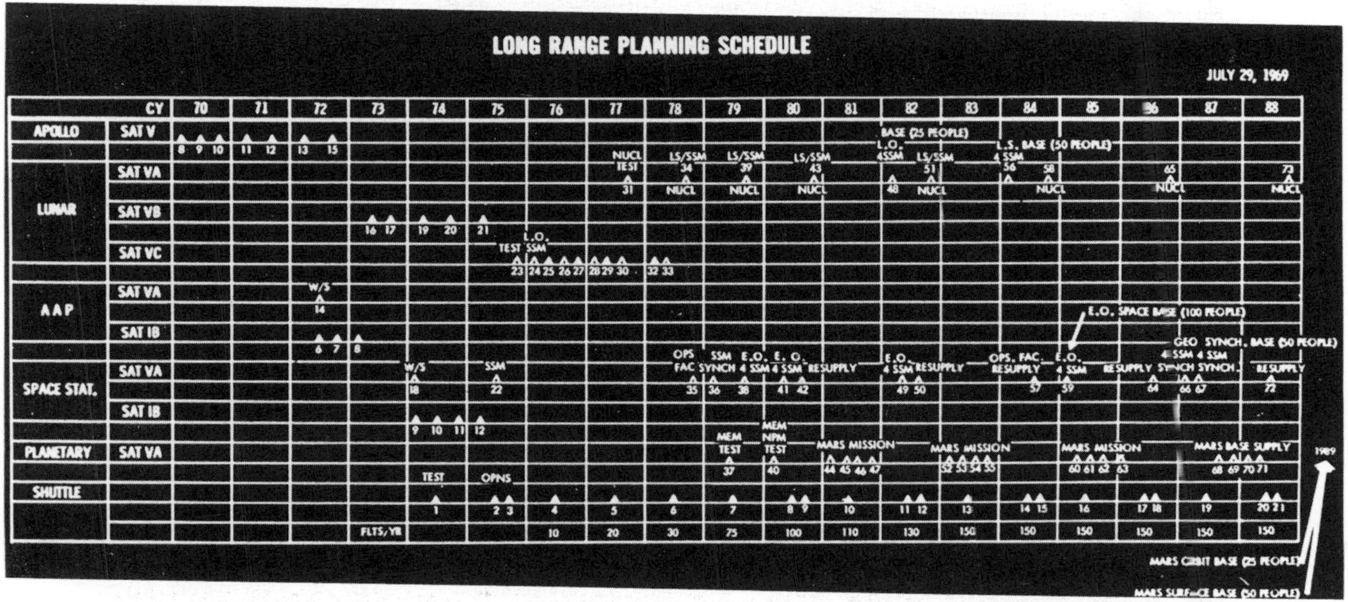

LONG RANGE PLANNING SCHEDULE

JULY 29, 1969

	CY	70	71	72	73	74	75	76	77	78	79	80	81	82	83	84	85	86	87	88
APOLLO	SAT V	8 9 10	11 12	13 15																
LUNAR	SAT VA								NUCL TEST 31	LS/SSM 34 NUCL	LS/SSM 39 NUCL	LS/SSM 43 NUCL		BASE (25 PEOPLE) L.O. 4SSM 48	LS/SSM 51 NUCL	L.S. BASE (50 PEOPLE) 4 SSM 56	58 NUCL	65 NUCL		73 NUCL
	SAT VB				16 17	19 20	21													
	SAT VC						TEST	L.O. SSM 23 24 25 26 27	28 29 30	32 33										
AAP	SAT VA			W/S 14																
	SAT IB			6 7	8															
SPACE STAT.	SAT VA					W/S 18	SSM 22			OPS FAC 35	SSM SYNCH 36 E.O. 4 SSM 38	E.O. 4 SSM 41 42	RESUPPLY	E.O. 4 SSM 49 50 RESUPPLY		OPS. FAC. RESUPPLY 57	E.O. 4 SSM 59 E.O. SPACE BASE (100 PEOPLE)	RESUPPLY 64	GEO SYNCH. BASE (50 PEOPLE) 4 SSM 4 SSM SYNCH SYNCH 66 67	RESUPPLY 72
	SAT IB					9 10 11	12													
PLANETARY	SAT VA										MEM TEST 37	MEM NPM TEST 40	MARS MISSION 44 45 46 47		MARS MISSION 52 53 54 55		MARS MISSION 60 61 62 63		MARS BASE SUPPLY 68 69 70 71	1989
SHUTTLE						TEST 1	OPNS 2 3	4	5	6	7	8 9	10	11 12	13	14 15	16	17 18	19	20 21
					FLTS/YR			10	20	30	75	100	110	130	150	150	150	150	150	150

MARS ORBIT BASE (25 PEOPLE)

MARS SURFACE BASE (50 PEOPLE)

MARS SURFACE EXCURSION
(1) DE-ORBIT
(2) ENTRY
(3) SHROUD & HEAT SHIELD JETTISON
(4) DESCENT BRAKING
(5) LANDING
(6) SURFACE OPERATIONS
(7) ASCENT
(8) STAGING
(9) RENDEZVOUS & DOCKING
MSFC-69-PD-SA 170

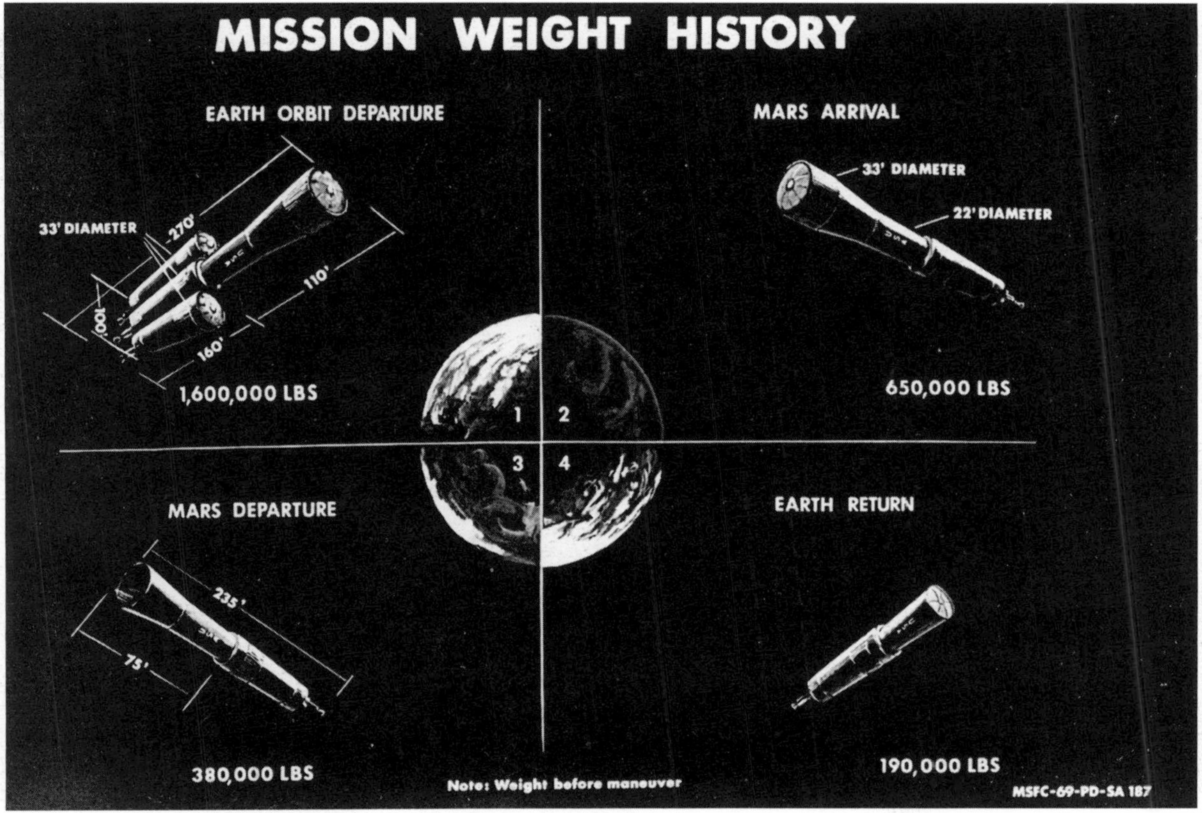
MISSION WEIGHT HISTORY
EARTH ORBIT DEPARTURE
33' DIAMETER
270'
110'
100'
160'
1,600,000 LBS
MARS ARRIVAL
33' DIAMETER
22' DIAMETER
650,000 LBS
1
2
3
4
MARS DEPARTURE
235'
75'
380,000 LBS
EARTH RETURN
190,000 LBS
Note: Weight before maneuver
MSFC-69-PD-SA 187

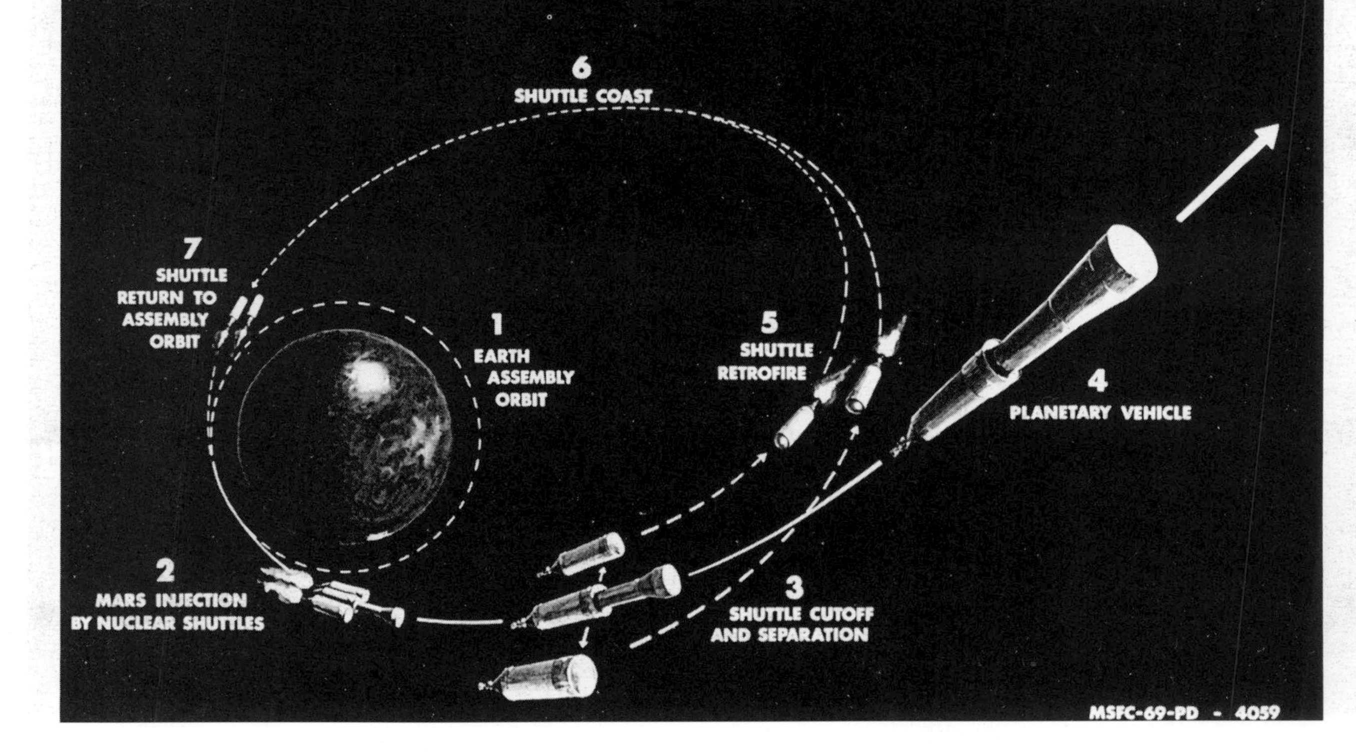
EARTH ORBIT DEPARTURE MANEUVERS
6
SHUTTLE COAST
7
SHUTTLE
RETURN TO
ASSEMBLY
ORBIT
1
EARTH
ASSEMBLY
ORBIT
5
SHUTTLE
RETROFIRE
4
PLANETARY VEHICLE
2
MARS INJECTION
BY NUCLEAR SHUTTLES
3
SHUTTLE CUTOFF
AND SEPARATION
MSFC-69-PD - 4059

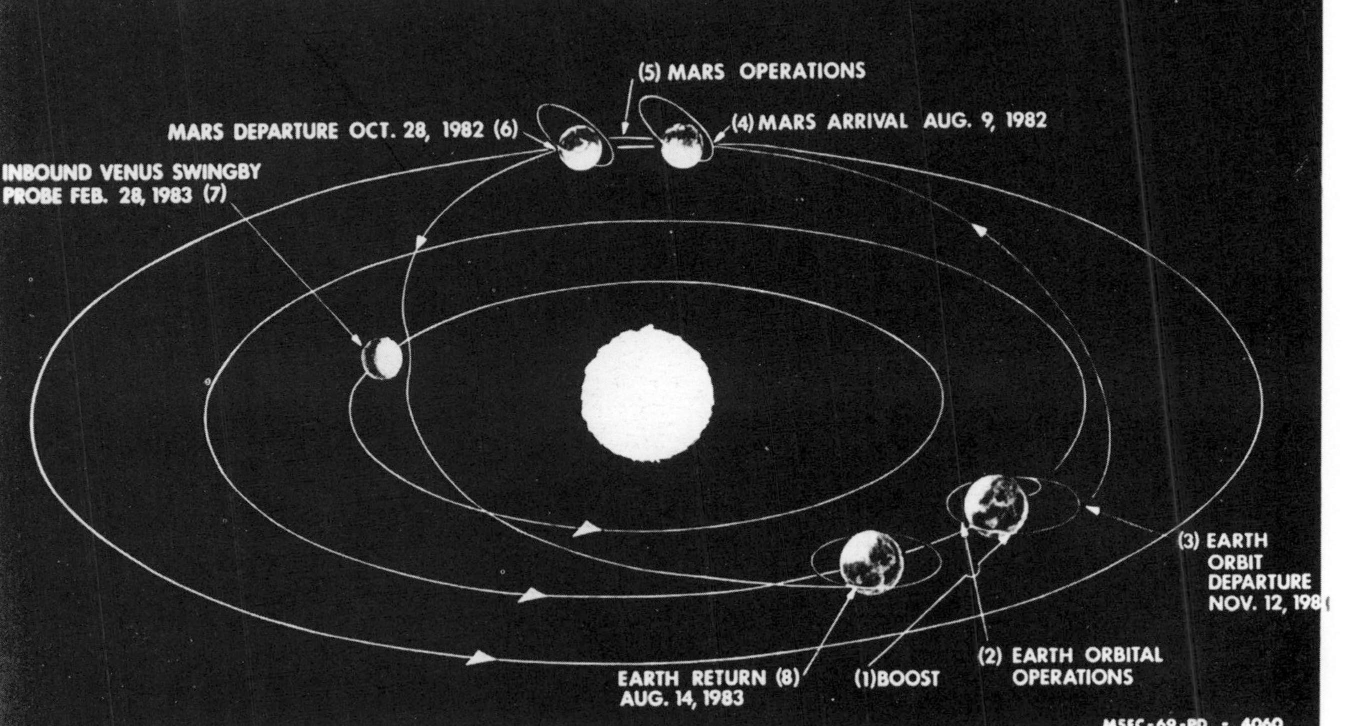
1981 MARS LANDING MISSION PROFILE
(5) MARS OPERATIONS
MARS DEPARTURE OCT. 28, 1982 (6)
(4) MARS ARRIVAL AUG. 9, 1982
INBOUND VENUS SWINGBY
PROBE FEB. 28, 1983 (7)
(3) EARTH
ORBIT
DEPARTURE
NOV. 12, 198
(2) EARTH ORBITAL
OPERATIONS
EARTH RETURN (8)
AUG. 14, 1983
(1)BOOST
MSFC-69-PD - 4060

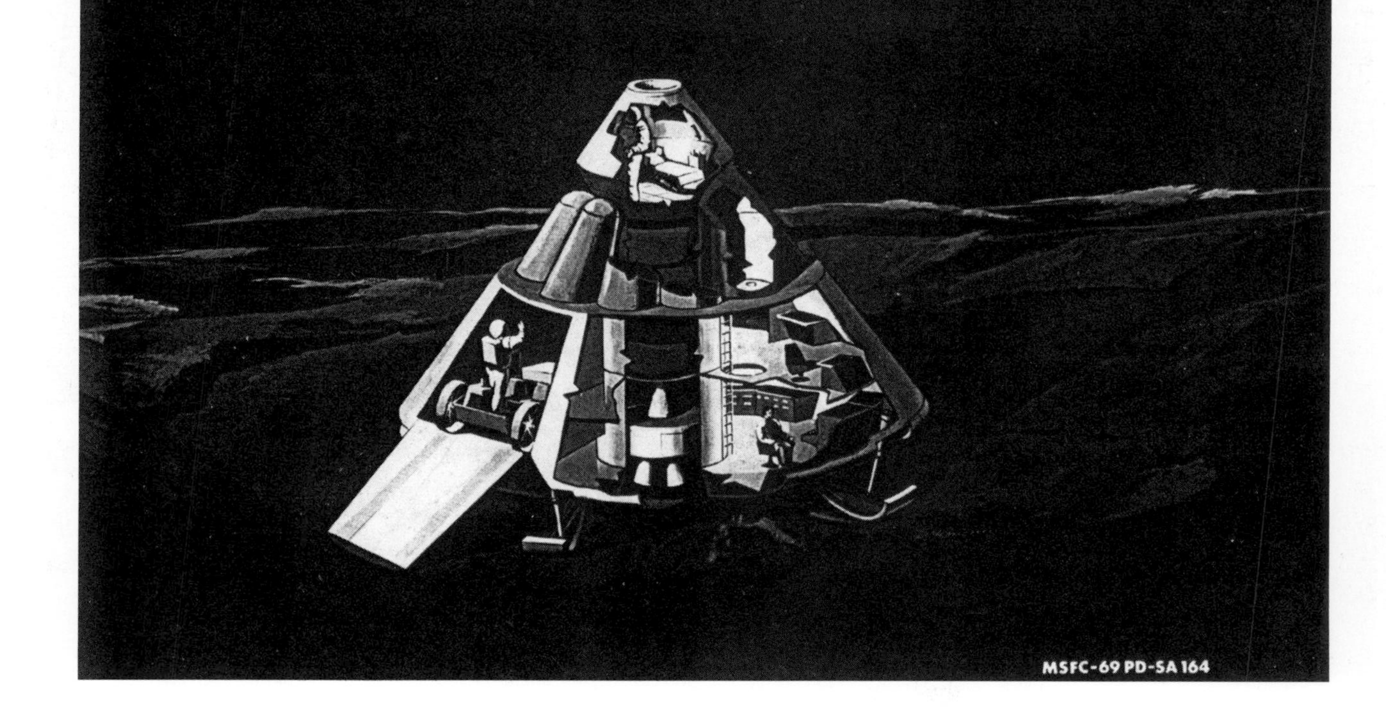
MARS EXCURSION MODULE CONFIGURATION
MSFC-69 PD-SA 164

USA
EARTH ORBIT DEPARTURE
6 MEN PER SHIP
MSFC-69-PD-SA-176

Appendix 3
The Armstrong Photo Mystery

Thirty years after the first moonlanding there is still speculation as to why there are no pictures of Neil Armstrong, the first man on the Moon – apart that is from the one taken by Armstrong himself, in which he is reflected in Aldrin's facepiece. It can be partially explained by a study of the Apollo 11 Lunar Surface Operations plan, reproduced here.

It was really left to Armstrong and Aldrin to make on-the-spot decisions; and in the event, Armstrong took many photographs, but when he attempted to hand over the camera to Aldrin to have himself photographed, Aldrin was too busy with other things; and when Aldrin did have the camera, he was apparently busy fulfilling commitments to photograph the experiments.

R TURNILL
1818

NATIONAL AERONAUTICS AND SPACE ADMINISTRATION

FINAL

APOLLO 11 LUNAR SURFACE OPERATIONS PLAN

PREPARED BY

LUNAR SURFACE OPERATIONS OFFICE
MISSION OPERATIONS BRANCH
FLIGHT CREW SUPPORT DIVISION

JUNE 27, 1969

MANNED SPACECRAFT CENTER
HOUSTON, TEXAS

CDR	LMP
	Photo area where bulk sample was collected
	Deploy ALSCC: (Deployment of the ALSCC will be delayed until the documented sample collection if behind in the timeline)
	a. Remove isolator latch pin and pivot cover b. Pull camera from MESA c. Place camera on secondary gear strut and exert pressure on camera cover. Pull the two skirt lanyards d. Rotate handle retaining latch e. Swing handle clockwise 150° and pull until fully extended f. Place camera on surface

ALSCC OPERATION

Close-up photographs will be taken by either crewman when time is available between or during other tasks. Several times within the EVA are suggested when it may be convenient for the crew to take photos. This is not a requirement to take photos nor does it prohibit them from obtaining photographs at other times which may be feasible.

In general the camera operation is:

a. Estimate position of object plane relative to camera bearing surface
b. Position camera over object (Describe object and location)

FLIGHT PLAN

CSM

CMP

SET UP CAMERA FOR TRACKING
EL/250/BW-BRKT
INT (f5.6,250,INF)

PITCH DOWN 172° TO HEADS DOWN FOR LUNAR SURFACE OBSERVATION, ORB RATE

R180, P282/44, Y0

113:00
113:19
113:30
113:32
113:38
REV 20
114:00

MSFN

LM

CDR

PRELIMINARY CHECKS
CK LM STATUS
CK LIGHTING VISIBILITY

REST
MONITOR AND PHOTOGRAPH LMP EGRESS

TV DEPLOYMENT
CAMERA EQPT FROM MESA
CARRY TV TO SITE
MOUNT TRIPOD, PANORAMA, POSITION FOR EVA
PHOTOGRAPH SWC
PHOTO BULK SAMPLE AREA

BULK SAMPLE COLLECTION
CAMERA ON MESA
PREPARE SRC
COLLECT ROCK FRAGMENTS AND LOOSE MATERIAL
WEIGH SAMPLE
PACK AND SEAL SRC, CONNECT TO LEC
REST

LM INSPECTION
INSPECT QUAD IV, +Y GEAR
EVAL TERRAIN, VISIBILITY
INSPECT QUAD III, -Z GEAR
PHOTO QUAD I, EASEP OFF LOADING
INSPECT, PHOTO -Y GEAR
PHOTO PANORAMA
TAKE CLOSEUP PHOTOS
EASEP DEPLOYMENT

LMP

STILL-CAMERA TO SURFACE
FINAL LM CK
EVA GO

INITIAL EVA
EGRESS
DESCEND TO SURFACE

ENVIRONMENT FAMILIARIZATION
CK BALANCE, STABILITY, REACH, WALKING, EMU

SWC DEPLOYMENT
DEPLOY SWC IN SUN

EVA AND ENIRON EVAL
EVAL EVA CAPABILITY AND EFFECTS
EVAL LIGHTING/VISIBILITY AND SURFACE CHARACTERISTICS
PHOTO PANORAMA

LM INSPECTION
PHOTO QUAD I, +Z GEAR
PHOTO BULK SAMPLE AREA
DEPLOY ALSCC
PHOTO QUAD IV, +Y GEAR
PHOTO PANORAMA
PHOTO QUAD III, -Z GEAR
CAMERA TO CDR

EASEP DEPLOYMENT
REMOVE EXPERIMENTS

MCC-H

0+30
EVA GO
0+40
0+50
1+00
1+10
1+20
1+30

MISSION	EDITION	DATE	TIME	DAY/REV	PAGE
APOLLO 11	FINAL	JULY 1, 1969	113:00 - 114:00	5/19-20	3-80

This page from the final Flight Plan shows the jobs allotted to the Commander (CDR) (Armstrong) and the LMP (Aldrin) as they started work on the lunar surface. On the left is shown what the CMP (Collins) is doing as he orbits above them. The plan shows what Mission Control *hopes* the astronauts will be able to do. In the event it was extraordinary how closely they kept to it all – a tribute to the long hours of rehearsal before the launch. (NASA)

Appendix 4
US Spaceflights Described in *The Moonlandings*

Spacecraft	Launch date	Astronauts	Flight time Days	Hrs	Min	Highlights
Mercury 3	5.5.61	Alan Shepard	00.	00.	15	1st American in space
Mercury 4	21.7.61	Virgil Grissom	00.	00.	16	Capsule sank
Mercury 6	20.2.62	John Glenn	00.	04.	55	1st American in orbit
Mercury 7	24.5.62	M. Scott Carpenter	00.	04.	56	Landed 402 km from target
Mercury 8	3.10.62	Walter Schirra	00.	09.	13	Landed 8 km from target
Mercury 9	15.5.63	L. Gordon Cooper	01.	10.	20	1st long flight by an American
Gemini 3	23.3.65	Virgil Grissom John Young	00.	04.	53	1st manned orbital manoeuvres
Gemini 4	3.6.65	James McDivitt Edward White	04.	01.	56	21-min 'spacewalk' (White)
Gemini 5	21.8.65	L. Gordon Cooper Charles Conrad	07.	22.	58	1st extended manned flight
Gemini 7	4.12.65	Frank Borman James Lovell	13.	18.	35	Longest US flight for 8 years
Gemini 6	15.12.65	Walter Schirra Thomas Stafford	01.	01.	51	RV to 1.8 m of Gemini 7
Gemini 8	16.3.66	Neil Armstrong David Scott	00.	10.	41	1st docking; emergency splashdown

Spacecraft	Launch date	Astronauts	Flight time Days	Hrs	Min	Highlights
Gemini 9	3.6.66	Thomas Stafford Eugene Cernan	03.	00.	21	2-h spacewalk (Ceman)
Gemini 10	18.7.66	John Young Michael Collins	02.	22.	47	RV with 2 targets; Agena package retrieved
Gemini 11	12.9.66	Charles Conrad Richard Gordon	02.	23.	17	RV and docking
Gemini 12	11.11.66	James Lovell Edwin Aldrin	03.	22.	34	Dockings; 3 spacewalks
Apollo 7	11.10.68	Walter Schirra Donn Eisele Walter Cunningham	10.	20.	09	1st manned Apollo flight
Apollo 8	21.12.68	Frank Borman James Lovell William Anders	06.	03.	00	1st manned flight around Moon
Apollo 9	3.3.69	James McDivitt David Scott Russell Schweickart	10.	01.	01	Docking with Lunar Module
Apollo 10	18.5.69	Thomas Stafford Eugene Cernan John Young	08.	00.	03	Descent to within 14 km of Moon
Apollo 11	16.7.69	Neil Armstrong Edwin Aldrin Michael Collins	08.	03.	18	Armstrong and Aldrin land on Moon; 20 kg samples
Apollo 12	14.11.69	Charles Conrad Richard Gordon Alan Bean	10.	04.	36	2 EVAs, total 7 h 39 min; 34 kg samples
Apollo 13	11.4.70	James Lovell John Swigert Fred Haise	05.	22.	55	Mission aborted following oxygen tank explosion
Apollo 14	31.1.71	Alan Shepard Stuart Roosa Edgar Mitchell	09.	00.	42	2 EVAs, total 9 h 26 min; 44 kg samples
Apollo 15	26.7.71	David Scott James Irwin Alfred Worden	12.	07.	12	3 EVAs, total 18 h 36 min; 78 kg samples

Spacecraft	Launch date	Astronauts	Flight time			Highlights
			Days	Hrs	Min	
Apollo 16	16.4.72	John Young Ken Mettingly Charles Duke	11.	01.	51	3 EVAs, total 20 h 14 min; 97.5 kg samples
Apollo 17	6.12.72	Eugene Cernan Ronald Evans Harrison Schmitt	12.	13.	51	3 EVAs, total 22 h 06 min; 113 kg samples
Skylab 1	14.5.73	Unmanned				Vibration damage during lift-off
Skylab 2	25.5.73	Charles Conrad Joseph Kerwin Paul Weitz	28.	00.	50	Exceeded Soviet duration record EVAs repaired damage
Skylab 3	28.7.73	Alan Bean Owen Garriott Jack Lousma	59.	11.	09	Rescue mission prepared but not needed
Skylab 4	16.11.73	Gerald Carr Edward Gibson William Pogue	84.	01.	15	Total Skylab EVAs 3 days, 5 h 48 min
ASTP	15.7.75	Thomas Stafford Vance Brand Donald Slayton	09.	01.	28	1st US–Soviet Joint Flight; 2 days docked activities
STS-92 Discovery	29.10.98	Curtis Brown + 5 John Glenn	08.	21.	43	Glenn, 77 became oldest human in space

Index

Page references in bold indicate illustrations.